FORMULATIONS, REGULATIONS, AND CHALLENGES OF NUTRACEUTICALS

AAP Advances in Nutraceuticals

FORMULATIONS, REGULATIONS, AND CHALLENGES OF NUTRACEUTICALS

Edited by

Tingirikari Jagan Mohan Rao, PhD
Rajesh K. Kesharwani, PhD, MTech
Raj K. Keservani, PhD, MPharm
Anil K. Sharma, PhD, MPharm

AAP | APPLE ACADEMIC PRESS

First edition published 2025

Apple Academic Press Inc.
1265 Goldenrod Circle, NE,
Palm Bay, FL 32905 USA

760 Laurentian Drive, Unit 19,
Burlington, ON L7N 0A4, CANADA

CRC Press
2385 NW Executive Center Drive,
Suite 320, Boca Raton FL 33431

4 Park Square, Milton Park,
Abingdon, Oxon, OX14 4RN UK

Library and Archives Canada Cataloguing in Publication

Title: Formulations, regulations, and challenges of nutraceuticals / edited by Tingirikari Jagan Mohan Rao, PhD, Rajesh K. Kesharwani, PhD, MTech, Raj K. Keservani, PhD, MPharm, Anil Kumar Sharma, PhD, MPharm.
Names: Rao, Tingirikari Jagan Mohan, editor. | Kesharwani, Rajesh Kumar, 1978- editor. | Keservani, Raj K., 1981- editor. | Sharma, Anil K., 1980- editor.
Series: AAP advances in nutraceuticals.
Description: First edition. | Series statement: AAP advances in nutraceuticals | Includes bibliographical references and index.
Identifiers: Canadiana (print) 20240311914 | Canadiana (ebook) 20240312031 | ISBN 9781774914564 (hardcover) | ISBN 9781774914571 (softcover) | ISBN 9781003412496 (ebook)
Subjects: LCSH: Functional foods—Safety regulations. | LCSH: Functional foods—Labeling.
Classification: LCC QP144.F85 F67 2024 | DDC 613.2—dc23

Library of Congress Cataloging-in-Publication Data

CIP data on file with US Library of Congress

ISBN: 978-1-77491-456-4 (hbk)
ISBN: 978-1-77491-457-1 (pbk)
ISBN: 978-1-00341-249-6 (ebk)

ABOUT THE BOOK SERIES: AAP ADVANCES IN NUTRACEUTICALS

Series Editor:
Raj K. Keservani, PhD, MPharm
Faculty of B. Pharmacy, CSM Group of Institutions, Allahabad, India

Rajesh K. Kesharwani, PhD
Nehru Gram Bharati (Deemed to be University), Prayagraj, India

Anil K. Sharma, PhD, MPharm
Department of Pharmacy, School of Medical and Allied Sciences,
GD Goenka University, Gurugram, India

In the modern era, mankind has witnessed a paradigm shift with respect to fundamental eating behavior. The lack of physical workouts and busy schedules at offices and in households have promoted the consumption of junk foods, which eventually results in numerous diseases and disorders since the nutritional content from fast food is inadequate. Plumpness and obesity have become a global health threat. The leading causes of death from most developing nations are noncommunicable diseases such as cardiovascular diseases, cancer, arthritis, osteoporosis, and liver toxicity. Patients suffering from such lifestyle ailments are a bit apprehensive towards prolonged use of costly modern therapeutics, which encourages instead the use of alternative approaches for management of such diseases and disorders.

The emerging sector of the nutraceutical industry encompasses products derived from nature, dietary supplements, and functional foods. Nutraceuticals are used for treatment and prevention of a broad range of diseases, such as the common cold, arthritis, sleep-related disorders, cancers, cardiovascular complications, metabolic disorders, and others.

The research on nutraceuticals is increasing day by day, considering the beneficial effects of food or food supplements in the management of diverse diseases. The issue of paramount concern is standardization and establishment of clinical efficacy of nutraceuticals, which is in fact a challenge for researchers around the globe.

This book series aims at realizing the significance of the variety of nutraceuticals for human well-being. The books in the series will also emphasize the role of dieticians and nutritionists for the prescription of judicious eating habits. The key food components such as carbohydrates, proteins, and lipids as well as micronutrients (vitamins, minerals) are demonstrated to maintain good health, obviating the need for medicines. Thus, nutraceuticals that are indeed derived from food ingredients are believed to be potential alternative therapeutics.

Asymptomatic diseases/disorders necessitate proper diagnosis. Stressed circumstances have been reported to cause weight loss. Workplace stress is a key variable in the genesis and propagation of metabolic disorders. The nuances of the underlying mechanisms are still to be deciphered. Poor lifestyle trends have been attributed along with the consumption of junk foods in several such instances. Sugar-rich carbonated beverages, the use of contraband drugs, and the consumption of liquor in excess are assigned as risk factors. This book series will cover advances and applications in addition to providing the basics of nutraceuticals.

The books in this series will be valuable resources for industry professionals to design and develop quality products for end users. In addition, the book series will be fortified by ideas of innovation, concept building, manufacturing aspects, quality control, and regulatory status of nutraceuticals.

Ostensibly, business has seen a rise in the nutraceutical consumption in recent years. Pure, safe, efficacious products are the need of the hour globally. There are a number of books on nutraceuticals in the market, yet these are only a few books presenting certain aspects of nutraceuticals therein. This book series aims to offer a holistic view of this promising strategy, which is picking up pace with time. This book series will be a unique endeavor to bring manufacturing, research and development, and marketing strategies under a single umbrella.

CURRENT BOOKS IN THE SERIES

- **Micronutrients and Macronutrients as Nutraceuticals**
  Editors: Prakash Chandra Gupta, PhD, Sayan Bhattacharyya, MBBS, MD, Nisha Sharma, PhD, MPharm, Rajesh K. Kesharwani, PhD, and Raj K. Keservani, PhD, MPharm

- **Formulations, Regulations, and Challenges of Nutraceuticals**
  Editors: Tingirikari Jagan Mohan Rao, PhD, Rajesh K. Kesharwani, PhD, Raj. K. Keservani, PhD, and Anil K. Sharma, PhD

- **Food Supplements and Dietary Fibers in Health and Disease**
  Editors: Bhushan R. Rane, PhD, MPharm, Raj K. Keservani, PhD, MPharm, Durgesh Singh, DPhil, Nayan A. Gujarathi, PhD, and Ashish S. Jain, PhD
- **Nutraceuticals in Cancer Prevention, Management, and Treatment**
  Editors: Raj K. Keservani, PhD, MPharm Bui Thanh Tung, PhD, Sippy Singh, DPhil, and Rajesh K. Kesharwani, PhD
- **Advances in Flavonoids for Human Health and Prevention of Diseases**
  Editors: Nisha Sharma, PhD, Deepika Saini, PhD, Rajesh K. Kesharwani, PhD, Prakash Chandra Gupta, and Raj. K. Keservani, PhD
- **The Flavonoids: Extraction and Applications**
  Editors: Deepika Saini, PhD, Rajesh K. Kesharwani, PhD, and Raj K. Keservani, PhD, MPharm
- **Immune-Boosting Nutraceuticals for Better Human Health: Novel Applications**
  Editors: Urmila Jarouliya, PhD, Raj K. Keservani, PhD, MPharm Rajesh K. Kesharwani, PhD, Virendra K. Patel, PhD, and Adi D. Bharti, MPharm
- **Applications of Functional Foods in Disease Prevention**
  Editors: Raj K. Keservani, PhD, MPharm, and Eknath D. Ahire, MSPharm
- **Plant Metabolites and Vegetables as Nutraceuticals**
  Editors: Raj K. Keservani, PhD, MPharm, Bui Thanh Tung, PhD, Rajesh K. Kesharwani, PhD, MPharm, and Eknath D. Ahire
- **Preventive and Therapeutic Role of Vitamins as Nutraceuticals**
  Editors: Khemchand R. Surana, Eknath D. Ahire, Raj K. Keservani, PhD, MPharm, and Rajesh K. Kesharwani, PhD
- **Nutraceuticals and Bone Health**
  Editors: Deepak Sharma, PhD, Madan Mohan Gupta, PhD, Anil K. Sharma, PhD, MPharm, Raj K. Keservani, PhD, MPharm, and Rajesh K. Kesharwani, PhD
- **Flavonoids as Nutraceuticals**
  Editors: Rajesh K. Kesharwani, PhD, MTech, Deepika Saini, PhD, Raj K. Keservani, PhD, MPharm, and Anil K. Sharma, PhD
- **Herbals as Nutraceuticals: Their Role in Healthcare**
  Editors: Raj K. Keservani, PhD, MPharm, Rajeshwar Kamal Kant Arya, PhD, and Rajesh K. Kesharwani, PhD

- **Nutraceutical Fruits: Overview and Disease Prevention**
  Editors: Raj K. Keservani, PhD, MPharm
- **Nutrigenomics and Nutraceuticals**
  Editors: Raj K. Keservani, PhD, MPharm, Eknath D. Ahire, Shubham J. Khainar, PhD, Sanjay J. Kshirsagar, PhD, and Rajesh K. Kesharwani, PhD
- **Nutraceuticals in Cardiac Health Management**
  Editors: Raj K. Keservani, PhD, MPharm, Rajesh K. Kesharwani, PhD, Praveen Kumar Jain, PhD, Bhushan R. Rane, PhD, and Nayan A. Gujarathi, PhD
- **Nutraceuticals for the Treatment and Prevention of Sexual Disorders**
  Editors: Raj K. Keservani, PhD, MPharm, Sharangouda J. Patil, PhD, and Ivan Aranha, PhD
- **Nutraceuticals in Obesity Management and Control**
  Editors: Raj K. Keservani, PhD, MPharm, and Alka Lohani, PhD
- **Antioxidants as Nutraceuticals**
  Editors: Nayan A. Gujarathi, PhD, Raj K. Keservani, PhD, MPharm, Rajesh K. Kesharwani, PhD, Bhushan R. Rane, PhD, and Yogeeta Sameer Goyal, PhD
- **The Nature of Nutraceuticals: History, Properties, Sources, and Nanotechnology**
  Editors: Rajesh K. Kesharwani, PhD, Prashant Kumar, PhD, and Raj K. Keservani, PhD, MPharm
- **Nutraceuticals in Insomnia and Sleep Problems**
  Editors: Raj K. Keservani, PhD, MPharm, Sayan Bhattacharyya, PhD, and Rajesh K. Kesharwani, PhD
- **Nutraceuticals in Respiratory and Pulmonary Diseases**
  Editors: Deepika Saini, PhD, Rajesh K. Kesharwani, PhD, and Raj K. Keservani, PhD, MPharm
- **Nutraceuticals in Arthritis and Psoriasis**
  Editors: Meenakshi Jaiswal, PhD, Raj K. Keservani, PhD, MPharm Rajesh K. Kesharwani, PhD, and Swati G. Talele, PhD

ABOUT THE EDITORS

Tingirikari Jagan Mohan Rao, PhD
Assistant Professor, Department of Biotechnology,
NIT Andhra Pradesh

Tingirikari Jagan Mohan Rao, PhD, is working as an Assistant Professor in the Department of Biotechnology, NIT Andhra Pradesh. He has research experience on synthesis and production of poly- and oligosaccharides. During his PhD, he worked on structural, biocompatible, and prebiotic potential of exopolysaccharides produced by lactic acid bacteria. While in his postdoctoral research in Brazil, he produced prebiotic oligosaccharides in orange juice by using immobilized and co-immobilized enzyme systems. While in South Korea, he worked on prebiotic properties of pectic polysaccharides using standard intestinal bacterial cultures and published his research findings in reputed international SCI-Journals. His significant contribution to scientific research can be reflected from his publications (17 papers, 6 books with citation more than 250 and h-index of 10).

Rajesh K. Kesharwani, PhD, MTech
Associate Professor, Department of Computer Application,
Nehru Gram Bharati (Deemed to be University), Prayagraj, India

Rajesh K. Kesharwani, PhD, MTech, is working as an Associate Professor, Department of Computer Application, Nehru Gram Bharati (Deemed to be University), Prayagraj, India. He has more than 11 years of research and nine years of teaching experience in various institutes of India, imparting bioinformatics and biotechnology education. He has received several awards, including the NASI-Swarna Jayanti Puruskar by The National Academy of Sciences of India. He has supervised one PhD, and more than 20 UG, PG students for their research work, and has authored over 49 peer-reviewed articles, 20 book chapters, and 14 edited books with international publishers (e.g., Springer Nature, IGI Global, Taylor & Francis, etc.). He has been a member of many scientific communities as well as a reviewer for many international journals. He has presented many papers at various national and international conferences. Dr. Kesharwani has received his PhD from the Indian Institute of Information Technology, Allahabad, and worked at NIT Warangal for two semesters. He has been a recipient of the Ministry of Human Resource Development (India) Fellowship and Senior Research Fellowship from the Indian Council of Medical Research, India. His research fields of interest are medical informatics, protein structure and function prediction, computer-aided drug designing, structural biology, drug delivery, cancer biology, nano-biotechnology, and biomedical sciences.

Raj K. Keservani, PhD, MPharm
Associate Professor, Faculty of B.Pharmacy, CSM Group of Institutions, Prayagraj, India

Raj K. Keservani, PhD, MPharm, is a Faculty of B. Pharmacy, CSM Group of Institutions, Allahabad, India. He has more than 12 years of academic (teaching) experience from various institutes of India in pharmaceutical education. He has supervised and more than 25 UG students for their research work, has published 31 peer-reviewed papers in the field of pharmaceutical sciences in national and international journals, one patents, 43 book chapters, three coauthored books, and 19 edited books. He is also active as a reviewer for several international scientific journals. Dr. Keservani graduated with a pharmacy degree from the Department of Pharmacy, Kumaun University, Nainital (Uttarakhand), India. He received his Master of Pharmacy (MPharm) (specialization in pharmaceutics) from the School of Pharmaceutical Sciences, Rajiv Gandhi Proudyogiki Vishwavidyalaya, Bhopal, India. Dr. Keservani is a life member of Society of Pharmaceutical Education and Research (SPER). His research interests include nutraceutical and functional foods, novel drug delivery systems (NDDS), transdermal drug delivery/drug delivery, health science, cancer biology, and neurobiology.

Anil K. Sharma, PhD, MPharm
Assistant Professor, School of Medical and Allied Sciences,
G D Goenka University, Gurugram, India

Anil K. Sharma, PhD, MPharm, is an expert in the area of pharmaceutics with a background in drug delivery. He has taught these subjects for nearly 10 years at universities such as the Delhi Institute of Pharmaceutical Sciences and Research, University of Delhi, and School of Medical and Allied Sciences, G D Goenka University, India. Prior to taking up his current role in 2018, Dr. Sharma served in academic positions such as lecturer (Pharmaceutics), Delhi Institute of Pharmaceutical Sciences and Research, University of Delhi, India. He has published over 28 peer-reviewed papers in the field of pharmaceutical sciences in both national and international journals as well many book chapters and several edited books. He holds a PhD (Pharmaceutical Sciences) from the University of Delhi; MPharmacy (Pharmaceutics) from the Rajiv Gandhi Proudyogiki Vishwavidyalaya; and BPharmacy from the University of Rajasthan.

CONTENTS

Contributors .... *xv*

Abbreviations .... *xix*

Preface .... *xxiii*

PART I: Overview on Nutraceutical Formulation and Challenges........ 1

1. **Overview of Formulation, Challenges, and Regulation for the Development of Nutraceuticals........ 3**
   Akash M. Bhoite, Utkarsha P. Patil, Harshada U. Bagul, Swati G. Talele, and Anil Jadhav

2. **Formulation and Challenges Involved in Developing Plant/Phytochemical-Based Nutraceuticals........ 29**
   Jijnasa Bordoloi, Jatin Kalita and Prasenjit Manna

PART II: Nutraceutical Formulations and Regulations........ 55

3. **Recent Developments in the Formulation of Nutraceuticals........ 57**
   Sibashish Kityania, Deepa Nath, Subrata Das, Rajat Nath, and Anupam Das Talukdar

4. **Future Trends in the Formulation of Nutraceuticals........ 81**
   Nidhi Parihar, Tulasi Naraharisetti, Prathamesh Patil, Sri Vathsan, and Deepak Bharadwaj PVP

5. **Major Factors Contributing to the Regulations of Nutraceuticals........ 105**
   Tesnim Arfi and Satyajit Saurabh

6. **Nutraceutical Regulation in Asian Countries........ 127**
   Navodit Kumar Singh, Richa Katiyar, Ranjana Kumari, Bishwambhar Mishra, Sanjeeb Kumar Mandal, and Sanjay Kumar

7. **Nutraceutical Regulation in European Countries........ 153**
   Satyajit Saurabh, Manali Singh, Shivani Bhutani, Sneha Srivastav, and Shilpi Kiran

8. **Regulatory Guidelines for the Development of Plant/Phytochemical-Based Nutraceuticals........ 171**
   Garima Sharma

9. **In Vitro and In Vivo Studies on Nutraceuticals........ 193**
   Joel Augustine, Raveena Ann Alex, and Jayanthi Abraham

10. Regulatory Framework for the Development of Nutraceuticals 213
Mansavi Bhardwaj

PART III: Opportunities and Challenges of Nutraceuticals 225

11. Opportunities and Challenges in the Development of Functional Foods and Nutraceuticals 227
Bhushan R. Rane, Akash J. Amkar, Vinit S. Patil, Prasad K. Vidhate, and Abhijit R. Patil

12. Nutraceuticals Manufacturing Challenges in Fast Growing Food Segment in India 255
Manish Kumar, Shiv Shankar Pandey, Pushpendra Kumar, Vimal Kumar Yadav, and Ajay Kumar Shukla

13. Challenges for Incorporation of Nutraceuticals in Food 285
Priya Sen, Preeti Khulbe, Manish Gupta, and Riya Thapa

Index 321

CONTRIBUTORS

Jayanthi Abraham
Microbial Biotechnology Laboratory, School of Biosciences and Technology, Vellore Institute of Technology, Vellore, Tamil Nadu, India

Raveena Ann Alex
Microbial Biotechnology Laboratory, School of Biosciences and Technology, Vellore Institute of Technology, Vellore, Tamil Nadu, India

Akash J. Amkar
Shri D.D. Vispute College of Pharmacy & Research Center, Panvel, Maharashtra, India

Tesnim Arfi
Department of Bioengineering and Biotechnology, Birla Institute of Technology, Mesra, Ranchi, Jharkhand, India

Joel Augustine
Microbial Biotechnology Laboratory, School of Biosciences and Technology, Vellore Institute of Technology, Vellore, Tamil Nadu, India

Harshada U. Bagul
Department of Pharmaceutics, Sandip Institute of Pharmaceutical Sciences, Nashik, Maharashtra, India

Mansavi Bhardwaj
School of Biotechnology, Shri Mata Vaishno Devi University, Katra, Jammu and Kashmir, India

Akash M. Bhoite
Department of Pharmaceutics, Sandip Institute of Pharmaceutical Sciences, Nashik, Maharashtra, India

Shivani Bhutani
Department of Biotechnology, Invertis University, Invertis Village, Bareilly-Lucknow National Highway, Bareilly, Uttar Pradesh, India

Jijnasa Bordoloi
MYAS-NIN Department of Sports Science, ICMR-National Institute of Nutrition, Hyderabad, Telangana, India

Subrata Das
Department of Life Science and Bioinformatics, Assam University, Silchar, India

Manish Gupta
Gyan Vihar School of pharmacy, Suresh Gyan Vihar University Jaipur, Gyan Vihar Marg, Jagatpura, Jaipur, Rajasthan, India

Anil Jadhav
Department of Pharmaceutics, Sandip Institute of Pharmaceutical Sciences, Nashik, Maharashtra, India

Jatin Kalita
Research Planning and Business Development Division, CSIR-North East Institute of Science and Technology, Jorhat, Assam, India
Center for Infectious Diseases, CSIR-North East Institute of Science and Technology, Jorhat, Assam, India

Richa Katiyar
Department of Chemical Engineering, IIT Delhi, New Delhi, India

Preeti Khulbe
Gyan Vihar School of pharmacy, Suresh Gyan Vihar University Jaipur, Gyan Vihar Marg, Jagatpura, Jaipur, Rajasthan, India

Shilpi Kiran
Department of Botany, Patna Women's College, Patna University, Patna, Bihar, India

Sibashish Kityania
Department of Life Science and Bioinformatics, Assam University, Silchar, India

Manish Kumar
Department of Pharmacy, Madhav University, Pindwara, Rajasthan, India

Pushpendra Kumar
Faculty of Pharmacy, Uttar Pradesh University of Medical Sciences Saifai Etawah, Uttar Pradesh, India

Sanjay Kumar
Department of Chemical Engineering, IIT Delhi, New Delhi, India

Ranjana Kumari
Department of Biotechnology, NIT Andhra Pradesh, Andhra Pradesh, India

Sanjeeb Kumar Mandal
Department of Biotechnology, Chaitanya Bharathi Institute of Technology, Hyderabad, India

Prasenjit Manna
Biological Sciences and Technology Division, CSIR-North East Institute of Science and Technology, Jorhat, Assam, India
Center for Infectious Diseases, CSIR-North East Institute of Science and Technology, Jorhat, Assam, India

Bishwambhar Mishra
Department of Biotechnology, Chaitanya Bharathi Institute of Technology, Hyderabad, India

Tulasi Naraharisetti
Department of Pharmacology & Toxicology, National Institute of Pharmaceutical Education and Research Guwahati, Assam, India

Deepa Nath
Department of Botany, Guru Charan College, Silchar, Assam, India

Rajat Nath
Department of Life Science and Bioinformatics, Assam University, Silchar, India

Shiv Shankar Pandey
Tahira Institute of Medical Sciences, Gorakhpur, Uttar Pradesh, India

Nidhi Parihar
Department of Pharmacology & Toxicology, National Institute of Pharmaceutical Education and Research Guwahati, Assam, India

Abhijit R. Patil
Shri D.D. Vispute, College of Pharmacy & Research Center, Panvel, Maharashtra, India

Prathamesh Patil
Department of Pharmacology & Toxicology, National Institute of Pharmaceutical Education and Research Guwahati, Assam, India

Utkarsha P. Patil
Department of Pharmaceutics, Sandip Institute of Pharmaceutical Sciences, Nashik, Maharashtra, India

Vinit S. Patil
Shri D.D. Vispute, College of Pharmacy& Research Center, Panvel, Maharashtra, India

Deepak Bharadwaj PVP
Department of Pharmacology & Toxicology, National Institute of Pharmaceutical Education and Research Guwahati, Assam, India

Bhushan R. Rane
Shri D.D. Vispute College of Pharmacy & Research Center, Panvel, Maharashtra, India

Satyajit Saurabh
DNA Fingerprinting Laboratory, Bihar State Seed and Organic Certification Agency, Patna, Bihar, India

Priya Sen
Gyan Vihar School of pharmacy, Suresh Gyan Vihar University Jaipur, Jagatpura, Jaipur, Rajasthan, India

Garima Sharma
Department of Botany, Shrimant Madhavrao Scindia Govt. Model Science College Gwalior, Jhansi Road Gwalior, MP, India

Ajay Kumar Shukla
Institute of Pharmacy, Dr. Ram Manohar Lohia Avadh University, Ayodhya, India

Navodit Kumar Singh
Department of Biochemical Engineering and Biotechnology, IIT Delhi, New Delhi, India

Manali Singh
Department of Biotechnology, Invertis University, Invertis Village, Bareilly, Uttar Pradesh, India

Sneha Srivastav
Department of Biotechnology, Invertis University, Invertis Village, Bareilly, Uttar Pradesh, India

Swati G. Talele
Department of Pharmaceutics, Sandip Institute of Pharmaceutical Sciences, Nashik, Maharashtra, India

Anupam Das Talukdar
Department of Life Science and Bioinformatics, Assam University, Silchar, India

Riya Thapa
Gyan Vihar School of Pharmacy, Suresh Gyan Vihar University Jaipur, Jagatpura, Jaipur, Rajasthan, India

Sri Vathsan
Department of Pharmacology & Toxicology, National Institute of Pharmaceutical Education and Research Guwahati, Assam, India

Prasad K. Vidhate
Shri D.D. Vispute College of Pharmacy & Research Center, Panvel, Maharashtra, India

Vimal Kumar Yadav
Institute of Pharmacy, Dr. Ram Manohar Lohia Avadh University, Ayodhya, India

ABBREVIATIONS

ABIAD	Association of Foods for Special Purposes and Congeners
AEC	ASEAN Economic Committee
ALA	arachidonic acid
AMD	age-related macular degeneration
APP	amyloid precursor protein
ARTG	Australian Register of Therapeutic Goods
ASEAN	Association of Southeast Asian Nations
AYUSH	Ayurveda, Yoga and Naturopathy, Unani, Siddha, and Homeopathy
BADS	biologically active dietary supplement
BAFS	biologically active food supplements
BAS	biologically active supplements
BCP	bioactive collagen peptide
CAA	Consumes Affairs Agency
CAGR	compound annual growth rate
CBEC	cross-border e-commerce
CD	cyclodextrin
CHCA	China Health Care Association
CII	Confederation of Indian Industry
CINP	chemotherapy-induced peripheral neuropathy
CLA	conjugated linolenic acid
CRS	Canadian regulatory system
CS	chondroitin sulphate
CVD	cardiovascular disease
DHA	docosahexaenoic acid
DRV	dietary reference values
DS	dietary supplements
DSHEA	Dietary Supplement Health Education Act
EFSA	European Food and Safety Authority
EGCG	epigallocatechin 3-gallate
EPA	eicosapentaenoic acid
EU	European Union
FA	folic acid
FBO	food business operator
FDA	Food and Drug Authority
FDR's	Food and Drug Regulations
FFC	Food with Function Claims

FHC	Foods with Health Claims
FIM	Foundation of Innovation Medicine
FNFC	Food with Nutrient Function Claims
FOSHU	Food for Specified Health Uses
FPIVS	Food Product Identity Verification System
FSA	Food Standard Agency
FSAN	Food Safety and Applied Nutrition
FSS	Food Safety and Standards
FSSA	Food Safety Standard and Standards Act
FSSAI	Food Safety and Standards Authority of India
FUFOSE	Functional Food Science in Europe
GABA	generate gamma-aminobutyric acid
GACP	Good Agricultural and Collection Practices
GHED	Global Health Expenditure Database
GHTF	Global Harmonization Task Force
GM	genetically modified
GMOs	genetically modified organisms
GMP	Good Manufacturing Practice
GOS	galacto-oligosaccharides
GRAS	Generally Recognized as Safe
GS	glucosamine sulphate
HA	hyaluronic acid
HADSA	Health Food and Dietary Supplements Association
HD	Huntington's disease
HFF	Health Functional Food
HFFA	Health Functional Food Association
HIF	hypoxia-inducible factor
HMP	herbal medicinal products
HPV	human papillomavirus
HUL	Hindustan Unilever
ICMR	Indian Council of Medical Research Association
ILSI	International Life Science Institute
INHDSA	Indian Health Food and Dietary Supplement
IRCH	International Regulatory Cooperation for Herbal Medicines
IS	Indian standard
JHFA	Japan Health Food Association
JHNFA	Japan Health Food and Nutrition Food Association
LDL	low-density lipoprotein
MFDS	Ministry of Food and Drug Safety
MHLW	Ministry of Health, Labor, and Welfare
MHPD	Marketed and Health Products Directorate
MOH	Ministry of Health
MP	milk permeate

MSM	methyl sulfonyl methane
NDA	nutrition, dietetic products, and allergies
NDDS	novel drug delivery system
NGP	next-generation probiotics
NHP	natural health product
NIN	National Institute of Nutrition
NLCs	nanostructured lipid carriers
NLEA	Health and Healthcare Labeling Act
NNMB	National Nutrition Monitoring Bureau
NPN	Natural Product Number
OA	osteoarthritis
OATP	organic anion transporting polypeptide
PASSCLAIM	process for the assessment of scientific support for claims
PCL	poly(ε-caprolactone)
PET	probiotics encapsulation technology
PMS	premenstrual syndrome
PUFA	polyunsaturated fatty acids
QC	quality control
RDA	recommended daily allowance
RDA	recommended dietary allowance
ROS	reactive oxygen species
SARS-CoV2	severe acute respiratory syndrome coronavirus two
SCF	Scientific Committee on Food
SFDA	State Food and Drug Administration
SFDA	Saudi Food and Drug Authority
SLNs	solid lipid nanoparticles
SPIP	single-pass intestinal perfusion
TCM	traditional Chinese medicines
TEWL	transepidermal water loss
TNF- α	tumor necrosis factor-α
TGA	Therapeutic Goods Administration
UL	upper level
USCHPA	US–China Health Products Association
USFDA	United States Food and Drug Administration
UV	ultraviolet
VCOs	voluntary consumer organizations
WADA	World Anti-Doping Agency
WHO	World Health Organization

PREFACE

The pharmaceutical products currently employed in disease treatment have several side effects on human health. These days with the increase in health consciousness, people believe that prevention of disease is always better than treatment. In addition, customers look for foods which are minimally processed and exhibit enriched nutritional benefits. Nutraceuticals are those food ingredients or dietary supplements which confer beneficial effect on human health without causing any side effects. Nutraceuticals are gaining great attention due to their safe, biocompatible, and nontoxic nature. As a result, the demand and consumer acceptance are increasing gradually for such products. In addition, as compared to traditional drugs, there are no minimal regulations on nutraceutical products which make them more market-friendly, thus, propelling their growth in global demand and wide consumer acceptance.

Since, nutraceutical ingredients have no defined regulations at the global level, it is not considered a medicine or a therapeutic agent. Some of the countries have devised their own guidelines for regulating the usage of nutraceuticals either as pharmaceutical ingredients or food supplements. Hence, the need of the hour is to evolve a common regulatory guideline so that companies can meet the growing demand of nutraceutical products. Nutraceutical ingredients have several limitations that affect their efficacy; some of the factors include solubility, shelf-life, low permeability, and fast metabolism. The stability of nutraceuticals greatly depends on the formulating material, processing conditions, and physicochemical properties of their matrix. Nutraceuticals are an emerging area and very less work has been done in this domain. It offers great scope for research and common regulatory guidelines are yet to be framed. There are very few publications which cover the above topics.

Hence, the current book, *Formulations, Challenges, and Regulations of Nutraceuticals*, focuses on the various novel micro, and nanocarriers being employed in the formulation, and delivery of nutraceutical ingredients to increase their efficacy, bioavailability, safety, and stability. It also highlights the current challenges and future strategies to be employed for the development of novel nutraceuticals and functional foods with enhanced health benefits. This book helps in addressing the challenges faced during the formulation and the current scenario of being employed for producing a stable and efficient nutraceutical formulation. The book discusses the role of various factors affecting the stability of nutraceuticals in food and gut environment by stressing the results generated from in vitro and in vivo studies.

Good manufacturing practices that should be followed for the development of nutraceuticals are also covered. In order to make products a commercial success, these should comply with all the regulatory guidelines issued by competent authorities of respective countries. Hence, this book discusses the numerous regulatory frameworks being employed in Asia and European regulatory agencies in commercialization of nutraceutical products produced by manufacturing companies. It also highlights the technological advancements and future prospects in the formulation of novel nutraceutical ingredients along with regulatory aspects to be followed for the successful growth of nutraceuticals. Thus, this book emphasizes the latest trends and technological advances, including the regulatory aspects, which will be useful for research scholars, scientists, and industries who are engaged in research and development activities of nutraceutical ingredients.

This book, *Formulations, Challenges, and Regulations of Nutraceuticals* comprises 13 chapters in three sections, which describe the formulation challenges during the manufacturing of nutraceuticals and its regulations worldwide.

PART I: OVERVIEW ON NUTRACEUTICAL FORMULATION AND CHALLENGES

The goals of nutraceutical regulation is focusing on safety and labeling with lesser importance, as compared to pharmaceuticals, on product claims and intended use which is discussed in Chapter 1, Overview on Formulation, Challenges, and Regulation for the Development of Nutraceutical, written by Swati G. Talele e al.

Chapter 2, Formulation and Challenges Involved in Developing Plant/Phytochemical-Based Nutraceuticals, written by Prasenjit Manna and associates, discusses the various formulations, health benefits, and the major challenges associated with plant-based nutraceuticals.

PART II: NUTRACEUTICAL FORMULATIONS AND REGULATIONS

The advance formulation strategies will direct to improve therapeutic safety and fulfill the market appeal of various nutraceutical products. Recent development in formulation of nutraceuticals with different techniques and prudent scientific validation will give direction to nutraceutical success in future which are discussed by Anupam Das Talukdar and colleagues in Chapter 3, Recent Developments in Formulation of Nutraceuticals.

Chapter 4, Future Trends in the Formulation of Nutraceuticals, written by Deepak Bharadwaj PVP and his associates, discusses the futuristic plans of nutraceutical formulations.

Chapter 5, Major Factors Contributing to the Regulations of Nutraceuticals, written by Satyajit Saurabh et al., provides an overview of different categories of nutraceuticals, their health applications, formulations, and factors contributing to the regulation of nutraceuticals.

Sanjay Kumar and his associates outline the regulatory process and regulations for international business developers looking to launch new nutraceutical goods into Asian markets in Chapter 6, Nutraceutical Regulation in Asian Countries.

Chapter 7, Nutraceutical Regulation in European Countries, written by Satyajit Saurabh and associates, discusses nutritional health applications and the addition of vitamins and minerals and other nutrients to the diet as two of the current proposals in the legal system, which will be key factors in the future promotion of "active food" in Europe.

Chapter 8, Regulatory Guidelines for the Development of Plant/Phytochemical-Based Nutraceuticals, written by Garima Sharma, discusses the various guidelines which regulates development of phytochemical-based nutraceutical products.

Chapter 9, In Vitro and In Vivo Studies on Nutraceuticals, written by Jayanthi Abraham et al., reveals the properties of nutraceuticals in preventing diseases.

Mansavi Bhardwaj provides a snapshot of regulatory framework for the development of nutraceuticals in India as well as abroad. This includes all the legal and necessary requirements for the product to be approved as a nutraceutical in Chapter 10, Regulatory Framework for the Development of Nutraceuticals.

PART III: OPPORTUNITIES AND CHALLENGES OF NUTRACEUTICALS

Functional foods and nutraceuticals are whole, fortified, enriched, or improved foods that, when consumed in sufficient quantities as a regular part of a varied diet, provide health benefits in addition to essential nutrients (such as vitamins and minerals). This topic is covered in Chapter 11, Opportunities and Challenge in the Development of Functional Foods and Nutraceuticals by Bhushan R. Rane et al.

The purpose, scope, and significance of using nutraceuticals are covered in Chapter 12 by Ajay Shukla et al., titled Nutraceuticals Manufacturing Challenges in Fast Growing Food Segment in India, which also discusses the difficulties facing the government in enforcing rules and regulations regarding the purity of active ingredients and rationalizing the use of nutraceuticals.

The problems with adding nutraceuticals to meals is covered in Chapter 13, Challenges for Incorporation of Nutraceuticals in Food, written by Preeti Khulbe et al.

PART I

Overview on Nutraceutical Formulation and Challenges

CHAPTER 1

OVERVIEW OF FORMULATION, CHALLENGES, AND REGULATION FOR THE DEVELOPMENT OF NUTRACEUTICAL

AKASH M. BHOITE, UTKARSHA P. PATIL, HARSHADA U. BAGUL, SWATI G. TALELE, and ANIL JADHAV

Department of Pharmaceutics, Sandip Institute of Pharmaceutical Sciences, Nashik, Maharashtra, India

ABSTRACT

Nutraceuticals being food with medicines have gained increasing interest due to their speculated safety. They have created an open platform for products that can assure novel remedies to health-related problems. This chapter focuses on different aspects of the use of nutraceuticals; it may be possible to reduce the chance of several adverse effects. This chapter reviews various types of nutraceuticals and their use in numerous fields, various formulations of nutraceuticals, such as powder, gel, and tablet, and advanced drug delivery systems such as liposomes and niosomes. Along with the properties, it covers the various types of nutraceutical formulations available and their significance, challenges faced during formulations of nutraceuticals, and regulations and authorities controlling them. The importance of nutraceuticals is spreading globally in terms of scientific services, legal aspects, marketing strategies for health promotion, and healthcare costs. The objectives of nutraceutical guidelines have been centered around safety and labeling with lesser importance, as compared with pharmaceuticals, on product claims, and intended use.

Formulations, Regulations, and Challenges of Nutraceuticals.
Tingirikari Jagan Mohan Rao, Rajesh K. Kesharwani, Raj K. Keservani, Anil Kumar Sharma (Eds.)

1.1 INTRODUCTION

Stephen DeFelice, MD, founder and chairman of the Foundation for Innovation in Medicine coined the term "nutraceutical" from "nutrition" and "pharmaceutical" in 1989. He defined a nutraceutical as a "food, or parts of a food, that provide medical or health benefits, including the prevention and treatment of disease." Nutraceuticals are also called medical foods, designer foods, functional foods, phytochemicals, and nutritional supplements, which comprise such everyday products as "bio" yoghurts and fortified breakfast cereals, as well as vitamins, herbal remedies, and even genetically/living modified foods and supplements.[1] Recently, nutraceuticals have drawn in interest attributable to their healthful, safety, and therapeutic effects.

1.2 CLASSIFICATION OF NUTRACEUTICALS[2–6]

1. Traditional nutraceutical
   - ➢ Chemical constituents
     a. Nutrients
     b. Herbals
     c. Phytochemical
     d. Polyunsaturated fatty acids
   - ➢ Probiotics and prebiotics
   - ➢ Nutraceutical enzymes
   - ➢ Functional foods
2. Non-traditional nutraceutical
   - ➢ Fortified nutraceuticals
   - ➢ Recombinant nutraceutical
3. Dietary supplements and fibers

1.2.1 TRADITIONAL NUTRACEUTICALS

Traditional nutraceuticals are food or supplement which are natural components. These nutraceuticals provide welfare beyond basic nutrition, such as lycopene in tomatoes, saponin in soys, calcium from milk. The following are the classification of traditional nutraceuticals.

1.2.1.1 CHEMICAL CONSTITUENTS

1. *Nutrient*
   Concurring to the Association of American Feed Control Officials "AFFCO," the definition could be a nourishing constituent in a shape and at a level that will offer assistance back the life of an animal. These are very accessible nutrients, such as minerals, amino acids, fatty acids, and the most frequently used nutrients are vitamins, minerals, and antioxidants.

   - Vitamins
     Vitamins play a very important part in day-to-day purposes like cell reproduction and growth but play a major role in the processing of energy in cells. Vitamins are grouped into two classes, fat-soluble vitamins and water-soluble vitamins. Fat-soluble vitamins are vitamin A, vitamin D, vitamin K, and vitamin E, while water-soluble vitamins contain vitamin B complex and vitamin C. The large source of vitamins is from plant and plant biotechnology.

TABLE 1.1 Vitamins: Their Source and Health Benefits.

Name of the vitamins	Sources of vitamin	Health benefits
Vitamin A	Dairy products, liver, fish, papaya	Nurture healthy skin and vision
Vitamin D	Salmon, milk, yolk, cod liver oil, cereals, dairy products	Provide proper calcium to the bones and make them strong
Vitamin K	Green leafy vegetables, kiwi, blackberries, figs, pomegranate, tomatoes	Important in blood clotting
Vitamin E	Plants oils, nuts, seeds, wheat germ	Back a safe framework and acts as an antioxidant
Vitamin B1	Peas, fruits (bananas and oranges), nuts, and liver	Essential for inflammation of nerves
Vitamin B2	Dairy milk, cheese, eggs, yoghurt, salmon	Maintain the body's energy supply
Vitamin B3	Liver, eggs, fish, dairy products, nuts, seed	Maintain proper blood fat level, maintain high blood pressure treat type 1 diabetes
Vitamin B5	Tomato, egg, milk, banana	Employ the proper working of the nervous system
Vitamin B6	Banana, tomato, chicken, or turkey	Convert protein into the energy
Vitamin C	Citrus fruits, potatoes, strawberries	For proper growth and repairing body tissues

- Minerals
  Minerals are the compounds that are used for providing the proper amount of elemental metal which possesses a distinctive function of the body and due to their improper presentation, various normal functions of the body get shuffled. Ca, I, Fe, Zn, and Mg are the elemental metals important for human welfare.

TABLE 1.2 Minerals: Their Sources and Health Benefits.

Minerals	Uses	Source
Phosphorus	Formation of bone and teeth	Chicken, organ meats, dairy products, grains, nuts
Iron	Normal cell production and generation of hemoglobin	Fruit, dates
Iodine	Proper thyroid functioning	Fish, dairy products, iodized salt
Calcium	Bone development and to make them strong, nails and teeth development, and releasing hormone	Milk and dairy products, green vegetables
Zinc	Cell growth wound healing	Beans, nuts, seafood, grain, dairy products
Magnesium	Regulates muscle and nerve functions and blood pressure	Nuts, seeds

- Amino acids
  The combination makes proteins called amino acids that carry various body functions. Lysine, histidine, threonine, methionine, valine, isoleucine leucine, phenylalanine, and tryptophan are the essential amino acids required for proper body functions, and due to their deficiency, depression, immunity problems, digestive problems, and fertility issues can occur. Good sources of amino acids include meat, eggs, soy, dairy products, and tofu.

2. *Herbals*

From ancient times herbals are used to cure acute and chronic diseases. These natural remedies in India are written, called Ayurveda. Various nutraceuticals are available in medicinal herbs of a special component. Flaxseed oil, garlic extract, ginger roots, and Aloe vera gel are examples.[40,41]

3. *Phytochemicals*

This word is a composition of two words phyto, means plants and chemicals means the substance which is derived from plant sources to increase the health of patients. These are obtained from plants and contain important components such as lutein and lycopene which can help in proper brain balancing also, they are found in fruits, vegetables, and eggs, for example, eugenol and β-carotene.

4. *Polyunsaturated fatty acids (PUFA)*

For all living organisms, fatty acids have crucial properties, omega-3 and omega-6 are very commonly used PUFA. The most common constituents of fatty acids are oils and fats which exist in animal fats, fish oil, seeds, coconuts, etc. Omega-3 PUFAs have been utilized as accessory therapy for cardiovascular disease, asthma, diabetes, etc. PUFAs are used as nutraceuticals because of their advantageous effects, and they also have low chemical stability due to the presence of leading oxygen.

1.2.1.2 PROBIOTICS AND PREBIOTICS

Havenaar and Huis in 't Veld introduced the definition of probiotics "as a viable mono- or mixed culture of bacteria which, when applied to animal or man, affects the utmost beneficially by improving the properties of the indigenous flora. Probiotics are the microbes that are advantageous to health and utilized in food mostly milk.

For example, *Lactobacillus, Bifidobacterium* spp. and *Streptococcus*; these bacteria are utilized as probiotic nutraceuticals, which provide health by affording immunological and digestive properties. Probiotics are short-chain carbohydrates that ameliorate the activity of fructooligosaccharides, insulin, etc.

1.2.1.3 NUTRACEUTICAL ENZYMES

Enzymes are the essential proteins in the body; the body cannot function without enzymes. These enzymes are obtained from plant, microbial, and animal sources; hypoglycemia, blood sugar disorders, digestive problems, and obesity can be cured by supplementing these nutraceutical enzymes.

For example, catalase, invertase-sucrose biodiastase, pectinase, pepsin, and lysozyme.

1.2.2 *DIETARY SUPPLEMENTS*

In the United States, the Dietary Supplement Health and Education Act (DSHEA) of 1994 defined the dietary supplement. It is a product taken orally that contains dietary ingredients, such as vitamins, minerals, amino acids, herbs, etc. These dietary supplements are only for supporting the body's function and not for disease treatment. All fruits, vegetables, dairy, products, and poultry products are dietary supplements.

1.2.3 *NON-TRADITIONAL NUTRACEUTICAL*

Biotechnology has allowed for the improvement in the quality of nutrition and the addition of artificial nutraceutical. Non-traditional nutraceuticals are hybrid plants or microorganisms which are originated by incorporating two genes to increase the component or to add a new component in them.

1.2.3.1 FORTIFIED NUTRACEUTICALS

To increase the efficacy and nutritional value of food, the addition of micronutrients like essential trace elements and vitamins takes place and these are called fortified nutraceuticals. These are also known as design food. Design food is a food that is fortified with calcium, cereals, folic acids, prebiotics, and probiotic-fortified milk. Design foods are used in the treatment of diarrhea, respiratory infection, etc.

1.2.3.2 RECOMBINANT NUTRACEUTICAL

Nutraceuticals prepared by using techniques such as fermentation and genetic engineering are called recombinant nutraceuticals. In these nutraceuticals, modification of genetic material takes place due to which the changes in the food properties take place. A variety of techniques are utilized such as genetic engineering and fermentation methods for the production of these nutraceuticals, for example, bread and cheese.

1.3 FORMULATION OF NUTRACEUTICALS

Nutraceuticals refer to foods having a medicinal effect on the health of human beings. It consists of food supplements, herbal products, probiotics and prebiotics, and medical foods meant for the prevention and treatment of diseases. Nutraceuticals may play an important role in future therapeutic developments. Nutrients in the form of small solids or droplets improve bioavailability. Nutraceuticals are a group of products that are more than food but less than pharmaceuticals.[35]

1.3.1 TABLET

Tablets with breakmarks were developed to allow tablets to be split in to two or four equal parts to help the ingestion of larger tablets and give consequences against orders. Nutraceutical tablets vary from pharmaceutical tablets only because of the formulas used in vitamin, mineral, and food supplement production. The most common multivitamin tablet forms can accommodate active ingredients and two to eight excipients including coating ingredients. These tablets are manufactured by three processes which include dry granulation, wet granulation, and direct compression. Fast-dissolving nutraceutical tablet contains clove.[7,9] Clove is a nutraceutical that is used as a powerful germicidal, antiseptic, antifungal, skin disorder, anti-inflammatory, and blood purifier, and in the preparation of this tablet, clove powder, lactose, sodium saccharine, cross povidone, magnesium stearate, and talc are used as ingredients.[30]

1.3.2 CAPSULE

These are the oral solid dosage form with hard and soft shells of various structures and capacities containing active ingredients, multi-minerals, probiotics, omega fatty acids, and amino acids these dietary supplements in which capsules are accessible as a dosage form.[8]

1.3.2.1 HARD CAPSULE

Hard capsules are especially applicable for vitamins and minerals also for plant powders and powdered extract. They comprise two prefabricated parts, the body is filled while a cap larger in diameter a fine grove assures that body and cap are tightly sealed. Different-sized capsule shells are made of gelatine and HPMC.

1.3.2.2 SOFT CAPSULE

Soft capsules are single-unit solid dosage forms comprised of liquid or semi-solid fill covered by a one-piece sealed elastic outer shell. Oils, emulsions, or other similar materials made up of paste are especially applicable to soft capsules. Soft gels put forward the prospect of delivering a liquid in solid oral dosage forms, such as the solution, suspension, or emulsion, and conduct the best absorption as compared with tablets or powder. From gelatine or form, a vegetarian source like carrageenan's, the soft capsule can be made chew-able capsule-one of the special types of soft gelatine capsules is a chewable capsule that the capsule shell as well as its content especially developed to be chewed. A combination of most vitamin C compounds and sweeteners assures a taste which is balanced between sweet and sour, for example, a vitamin E capsule is used to treat and prevent the vitamin E deficiencies like anemia, hair loss, muscle weakness, dry skin, leg cramps, cancer blindness, immune impairment, and fertility problems. Beginning from head-to-toe, vitamin E oil is very beneficial to the body such as for nail growth; it is used as overnight cream by mixing vitamin E oil with regular night cream, also used as antiwrinkle cream; it also provides relief from sunburn.

1.3.3 GELS

These are the semi-solid dosage form, the most commonly used gel in the nutraceutical is *Aloe vera* gel. *Aloe vera* constituents are used as a dietary nutraceutical and possess therapeutic potential. This gel is used as skin protective (care gel) largely for dryness of skin, sunburn, moisturizing, wound healing, etc. The gel contains amino acids, minerals, vitamins, enzymes, polysaccharides, and biological stimulators.[10–12]

1.3.4 SYRUP

The syrup is a concentrated solution of sugar in water or other aqueous liquid. The syrup is probably given the macronutrients like minerals and vitamins and phytochemicals mostly phenolics, antioxidants, amino acids, etc.[13] Maple syrup – broadly maple syrup is used to produce food products. The significant carbohydrate found in maple syrup is sucrose, a complex carbohydrate. It also contains minerals (K, Ca, Mg, Na, Al. Zn, Fe, etc.) vitamins, amino acids, and phytohormones.[44]

1.3.5 POWDER

A combination of finely divided active ingredients or other ingredients in dry form is called powder which can be used either internally or can be used for external applications.

1.3.5.1 PROTEIN POWDER

The four main protein sources are wheat, egg, soy, and rice. Protein powders are prepared into powder form which can be reassembled into liquid or shakes. Protein powder is utilized for ameliorating athletic performance; it is used as a food supplement, for replacing or supplementing milk-based infant formulas, in treating protein allergy, asthma, high cholesterol, obesity, weight loss, and for preventing allergies. For example, L-carnitine + co-enzymeQ10 + zinc + lycopene + astaxanthin + folic acid powder. In every cell of the body, coenzyme Q10 is found which acts like a vitamin, which is used by the cells to produce energy for maintaining and growing cells. Astaxanthin focuses on the development of quality and fertilizing the semen, also decreasing oxidative stress. L-carnitine, an amino acid, decreases fatigue, produces energy, and is important for a healthy heart and proper brain functioning. Lycopene provides nutrients that are important for good health.[47]

1.3.6 ESSENTIAL OILS

Essential oils are a combination of volatile and natural substances recognized and specified by a strong odor, produced by aromatic plants also called volatile oils. These are mainly a mixture of terpenes and phenylpropenes. These oils are mainly used as antioxidants and exhibit antimicrobial activity.

1.3.7 STICK

Stick is a new and very modish choice to tablets and capsules in the field of nutraceutical as dietary supplements. Sticks can carry approx. 1–10 g powder or granules depending on the concentration. These powder sticks are consumed according to taste, that is, either directly by dissolving in water by stirring into yoghurt or mixing into many extra dishes.

For example, nutraceutical chewing candy formulations are also developed based on fermented milk permeate, psyllium husk, and apple by-products. For the preparation of this nutraceutical chewing candy, agar, milk, permeate, psyllium husk, apple by-product, ascorbic acid, and xylitol are used. Nutraceutical candy is used as a health supplement. Milk permeate (MP) is a by-product that comes from the dairy industry. MP is used against pathogenic strains. Psyllium husk is obtained from natural hydrocolloids which reduces the possibility of coronary heart disease. By the addition of supplementary natural ingredients such as by-products of apple juice production, the functional use of fermented MP can be ameliorated.[30]

1.4 NOVEL DRUG DELIVERY SYSTEM OF NUTRACEUTICAL

Nowadays there is an advancement in the nutraceutical delivery system. Due to low solubility, poor permeability and fast metabolism, short half-life, and many such problems, most nutraceuticals suffer from improper formulation. The predilection of people to eat healthy food products as well as nutraceuticals prevent many diseases. Compelled scientists and researchers to look for an efficient delivery system.[14]

1.4.1 LIPOSOMES

Liposomes are spherical vesicles composed of phospholipids comprising lipid bilayers. They are formulated using cholesterol and a natural phospholipid called liposomes. The application of liposome vesicles on the skin has many advantages over conventional dosage forms. Liposomes are used in the preparation of gels, creams, lotions, etc. to treat skin disorders.

Buccal liposomal formulation using silymarin reveals the hepatoprotective action of silymarin with enhanced bioavailability and the best pharmacological response.[15]

For example, In oral supplementation with liposomal glutathione (GSH), glutathione is the antioxidant condemnatory regulator of oxidative stress. It is very important to maintain proper glutathione levels if it gets decreased; it can increase the risk of diseases like cancer, cardiovascular disorder, arthritis, and diabetes. By administering liposomal GSH, the level of GSH in the body can be maintained. Enhanced liposomal formulation of glutathione protects the drug against oxidation and degradation. This liposomal formulation of GSH also maintains immune function.[16,17]

1.4.2 NIOSOMES

Niosomes are nonionic surfactant-based multi- or unilamellar vesicles in which an aqueous solution of solute is enclosed by a membrane from surfactant macromolecules as a bilayer. Both lipophilic as well as hydrophilic nutraceuticals are protected in vivo by niosome vehicles. Novel encapsulation of lycopene in niosome and the evaluation of its anticancer activity have also been performed.

The red-colored carotenoid pigment present in ripe tomatoes is lycopene. It has gained appreciable focus for its health benefits. By reducing DNA damage and by lowering prostate-specific antigens, lycopene prevents cancer. For the extraction of lycopene, span 60, cholesterol, chloroform, n-hexane, ethanol, and acetone are used. Lycopene is a molecule that is more receptive to physical and chemical factors. These are utilized as dietary antioxidants with free-radical scavenging activity and play a role in human health. Lycopene is the care material, and it is enclosed by a surfactant membrane.[18]

1.4.3 NANOEMULSION

Nanoemulsion is the nano-sized emulsions in which two immiscible liquids are mixed to form a single phase with the help of an emulsifying agent. The size of the nanoemulsion droplet is about 100 nm. Nanoemulsion mediates the bioactives to the target site for a longer period and maintains blood plasma concentration, that is, bioavailability. Nanoemulsion is prepared by using olive oil, corn oil, flan seed oil, etc. For example, nanoemulsion gel of nutraceutical co-enzyme Q10.

Co-enzyme Q10 is poorly soluble in water, so, to overcome this problem, nanoemulsion gel of co-enzyme Q10 is formulated to improve and ameliorate its solubility and skin permeability. This nanoemulsion is prepared by using isopropyl myristate as oil, tween 80, transductal HP, surfactant, water, Carbopol as gelling agents and drugs. In this formulation, the low water-soluble and low permeable drug is added in an aqueous liquid of gelling agent liquid. Q10 acts as an antioxidant including in antiaging of the skin, skin antiwrinkle, etc. The nanoemulsion gel is a topical gel in which properly weighed co-enzyme Q10 is mixed in proper quantities of oil, surfactant, and co-surfactant to form the emulgel.[19]

1.4.4 PHYTOSOMES

These are the combination of phospholipid and biologically active ingredients. The phytosomal oral preparation of curcumin by using curcumin and phospholipid combination also can increase the bioavailability by elevating the antioxidant activity. Also, hawthorn (flavonoid) an oral phytosomal preparation can be used as a cardioprotective and antihypertensive.

1.4.5 NANOPARTICLE

Nanoparticle has three nanoscale dimensions and which range approximately from 1 to 100 nm. Nanoparticles can be made up of materials such as polymers, proteins that have biodegradability, bioavailability, and the possibility of encapsulating hydrophobic compounds, so researchers captivate to use this in the design of nanoparticle curcumin which is a hydrophobic polyphenol with antioxidant and anti-inflammatory activities. *Resveratrol* is a nutraceutical that shows anticancer activity against MCF-7, when encapsulated into a nanoparticle, it further reveals the development of nanoparticle,[42] for example, a curcumin nanoparticle.

Curcumin is a composite obtained from the plant *Curcuma longa* species utilized as an antioxidant, anti-inflammatory, and antitumor activity for the proper delivery of curcumin to specific sites the phospholipid vesicle or lipid nanospheres are formulated for proper delivery of curcumin by IV. Curcumin is hydrophobic, that is, it is solubilized in a hydrophobic solvent to form nanoparticle dispersion and this preparation has a great therapeutic effect as an antioxidant and shows anti-inflammatory activity.[20]

1.4.6 NANOSPHERES AND NANOCAPSULES

As the name indicates, both nanosphere and nanocapsules have a diameter of less than 1 μm. Nanocapsules have a polymeric membrane in which active material gets encapsulated. Nanosphere has a polymeric matrix in which active ingredients get dispersed. β-Carotene which is an antioxidant that has been encapsulated and has been used in slow-release formulations.

Solid–lipid nanoparticles (SLN) are lipid-based nanocarrier systems extensively applied in the nutraceutical field. Solid–lipid nanoparticles are a mixture of solid–lipids, and surfactants SLN can carry mainly lipophilic nutraceuticals like a fat-soluble vitamin (vitamin A). Turmeric extract is also encapsulated into SLN which has anticancer, antioxidant, and antimicrobial activities.[39] For example, nanoscale delivery of resveratrol.

Resveratrol is known for its stability, biocompatibility, and intestinal permeability over $CaCO_2$ cell monolayer in its free form or get encapsulated into SLN particles. This preparation shields resveratrol from photodegradation. This preparation is utilized as a nutraceutical as an oral administration for resveratrol supplementation.[14]

1.5 FORMULATIONCHALLENGES IN NUTRACEUTICALS[8,9,21]

1.5.1 SOLID DOSAGE FORM

1.5.1.1 STABILITY

Stability means the capacity of dosage form in formulated drugs which does not degrade and demeanor dosage form, the dosage form remains in the confirmed specification and ensures identity, quality, and purity in a self-period. Solid dosage form contains tablets and capsules that have challenges related to their ingredient interactions and stability. In nutraceutical tablets or solids, the dosage form contains active ingredients. About 70–90% of complexes contain active compounds and it is a challenge to make the formulation of solid dosage form by considering its stability. Vitamin E and β-carotene are nutraceutical antioxidants that can contribute hydrogen atoms or electrons to form stable radical intermediates and can prevent the oxidation of many biological components and make the formulation stable. Due to the presence of moisture contained in tablets or capsules, microbial growth takes place. Hence, the stability of the dosage form is disturbed because of this microbial contamination. The therapeutic activity of the nutraceutical formulation gets decreased and these are the stability-related challenges to the formulator for tablets and capsules.[48,49]

1.5.1.2 SOLUBILITY, DISSOLUTION, AND DISINTEGRATION

Solubility is another major challenge in the formulation of the solid dosage form. Some nutraceuticals, such as carotenoids, some vitamins, phytosterols, lutein, etc. are not soluble in water, hence, this is another provocation to the formulator. Other challenges are proper disintegration and dissolution of nutraceutical tablets or capsules. For example, there are many ingredients in nutraceutical tablets, when compared with pharmaceutical tablets, nutraceutical tablets required more compression force to form harder tablets, which has difficulty related to disintegration and dissolution. Due to these difficulties, tablets cannot provide any therapeutic effect, for example, multivitamin tablets.[8,9]

1.5.1.3 NATURE OF NUTRACEUTICAL INGREDIENTS

The nature of nutraceutical ingredients conducts provocation related to uniformity, moisture, sensitivity, compressibility, particle size, flowability, ingredient interaction, quality control, etc., for example, nutraceutical ingredients show different natures, some are hydrophobic and some are hydrophilic. Some are in granular form while some are in fine powder form. So, it is difficult to blend the ingredients. Ingredients are particles of many different sizes. Particles have different properties. Due to the flow problems, problems with capping, sticking, mottling, and consistency of substance may result from segregation. For example, multivitamin tablets contain a variety of particle sizes and minerals with different sizes can cause this problem.

1.5.1.4 DESIGN OF DOSAGE FORM

Another important challenge of the solid dosage form is the careful design, that is, its shape, size, color, texture, and consideration are also very important in the formulation of the solid dosage form. The instability of many ingredients present in the nutraceutical tablets, which are concentrated in a very little place in a large number of excipients presents a new challenge in formulating nutraceutical solid dosage forms. The lesser excipients and discrepancy of activity in some formulations make it difficult to achieve formulation outcomes, such as disintegration time, hardness, and friability. Cautious design of solid dosage form by utilizing colors such as brown and gray by giving the appearance and texture of gummies and lozenges which can make it difficult.[43]

1.5.1.5 ADDITION OF NATURAL INGREDIENTS

It is very unrefined, abrasive, and hard to add a new ingredient in a dosage form that is subjected to damage, and it is challenging to formulate by adding a new ingredient.

1.5.2 LIQUID DOSAGE FORM

1.5.2.1 STABILITY

Very most important factors considered in the stability of liquid dosage form are pH, aqueous and organic solubility and compatibility, ingredients compatibility, etc. For proper safety, efficiency, and therapeutic activity of dosage form, stability plays a very important role for the formulator to make a stable form of the liquid dosage form and is a key challenge. Due to the presence of an aqueous vehicle, the increase in microbial contamination is a risk factor, and considering this aspect, the formulator has the challenge to formulate the liquid dosage form.

For example, the oral delivery of probiotics is embarrassed by the low instability of bacteria in GIT and subsequent loss of viability under the effect of high acidity and bile salt concentration.

1.5.2.2 SOLUBILITY

Many nutraceutical ingredients are insoluble in the aqueous vehicle due to improper mixing of active ingredients into the vehicle to face the efficiency, stability, and pharmacological issues. So, the solubility of nutraceuticals is another challenge faced by the formulator. Examples include carotenoids, water-insoluble vitamins.

1.5.2.3 BIOAVAILABILITY OF INGREDIENTS

The bioavailability of the ingredients at the site of action is very important to exhibit an essential therapeutic effect. Good bioavailability and permeability of these ingredients from GIT are necessary. So, formulating the products by taking into consideration of bioavailability and permeability is a great challenge. For example, the bioavailability of lipophilic antioxidant coenzyme Q10 is challenged due to its (1) low aqueous solubility and dissolution rate and (2) due to its large molecular weight, permeability is limited.

1.5.2.4 INTERACTION

It is possible that the inactive ingredients may interact with each other. Due to this instability getting shuffled, the risk of change in the structure of active compounds is also possible. Also, the interaction with gastrointestinal fluid can decrease the pharmacological activity of the drug.

1.5.3 NOVEL DRUG DELIVERY SYSTEM

1.5.3.1 ENCAPSULATION

Capsulation of active components (i.e., nutraceutical) within another inactive ingredient is in the form of a matrix or wall material. But the encapsulation process is another challenging one because it is very difficult to entrap the active ingredient within the matrix or wall material. It is very strenuous to incorporate the active material. Encapsulation of curcumin which is a commonly used dietary supplement within nanoparticles by considering its stability and solubility is an example.[45]

1.5.3.1.1 *Stability and Solubility*

The stability of nutraceuticals is also a substantial thing that is challenging in formulations, for example, curcumin is unstable in all aqueous solutions.

1.6 REGULATIONS OF NUTRACEUTICAL

The goal of nutraceutical regulation is focused on safety and labeling with lesser importance as compared with pharmaceuticals, product claims, and intended use. This is obtained through Good Manufacturing Practice (GMP) regulations and a new expansion in implementation. The nutraceutical guidelines present an excellent test for the globalization of nutraceuticals.[36]

1.6.1 REGULATORY AGENCIES FOR NUTRACEUTICAL

1.6.1.1 JAPAN

After the USA, Japan is in second place in the nutraceutical market. Dietary supplements and natural nutraceuticals from Japan are chosen as "food with health claims," and under this, there are three distinctive asserts: (1) food with nutrient function claims (FNFC) which mainly consists of 12 vitamins and 5 minerals, (2) food for specified health used (FOSHU), and (3) food with function claims (FFC).

The main regulatory body for drug and health supplements is "The Ministry of Health, Labor and Welfare" (MHLW). It continues to regulate r medicines and dietary supplements.

1.6.1.1.1 *Food for Specified Health Use (FOSHU)*

Japan became the first country by issuing FOSHU to modulate food supplements. It was introduced as a regulatory system in 1991.

1.6.1.1.2 *Food with Function Claims (FFC)*

It was brought in 2015 to classify dietary supplements for health promotion and disease prevention apart from FOSHU. For approval purposes, FFC products should be submitted to the secretary-general of the "consumes affairs agency" (CAA).

The other regulatory authorities involved in the regulations of neutraceutical and food products are Food Safety Commission, Pharmaceutical Affairs and Food Sanitation Council, Japan Health Food Association (JHFA), and Japan Health Food and Nutrition Food Association (JHNFA).

1.6.1.2 CHINA

It is the third largest nutraceutical market in the world after the US and Japan. Previously, the Ministry of Health (MOH) controlled health food registration approvals in June 2003, the State Food and Drug Administration (SFDA) of China took this responsibility from the MOH. Chinese regulation depends more on traditional Chinese medicines (TCM) also called traditional Chinese herbal medicines. China exports about 65% of rare materials for food product manufacturing. The SFDA is the controlling head of dietary supplements and issues registrations, while the MOH is responsible for new compliance of novel food constituents. The export and import are controlled by the Authority Administration of Quality Supervision Inspection and Quarantine (AQSIQ). It is also responsible for the national quality entry–exit health quarantine, import–export food safety.[22,26]

In China, other bodies associated with the regulation of nutraceuticals are the China Health Care Association (CHCA) and the US–China Health Products Association (USCHPA).

1.6.1.3 INDIA[22,23,27]

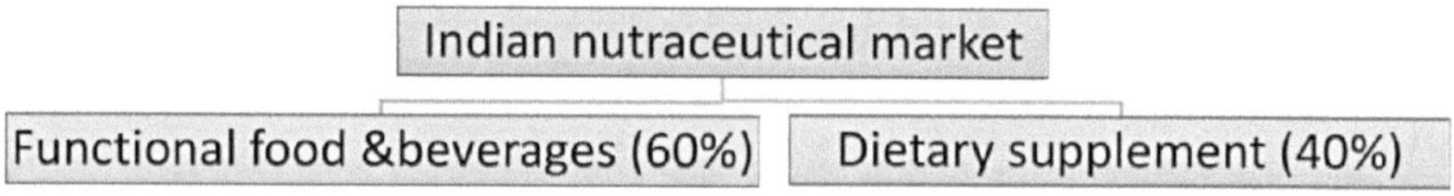

The Indian nutraceutical market is sectioned into two core groups:

1. Functional food and beverages
   Fortified food, fortified juices, energy drinks, sports drinks
2. Dietary supplements
   This subdivision includes vitamins, minerals, herbal, protein supplements, and Chyawanprash

   - The Food Safety Standard and Standards Act (FSSA) 2006
     FSSA regulates the manufacturing of food products and other nutraceuticals, storage of food products and distribution, sale, and import of food products. Its rules and regulations came in 2011 according to Sec. 22 (1) of FSSA. Nutraceuticals are food items that are processed especially so that they add dietary requirements to the composition of food. They can contain plants/botanicals (in the form of powders, concentrate, minerals/vitamins/proteins.
   - The Food Safety Standards Authority India (FSSAI) authority approves the registration of food products.
   - The other regulatory body in India, that is, DSHEA. It is the governing body that cliques the rules and regulations for nutraceutical products in the market. It was approved in 1994.
   - Indian Pharmacopoeia
     It decides the standards for the safety and quality of nutraceutical ingredients. For example, plant extracts and phytochemicals.
   - Federation of Indian Chambers and Commerce of Industry (FICCI)

It checks and regulates the validation of products and claims which can meet consumers' demands. The department under FDA Centre for Food Safety and Applied Nutrition (FSAN) is responsible for promising and protecting the public's health by confirming that nation's food supply is safe and wholesome.

- National Institute of Nutrition (NIN)-It mainly focuses on studies related to protein-energy malnutrition, procedures of organization, and deterrence of nutritional difficulties.

  It studies the drug nutrition interactions, such as drug metabolism, toxicity, and identity. Naturally occurring food ingredients which are rich in antioxidants are associated with the regulatory body and come under Food and Drug Toxicology Research Centre (FDTRC). The other Indian regulatory bodies are the Health Food and Dietary Supplements Association (HADSA), the National Nutrition Monitoring Bureau (NNMB), the Indian Health Food and Dietary Supplement (INHDSA), the Indian Council of Medical Research Association (ICMR).

1.6.1.4 UNITED STATES

In the United States, nutraceuticals are termed dietary supplements. The US Dietary Supplements Health and Education Act (DSHEA). It is regulatory in it for a dietary supplement as food through the Food and Drug Administration in 1994. Dietary supplements are a product that encompasses one or additional dietary constituents,[23,25,28]

1. Vitamins
2. Minerals
3. Herbs/Botanicals
4. Amino acids
5. Metabolites, extracts

1.6.1.4.1 US FDA Modernization Act 1997

This Act regulates the supplements data which is to be marketed according to health claims and nutrition content claims.

Other regulatory bodies are the Food and Drug Administration (FDA) and the US Department of Agriculture Federal Trade Commission.

All the regulatory bodies evaluate, investigate, regulate, and inspect the nutraceutical materials as ingredients to be used in dietary supplements.

1.6.1.5 RUSSIA

Russian Framework includes two types of documents for food as dietary use and functional foods.[23,24,27]

1. Russian Federation food security.
2. Russian Federation Policy Framework on healthy public nutrition.

The Ministry for Health and Social Development maintains the nutraceutical under the title of biologically active dietary supplement (BADS)

1.6.1.6 CANADA

- The Food and Drug Authority (FDA) of the state set regulations for the maintenance of dietary supplements. Food and Drugs Act came into effect on January 1, 2004.[22]
- Natural Health Product Regulations which sets up the requirements for efficient and quality reviews to provide natural product numbers (NPN).
- Canadian Food Inspection Agency: It regulates the inspection and analysis of products regulated under the Food and Drug Regulations (FDRs).
- Nutraceuticals are considered "natural health products" (NHP's) and are under the authority of the "Natural and non-prescription health products directorate"(NNHPD)
- The Marketed and Health Products Directorate (MHPD): Works toward safety trends and are concerned about all the marketed health products regulations.

1.6.1.7 BRAZIL

Dietary supplements in Brazil include vitamins, minerals, bioactive substances, novel food, and probiotics.[28,23]

The following are the institutes involved in the registration process:

1. Ministry of Health
2. Brazilian Health Surveillance Agency (ANVISA): It coordinates, supervises, and controls the activities which are related to registration, information, inspection, and standard establishments.

3. General Food Management: It is responsible for supplementary assistance to the state and the Municipal Health Surveillance Agencies are involved in the development and implementation of health and foods.
4. National Policy of Integrative and Complementary Practice (PNPIC) in the Unified Health System (SUS): Mainly involved in the research and the use of medicinal plants and herbal medicines according to the quality, safety, and efficacy norms.

Other regulatory committees include the Brazilian Association of Foods for Special Purposes and Congeners (ABIAD) and Committee for the Scientific and Technical Assessment of Functional and New Foods (CTCAF). It is responsible for scientific advice regarding supplements, Ministerio da Agricultura, Pecuariae Abastecimento (MAPA).

1.6.1.8 UK (UNITED KINGDOM)

1. Food Standard Agency (FSA):
   - It is the non-ministerial government department of the government of the UK.
   - It is responsible for the protection of public health concerning foods.
   - The authority Medicines and Healthcare Products Regulatory Agency (MHRA) ensures and regulates guidelines for safe level intake of vitamins and minerals in nutraceuticals.

1.6.1.9 EUROPEAN UNION (EU)[25,27]

- The European Food and Safety Authority (EFSA) which regulates Food supplements under these are defined as "concentrated sources of nutrients and other substances which prove to be beneficial in showing nutritional effect."
- In January 1997, Regulations [EC] no. 258/97 on novel food and novel food ingredients was accepted and passed into effect on May 15, 1997.
- In Europe, dietary supplements can be categorized as follows:

1. Supplements containing active ingredients (vitamins and minerals).
2. Supplements containing water- and fat-soluble vitamins, minerals, and other non-botanical actives (e.g., glucosamine).
3. Supplements containing herbal ingredients and non-herbal ingredients.
4. Supplements containing only herbals/botanicals.

1.6.1.10 AUSTRALIA[22,23]

- Nutraceuticals that fall under the broad category are called "THERAPEUTIC GOODS"
- The therapeutic goods regulation comes under the Therapeutic Goods Act 1989.
- The nutraceuticals/medicines are categorized into two subunits:

  1. Low-risk medicines: are enumerated on the Australian Register of Therapeutic Goods (ARTG).
  2. High-risk medicines: These must be registered on ARTG.

- The Department of Health and Aging is another authority involved in the process of nutraceutical supplements.

1.7 CONCLUSIONS

Nutraceuticals have all the essential nutrition required in a healthy diet for the human body. From the above study, it is clear that the demand and use of nutraceuticals is uprising. But along with it, there come significant problems too. For the treatment of several diseases, there is no clear clinical evidence regarding the use of nutraceuticals as a remedy. Although scientific researches and clinical trials continue to add more to the nutraceutical industry. Several formulation methods involved do not have appropriate results including dosage forms like tablets. The regulatory bodies controlling the regulations and guidelines for the development, production, sales, manufacture, distribution, and import of dietary supplements are still in the way of evolving. Regardless of these factors, nutraceuticals have become a novel boost in the field of pharmaceuticals, and are constantly illuminating new approaches to make nutrition more handy and accessible.

KEYWORDS

- **functional foods**
- **phytochemicals**
- **herbals**
- **vitamins and regulation**

REFERENCES

1. Jha, S. K.; Roy, P.; Chakrabarty, S. Nutraceuticals with Pharmaceutics: Its Importance and Their Applications. *Int. J. Drug Dev. Res.* **2021,** *13*, 1–10.
2. Rajam, R. P.; Murugeasan, G.; Vadivelkumar, K.; Menachisundaram, V. Nutraceuticals—A Review. *World J. Pharma. Res. 8* (11), 1354–1374.
3. Chaudhari, S. P.; Powar, P. V.; Pratapwar, M. N. Nutraceuticals: A Review. *World J. Pharm. Pharma. Sci. 6* (8), 681–739.
4. Chauhan, B.; Kumar, G.; Kalam, N.; Ansari, S. H. Current Concepts and Prospects of Herbal Nutraceutical: A Review. *J. Adv. Pharma. Technol. Res.* **2013,** *4*, 4–8.
5. AlAli, M.; Alqubaisy, M.; Aljaafri,M. N.; AlAli, A. O.; Baquais, L.; Molouki, A.; Abushelaibi,A.; Lai, K-S.; Lim, S. H. E. Nutraceutical: Transformation of Conventional Foods into Health Promotors/Diseases Prevents and Safety Considerations, 1–28.
6. Singh, J.; Sinha, S. Classification, Regulatory Acts and Applications of Nutraceuticals for Health. *Int. J. Pharm. Biol. Sci.* **2012,** *2*, 177–187.
7. Nagaich, U.; Pal, A. K.; Bharti, C.; Gulati, N. Formulation and Evaluation of Nutraceutical Tablet Using Herbal Drugs Delivery and Therapeutics **2014,** *4* (2), 47–51.
8. Singh, S.; Desai, I.; Noronha, S. Tablet & Capsules—The Go to Dietary Supplement Dosage Form, Oct 2018; pp 34–66.
9. Blanchard, R.; Higgins, T.; Holland, I. Overcoming Manufacturing Challenges of Nutraceutical Tablets.
10. Khan, A. W.; Kotta, S.; Ansari, S. H.; Sharma, R. K.; Kumar, A.; Ali, A. Formulation Development, Optimization and Evaluation of Aloe Vera Gel for Wound Healing **2013,** *9*, 36.
11. Upadhyay, R. K. Nutraceutical, Therapeutic and Pharmacological Potential of Aloe Vera: A Review. Jan–Mar, **2018,** *12* (1).
12. Tamasi, G.; Pardini, A.; Bonechi, C.; Donati, A.; Pessina, F.; Marcolono, P.; Gamberucci, A.; Leone, G.; Consumi, M.; Manani, A.; Rossi, C. Characterization of Nutraceutical Components in Tomato Pulp, Skin and Ocular Gel, 2019.
13. Zhang, Y.; Yuan, T.; Li, L.; Nahar, P.; Slitt, A.; Seeram, N. P. Chemical Compositional and Biological and Safety Studies of a Novel Maple Syrup Derived Extract for Nutraceutical Application. *J. Agric. Food Chem.* **2014,** 6687–6698.

14. Helal, N. A.; Eassa, H. A.; Amer, A. M.; Eltokhy, M. A.; Edafigha, I.; Nounou, M. I. Nutraceuticals Novel Formulations: The Good, the Bad, the Unknown and Patents Involved. Jun **2019,** *13* (2), 105–156.
15. Khan, M. G. The Novel Drug Delivery System. *World J. Pharm. Pharma. Sci. 6* (7), 477–487.
16. Maghimipour, E.; Aghel, N.; Mahmoudabadi, A. Z.; Ramezani, Z.; Handali, S. Preparation and Characterization of Liposomes Containing Essential Oil of *Eucalyptus cumaldulensis* Leaf. *Jundishpur J. Nat. Pharma Prod.* **2012,** *7* (3), 117–122.
17. Sinha, R.; Sinha, I.; Calcagnotto, A.; Trushin, N.; Haley, J. S. Schell, T.D.; Richie, J. P., Jr. Oral Supplementation with Liposomal Glutathione Elevates Body Stores of Glutathione and Markes of Immune Functions.
18. Sharma, P. K.; Saxena, P.; Jaswant, A.; Chalamainah, M.; Tekade, K. R.; Balasubramaniam A. Novel Encapsulation of Lycopene in CS and Assessment of Its Anti-Cancer Activity. *J. Bioequivalence Bioavail. 8* (5), 224–232.
19. El-Leithy, E. S.; Makky, A. M.; Khattab, A. M.; Hussein, D. G. Nanoemulsionel of Nutraceutical Co-enzyme Q10 as an Alternative to the Conventional Topical Delivery System to Enhance Skin Permeability and Anti-Wrinkle Efficiency.
20. Vergallo, C. Nutraceutical Vegetable Oil Nanoformulations for Prevention and Management of Disease. *Nanomaterials* **2020.**
21. Radheshyam, A. Review on Nutraceuticals Challenges in Formulation and Its Regulatory Aspects. *J. Pharma. Res.*.
22. Sharma, A.; Kumar, P.; Sharma, P.; Shrivastav, B. A Comparative Study of Regulatory Registration Procedure of Nutraceuticals in India, Canada and Australia. *Int. J. Pharma. Qual. Assurance* **2013,** *4* (04), 61–66.
23. Chaudhari, S. P.; Powar, P. V.; Pratapwar, M. N. Nutraceuticals: A Review. World J. Pharma. Pharma. Sci. *6* (8), 681–739.
24. Tutelyan, V. A.; Petrenko, A. S.; Smirnova, E. A.; Sukhanov, B. P.; Kutyshenko, A. V. Chapter 68—Regulatory Aspects of Nutraceuticals: Russian Perspect. In *Nutraceuticals*; 2016; pp 959–970.
25. Tedesco, D. E. A. Regulatory Guidelines for Nutraceuticals in the European Union. *Nutraceuticals in Veterinary Medicine*; pp 793–805.
26. Yang, M. S. Chapter 65—Regulatory Aspects of Nutraceutical: Chinese Perspective. In *Nutraceutical Safety, Efficacy, Toxicity*; 2016; pp 947–957.
27. FSA: Wikipedia, website= www.food.gov.uk
28. de Figueiredo Toledo, M. C. Lajolo, F. M. Supplements and Functional Foods Legislation in Brazil. In *Nutraceutical and Functional Food Regulation in the United States and Around the World Food Science and Technology*; 2008; pp 349–364.
29. Shashank, B. Capacity Building in the Indian Food Industry: Opportunities and Challenges. In: *Proceedings of the Indian Symposium, CFTRI Mysore (India)*, June 23–25, 2006.
30. Zokaityte, E.; Siriakovite, K.; Starkute, V.; Zavistanaviciute, P.; Lele, V.; Mozuriene, E.; Klusaite, D.; Viskelis, P.; Ruibys, R.; Guinea, R. P. F.; Burtkiene, E. Characteristics of Nutraceuticals Chewing Candy Formulations Based on Fermented Milk Permeate, Psyllium Husk and Apple By-products, Foods **2021,** *10* (4), 777.
31. Tiwari, R. K.; Singh, L.; Sharma, V.; Singh, P. Formulation, Development of Fast Dissolving Tablet of Clove—The Best Nutraceutical Analgesic Tablet. *Asian Food Sci. J.* **2018,** *1* (3), 1–7.

32. Pounikar, Y.; Jain, P.; Khurana, N.; Omray, I. K.; Patil, S. Formulation and Characterization of Aloe Vera Cosmetic Herbal Hydrogel. *Int. J. Pharm. Pharma. Sci.* **2012,** *4* (4), 85–86.
33. Amol, K.; Pratibha, P. Novel Drug Delivery System in Herbals. *Int. J. Pharm. Chem-Bio. Sci.* **2014,** *4* (4), 910–930.
34. Dillard, C. J.; German, J. B. Phytochemicals: Nutraceuticals and Human Health. *J. Sci. Food Agric.* **2000,** 1744–1756.
35. Chauhan, B.; Kumar, G.; Kalam, N.; Ansari, S. H. Current Concepts and Prospects of Herbal Nutraceutical: A Review. *J. Adv. Pharm. Technol. Res.* **2013,** *4*, 4–8 [Pubmed].
36. Palthur, M. P.; Palthur, S. S. S.; Chitta, S. K. Nutraceuticals; Concepts and Regulatory Scenario. *Int. J. Pharm. Pharm. Sci.* **2010,** *2*, 14–20.
37. https://www.fda.gov/for consumers/consumerupdates/ucm050803.htm
38. Khar, R. K.; Vyas, S. P.; Ahmed, F. J.; Jain, G. K. *The Theory and Practise of Industrial Pharmacy*, 4th ed.; Lachman's/Liberman's CBS Distributors, 2013; pp 872, 902, 905, 943.
39. Feng, J.; Wu, S.; Wang, H.; Liu, S. Improved Bioavailability of Curcumin in Ovalbumin-Dextran Nano Gel Prepared by Maillard Reaction. *J. Funct. Foods* **2016,** *27*, 55–68 [crosslef].
40. Grinlog, D.; Reynolds, T. The Aloe Vera Phenomenon: A Review of the Properties and Modern Uses of Parenchyma Gel. *J. Ethnopharmacol.* **1986,** *16*, 117–151.
41. Haddad, P.; Amouzgar-Hashemi, F.; Samsami, S.; Chinichain, S.; Oghabian M. A. Aloe Vera for Prevention of Radiation-Induced Dermatitis a Self-Controlled Clinical Trial. *Curr. Oncol.* **2013,** *20*, e345–e348.
42. Acosta, E. Bioavailability of Nanoparticles in Nutrient and Nutraceutical Delivery. *Curr. Opin. Colloid Interface Sci.* **2009,** *14*, 2–15.
43. Lalancette, K. G.; Pichette, A.; Legault, J. Sensitive cell-Based Assay Using DCFH Oxidation for the Determination of Pro and Antioxidant Properties of Compound and Mixtures; Analysis of Fruit and Vegetable Juice. *Foodchem* **2008,** *115*, 720–726.
44. Perkins, T. D.; Van den Berg, A. K. Maple Syrup- Production, Composition, Chemistry and Secondary Characteristics. *Adv. Food Nutr. Res.* **2009,** *56*, 101–143.
45. Nasri, H.; Baradaran, A; Shirzad, H.; Rafieian Kopaei, M. New Concepts in Nutraceuticals as an Alternative for Pharmaceuticals. *Int. J. Prev. Med.* **2014,** *5*, 1487.
46. Essien, J. P.; Akpan, E. J. Antifungal Activity of Ethanolic Leaf Extract of *Eucalyptus camaldulensis* Dehn., Against Ringworm Pathogens. *Global J. Pure Appl. Sci.* **2014,** *10*, 37–41.
47. Clinton, S. K. Lycopene: Chemistry, Biology and Implications for Human Health and Diseases. *Nutr. Rev.* **1998,** *56*, 35.
48. Augustin, M. A.; Sanguansri, L. Challenges and Solution to Incorporation of Nutraceuticals. *Foods*, **2015,** *6*, 463–477.
49. Kesarvani, R. K.; Kesharwani, R. K.; Sharma, A. K.; Gautam, S. P.; Verma, S. K. Nutraceutical Formulations and Challenges. *Dev. New Funct. Food Nutraceutical Prod.*, 167–177

CHAPTER 2

FORMULATION AND CHALLENGES INVOLVED IN DEVELOPING PLANT/PHYTOCHEMICAL-BASED NUTRACEUTICALS

JIJNASA BORDOLOI[1], JATIN KALITA[3,4], and PRASENJIT MANNA[2,4]

[1]*MYAS-NIN Department of Sports Science, ICMR-National Institute of Nutrition, Hyderabad, Telangana, India*

[2]*Biological Sciences and Technology Division, CSIR-North East Institute of Science and Technology, Jorhat, Assam, India*

[3]*Research Planning and Business Development Division, CSIR-North East Institute of Science and Technology, Jorhat, Assam, India*

[4]*Center for Infectious Diseases, CSIR-North East Institute of Science and Technology, Jorhat, Assam, India*

ABSTRACT

Plant-based nutraceuticals are natural bioactive products with therapeutic as well as nutritional values that hold immense health benefits against various ailments. These substances might encompass natural products, medicinal herbs, dietary supplements, and fortified foods. These nutraceuticals are based on varieties of phytonutrients, such as vitamins, minerals, indigestible fiber, essential fatty acids, essential amino acids, and phytochemicals, such as terpenoids, phenolic compounds, and nitrogen-containing compounds. Various scientific studies have claimed that phytonutrients, whether used alone or in combination, play significant roles in human health against

Formulations, Regulations, and Challenges of Nutraceuticals.
Tingirikari Jagan Mohan Rao, Rajesh K. Kesharwani, Raj K. Keservani, Anil Kumar Sharma (Eds.)

different medical conditions, such as pain, inflammation, allergy, bacterial and fungal infections, constipation, aging, diabetes, cancer, dyslipidemia, blood pressure, and heart disease. The use of plants for nutraceutical purposes is rapidly increasing due to their tremendous therapeutic benefits and the growing inclination of the public toward a vegan diet. However, there is an emerging public concern regarding the safety and efficacy of these products which primarily includes misidentification of the accurate source of raw materials, impurity of the products, presence of other active compounds, lack of proper experimental evidence, limited knowledge about the permissible limit, false advertising, mixing with illegal drugs to improve efficacy, and contamination with various substances. This chapter discusses the various formulations, health benefits, and the major challenges associated with plant-based nutraceuticals.

2.1 INTRODUCTION

The lifestyle patterns of people all over the world are changing drastically over the last few decades with the rapid development of science and technology. The major lifestyle changes were observed in terms of substandard dietary habits and absence of physical activity that resulted in the incidences of a variety of lifestyle disorders including cardiovascular diseases, type 2 diabetes, obesity, hepatic dysfunction, hypertension, etc. The increased demands for healthcare have drastically raised the cost of the amenities associated with disease diagnosis and treatment. Hence, natural food products with pharmacological properties are gaining tremendous attention in recent times as an advantageous alternative to the modern healthcare system to improve the health status of people worldwide.

Nutraceuticals have been shown to hold enormous potential in the prevention and management of the aforementioned disorders along with other ailments, such as allergies, infections, pain, inflammation, and aging[1] Nutraceuticals are natural food products with nutritional as well as pharmaceutical properties, which are distinct from synthetic pharmaceutical drugs. They have advantages over their modern synthetic counterparts in easy availability, affordability, and absence of side effects. Besides, they can be easily included in the diet to maintain a healthy way of living thus preventing the disease onset.

The term nutraceutical was coined by Stephen De Felice who defined nutraceutical as any food or a part of food that provides medical or health benefits, including the prevention and treatment of disease.[2] Plant-based nutraceuticals are food products obtained from plant sources that are rich sources of essential phytonutrients, such as vitamins, minerals, fibers, polyunsaturated fatty acids (PUFA), amino acids, terpenoids, phenolics, and nitrogen-containing compounds.[1,3] Plant-based natural food products, such as green tea, garlic, ginger, oats, avocado, spinach, lettuce, sweet potato, apple, almonds, walnuts, dates, broccoli, flax seeds, and chia seeds and medicinal herbs such as tulsi, neem, cinnamon, coriander, turmeric, fenugreek, brahmi, lemongrass, amla, clove, ashwagandha, giloy, and ajwain offer tremendous health benefits against different ailments due to their rich content of wide varieties of phytonutrients.

The nutraceutical industry is growing very fast due to the rising awareness among people toward health and nutrition. The global nutraceutical industry was worth around US$ 182.6 billion in 2015 and is projected to expand at an annual growth rate of 7.3% until 2021.[4] The Indian nutraceuticals industry is also considered one of the rapidly growing markets in the Asia-Pacific region.[4] The Indian Nutraceuticals market was estimated at around $4 billion in 2017 and is expected to grow at an annual growth rate of whooping 21% to $ 10 billion in 2022.[1] However, various concerns related to safety and efficacy are associated with these plant-based nutraceuticals, such as misidentification of the raw materials, impurity of the products, absence of scientific evidence, undefined dietary dosage limit, improper labeling, and mixing with banned drugs to improve efficacy.[5] In India, FSS (Food Safety and Standards) Act, 2006 and FSSAI (Food Safety and Standards Authority of India) regulations, 2016 were passed by the government to resolve the aforementioned challenges associated with the nutraceutical products.[6,7] This chapter discusses the different formulations, therapeutic significance, challenges, and regulations associated with plant/phytochemical-based nutraceuticals.

2.2 PRINCIPAL FORMULATIONS AND PHARMACOLOGICAL SIGNIFICANCE OF PLANT-BASED NUTRACEUTICALS

Plant/phytochemical-based nutraceuticals could be primarily available in the following formulations:

a. Natural plant products
b. Medicinal herbs
c. Dietary supplements
d. Fortified foods

2.2.1 NATURAL PRODUCTS

Natural plant products such as fruits, vegetables, millets, lentils, cereal grains, seeds, and nuts constitute an important group of plant-based nutraceuticals which primarily comprise nearly all essential nutrients, such as vitamins, minerals, dietary fiber, essential fatty acids, and phytochemicals like terpenoids, phenolic compounds, and nitrogen-containing compounds.[1,3] The health benefits offered by these phytonutrients have been discussed below.

2.2.1.1 VITAMINS

Vitamins constitute a complex organic group of essential micronutrients which are vital for the proper growth and development of organisms. They participate in diverse crucial biochemical functions, and hence, deficiency of these micronutrients may result in the occurrence of several health disorders.[1] Vitamins are broadly classified as fat-soluble and water-soluble vitamins. Vitamin A, vitamin D, vitamin E, and vitamin K are the fat-soluble vitamins, whereas vitamin B complex (vitamin B_1, vitamin B_2, vitamin B_3, vitamin B_5, vitamin B_6, vitamin B_7, vitamin B_9, and vitamin B_{12}) and vitamin C are the water-soluble vitamins.[3] Except for vitamin D, all the vitamins must be obtained from plant or animal-based diet. Every individual must meet the daily dietary reference values of the vitamins to prevent incidences of vitamin deficiency-associated disorders. Vitamin A deficiency is associated with night blindness and xerophthalmia, vitamin D with rickets (children) and osteomalacia (adults), and vitamin K with coagulation complications.[8] Similarly, vitamin C deficiency is linked with the incidence of scurvy, vitamin B_1 with beriberi, vitamin B_3 with pellagra, vitamin B_9 with megaloblastic anemia, and vitamin B_{12} with pernicious anemia.[8] Besides, vitamin C and vitamin E play a crucial role as natural antioxidants which are vital for warding off several oxidative stress-linked disorders.[3]

2.2.1.2 MINERALS

Minerals are a group of essential elements which are required for the proper functioning of various biological processes and metabolic reactions in the body. Minerals, as cannot be synthesized by living organisms, must be obtained from the plant as well as animal-based diet. Minerals are categorized into two major classes: macrominerals comprising calcium (Ca), magnesium (Mg), phosphorus (P), sulfur (S), potassium (K), sodium (Na), and microminerals or trace elements comprising copper (Cu), iodine (I), iron (Fe), manganese (Mn), selenium (Se), zinc (Zn), cobalt (Co), fluoride (F^-). Trace elements are needed in smaller quantities compared with the macrominerals.[9] These micronutrients are needed for various biological functions, such as bone formation, DNA, RNA, and protein synthesis, functioning as electrolytes, formation of hemoglobin, synthesis of thyroid hormones, and as essential co-factors for varieties of enzymes, such as matrix metalloproteinases (Zn), liver alcohol dehydrogenase (Zn), carbonic anhydrase (Zn), superoxide dismutase (Mn), cytochrome c oxidase (Cu), xanthine oxidase (Mo), aldehyde oxidase (Mo), sulfite oxidase (Mo), and glutathione peroxidase (Se).[3,9,10] The deficiency of these essential elements, hence, might result in the occurrence of several health complications.

2.2.1.3 DIETARY FIBER

Dietary fiber is the portion of plant-based food that cannot be completely digested by human digestive enzymes. It has diverse chemical composition principally comprising cellulose, hemicellulose, lignin, pectin, and β-glucan.[11] A diet rich in fiber is essential for supporting bowel health, relieving constipation, lowering LDL cholesterol, controlling hyperglycemia and obesity.[11–20]

2.2.1.4 POLYUNSATURATED FATTY ACID

Polyunsaturated fatty acids (PUFA) are a broad group of fatty acids with more than one double bond in their structure.[21] Omega-3 PUFA and omega-6 PUFA represent two major classes of essential fatty acids.[21] Alpha-linolenic acid (ALA, 18:3n-3) and alpha-linoleic acid (LA, 18:2n-6) are the parent compounds of omega-3 PUFA and omega-6 PUFA, respectively which must be primarily obtained from a plant-based diet.[21] ALA can be converted into longer chain derivatives, such as eicosapentaenoic acid (EPA, 20:5n-3) and docosahexaenoic acid (DHA, 22:6n-3) inside the body.[21] Likewise, LA can be converted into 20C derivative arachidonic acid (AA, 20:4n-6) inside the body.[21] However, these longer chain PUFAs (EPA, DHA, and AA) can also be directly obtained from an animal-based diet.[21] These essential fatty acids are associated with a lower risk of incidences of cardiovascular disease, cancer, inflammation, nonalcoholic fatty liver disease (NAFLD), and autoimmune diseases.[22–30] However, it is extremely crucial to maintain a proper ratio between the dietary intake of omega-3 PUFA and omega-6 PUFA for the management of the aforementioned disorders; preferably a low ratio of omega-6/omega-3 PUFA.[31] Table 2.1 demonstrates some important natural plant products and their nutrient constituents.

TABLE 2.1 Some Important Phytonutrients and Their Natural Plant Sources.

S. No.	Phytonutrients	Natural products
1	Vitamin A	Carrot, sweet potato, kale, apricot, broccoli, mango, tomato, pumpkin, papaya, spinach, lettuce, squash, grapefruit, peach, green peas, apricot[32]
2	Vitamin E	Avocado, almond, sunflower seeds, peanuts, spinach, broccoli[33]
3	Vitamin K	Broccoli, kale, spinach, lettuce, soybean, cabbage, green peas, avocado, cucumber, cauliflower[34]
4	Vitamin C	Guava, orange, lemon, kiwi, broccoli, strawberry, gooseberry, spinach, cauliflower, tomato, papaya, cabbage, pineapple, black currant[3]
5	Vitamin B_1 (Thiamine)	Rice, wheat, lentils like moong, masoor, chana, spinach, lettuce, fenugreek, cabbage[35]
6	Vitamin B_2 (Riboflavin)	Spinach, broccoli, barley, bajra, jowar, walnut, almond, sesame seeds, sunflower seeds[35]
7	Vitamin B_3 (Niacin)	Whole wheat, groundnut, sesame seeds, peanuts, brown rice, avocado, potato, green peas, almond[35]
8	Vitamin B_5 (Pantothenic acid)	Avocado, broccoli, kale, potatoes, dates, sweet potato, tomato, broccoli, banana, guava[36]
9	Vitamin B_6 (Pyridoxine)	Potato, pistachio, sunflower seeds, sweet potato, sesame seeds, avocado, chickpeas, banana[35]
10	Vitamin B_7 (Biotin)	Avocado, sweet potato, banana, sunflower seeds, broccoli[37]
11	Vitamin B_9 (Folic acid)	Spinach, fenugreek, beet, legumes, orange, lime, walnut, flax seeds, wheat, papaya, banana, avocado[38]
12	Calcium	Almonds, green peas, orange, kale, spinach, chia seeds, sesame, flax seeds, broccoli, fig, orange[39]
13	Magnesium	Avocado, almonds, cashews, legume, chia seeds, flax seeds, wheat, oats, spinach, banana[40]
14	Phosphorus	Sunflower seeds, pumpkin seeds, almond, cashews, wheat, rice, oats, lentils, pumpkin seeds, chia seeds, flax seeds[41]
15	Sulfur	Onion, garlic, cauliflower, broccoli, cabbage[42]
16	Potassium	Banana, avocado, sweet potato, spinach, watermelon, beet[43]

TABLE 2.1 *(Continued)*

S. No.	Phytonutrients	Natural products
17	Sodium	Cumin seeds, beetroot leaves, spinach, fenugreek leaves and seeds, parsley, carrot[44]
18	Copper	Almonds, cashews, sesame seeds, sweet potato, avocado, chickpea, sunflower seeds, walnut[45]
19	Iodine	Strawberry, cranberry, banana, seaweeds like kelp[46]
20	Iron	Lentil, Apricot, spinach, kale, tomato, potato, cashews, soybean, red kidney beans, black-eyed peas, chickpeas, pumpkin seeds, sesame seeds, flax seeds[47]
21	Manganese	Sweet potato, brown rice, chickpeas, spinach, pineapple, potato, almond[48]
22	Selenium	Brown rice, cashews, sunflower seeds, brazil nut, oats, spinach[49]
23	Zinc	Lentils, chickpeas, cashews, pumpkin seeds, oats, quinoa, green peas, spinach, chia seeds, flax seeds[50]
24	Dietary fiber	Oats, almonds, corn, chia seeds, banana, broccoli, avocado, kidney beans, lentils, chickpeas[51]
25	Omega-3 PUFA (ALA)	Seaweed, spirulina, chia seeds, flax seeds, walnut[52]
26	Omega-6 PUFA (LA)	Sunflower seeds, soybean oil, canola oil, almond, cashews, walnut[52]

2.2.1.5 PHYTOCHEMICALS

Phytochemicals are secondary metabolites produced by plants via several metabolic pathways.[53] They are non-nutritive compounds produced by a diverse group of plants primarily to protect themselves from herbivory.[53] Various studies have reported the prophylactic effects of phytochemicals against several disorders, such as inflammation, cancer, type 2 diabetes mellitus, heart diseases, hypertension.[54] Besides, phytochemicals such as carotenoids and flavonoids play a crucial role as natural antioxidants. Based on the chemical structure, phytochemicals are divided into three major categories, viz. terpenoids, phenolic compounds, and nitrogen-containing compounds which are further classified into several subgroups as demonstrated in Figure 2.1.[55] Among the phytochemicals, carotenoids, phytosterols, limonoids, flavonoids, and alkaloids offer specific pharmacological benefits toward human health.[55]

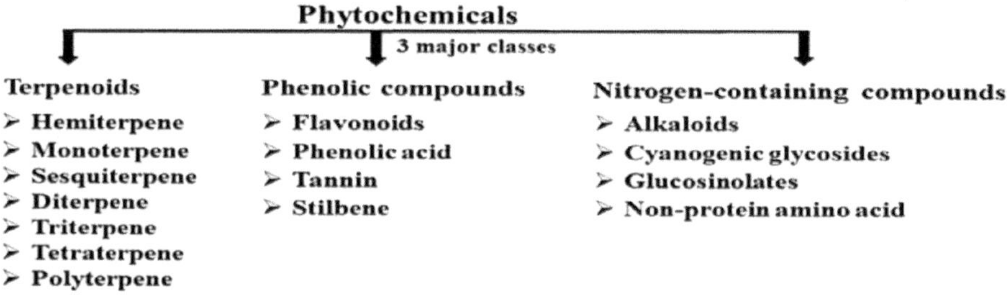

FIGURE 2.1 Classification of phytochemicals. Phytochemicals are categorized into three major classes, viz. terpenes, phenolic compounds, and nitrogen-containing compounds which are further divided into various subgroups based on their chemical structure.

2.2.1.5.1 *Carotenoids*

Carotenoids constitute a family of more than 500 fat-soluble terpene compounds which have proven health benefits against several health disorders.[32] The widely studied carotenoids are α-carotene, β-carotene, lycopene, zeaxanthin, β-cryptoxanthin, and lutein.[32,56] Various studies have reported that carotenoids hold beneficial effects against uterine, prostate, breast, stomach, colorectal, lung, and digestive tract cancers.[54,57–60] Besides, carotenoids are widely known to function as strong antioxidants providing cellular protection against oxidative stress.[61–63] Lycopene, which is found in red-colored fruits like tomato and watermelon, is recognized as the most effective antioxidant in the carotenoid family.[64] Carotenoids are also strongly associated with a lower risk of incidences of eye diseases like age-related macular degeneration (AMD).[65] β-Cryptoxanthin, another important carotenoid, functions as a precursor for vitamin A, which is an essential nutrient for normal eye function, immunity, cellular growth, and differentiation.[64]

2.2.1.5.2 *Phytosterols*

Phytosterols constitute a large group of fat-soluble compounds under the terpene subclass that structurally resembles animal cholesterol. The most common phytosterols available in the human diet are sitosterol, campesterol, and stigmasterol.[54] Studies have shown that phytosterols have anticarcinogenic properties against colon, breast, and prostate cancer.[66,67] Besides, phytosterols also have anti-inflammatory and cholesterol-lowering activity.[68,69]

2.2.1.5.3 Limonoids

Limonoids are a large group of triterpenoids primarily found in citrus fruits. Several studies have reported the anticancer, antioxidant, anti-inflammatory, neuroprotective, antimalarial, antiviral activities associated with different limonoid compounds.[70–84]

2.2.1.5.4 Phenolic Compounds

Phenolic compounds constitute the largest group of plant phytochemicals comprising more than 8000 compounds, which are broadly distributed in different plant species. Phenolic compounds are widely recognized for their strong antioxidant potential as well as their association with the lower risk of incidences of cardiovascular diseases, cancer, liver dysfunction, obesity, type 2 diabetes, etc.[85] Phenolic compounds are primarily categorized into flavonoids and non-flavonoids, such as phenolic acid, lignin, and stilbene.[86] Flavonoids are further classified into flavone, flavanone, flavonol, flavan-3-ol, isoflavone, chalcone, and anthocyanin.[86] Quercetin (flavonol), kaempferol (flavonol), catechin (flavan-3-ol), resveratrol (stilbene), rutin (flavonol), daidzein (isoflavone), and anthocyanin are some of the most studied phenolic compounds and are widely recognized for their anti-inflammatory, antibacterial, anticancer, hypolipidemic, hypoglycemic, neuroprotective, antioxidant, and heart-protective properties.[86–115]

2.2.1.5.5 Alkaloids

Alkaloids are nitrogen-containing phytochemicals that offer multiple medicinal benefits to human health. Alkaloids such as berberine, morphine, montanine, salsoline, galantamine, piperine, nicotine, caffeine, cocaine, harmine, vincristine, and vinblastine have shown tremendous potential in the treatment of various health disorders.[116] They possess potent analgesic (e.g., morphine), hypoglycemic (e.g., piperine), anticancerous (e.g., berberine, vincristine, and vinblastine), antibacterial (e.g., ciprofloxacin), anti-inflammatory (e.g., berberine, piperine, and caffeine), antioxidant (e.g., berberine, caffeine), and antiarrhythmic (e.g., quinidine) activities due to which they have been extensively utilized in modern medicines.[116–118] Besides, different alkaloids such as cocaine, caffeine, nicotine, arecoline, lobeline, galantamine, and morphine exert numerous neuroprotective activities against several diseases like epilepsy, dementia, depression, anxiety, schizophrenia, and alzheimer's disease.[119–128] Table 2.2 demonstrates some important phytochemicals and their plant sources.

TABLE 2.2 Some Important Phytochemicals and Their Plant Sources.

S. No.	Phytochemicals	Plant sources
1	β-Carotene	Apricot, carrot, spinach, sweet potato, kale, pumpkin[64]
2	Zeaxanthin/Lutein	Kale, spinach, broccoli, lettuce, green peas, corn, Brussel sprouts[32]
3	Lycopene	Tomato, guava, watermelon[32]
4	Phytosterol	Corn oil, sunflower oil, almonds, olive oil, soybean, sesame oil[69]
7	Limonoids	Citrus fruit[129]
8	Quercetin	Blueberry, blackberry, grape, tea, onion, tomato, apple[130]
9	Kaempferol	Apple, grapes, tomato, green tea, potato, onion, broccoli, squash, cucumber, lettuce, peaches, blackberry, raspberry, spinach[131]
10	Catechin	Green tea, apple, grape, pear, cherry[130]
11	Resveratrol	Grapes, peanuts, blueberry[132]
12	Rutin	Green tea, grapes, apple, citrus fruit, peach[131]
13	Daidzein	Soybean[133]
14	Anthocyanin	Cranberries, black currants, blackberry, red grapes, raspberry, strawberry, blueberry[131]
15	Alkaloids	Indian barberry (berberine), opium poppy (morphine), cocoa tree (caffeine), tobacco (nicotine), pink periwinkle (vincristine, vinblastine), black pepper (piperine), long pepper (piperine)[116]

2.2.2 MEDICINAL HERBS

Plants serve as an excellent source of phytochemicals with tremendous therapeutic benefits against various disorders. Medicinal herbs have been a vital part of the traditional healing system in many parts of the world due to their easy availability and affordability compared with modern synthetic medicines. These traditional herbal nutraceuticals are often used to provide basic as well as reliable health services, especially to people living in isolated areas where advanced health care services are not easily accessible and also to underprivileged people for whom folk medicines serve as the only affordable and alternate remedy. The global interest in the utilization of medicinal herbs in nutraceuticals is growing rapidly as many plant-based medicines, such as aspirin, digoxin, quinine, morphine, artemisinin, taxol, vincristine, vinblastine, etc. have been successfully employed for the treatment of several illnesses. Table 2.3 represents some common medicinal plants used as traditional herbal nutraceuticals against various ailments.

TABLE 2.3 Medicinal Herbs and Their Use as Nutraceuticals.

S. No.	Medicinal herb species	Common name	Medicinal purpose
1	*Aegle marmelos*	Golden apple	Diarrhea, type 2 diabetes, dysentery, cancer, ulcer, respiratory distress[134]
2	*Allium sativum*	Garlic	Low blood pressure, inflammation, high cholesterol, Cancer[135]
3	*Aloe vera*	Ghritkumari	Inflammation, skin allergy, wound, skin burn, constipation[136]
4	*Andrographis paniculata*	Kalmegh	Common cold, indigestion, diarrhea, osteoarthritis, intestinal worm[137]
5	*Azadirachta indica*	Neem	Inflammation, hyperglycemia, ulcer, cancer, type 2 diabetes[138]
6	*Cassia senna*	Senna	Constipation[139]
7	*Catharanthus roseus*	Pink periwinkle	Type 2 diabetes, cancer, hypertension, wound[140]
8	*Cinnamomum* sp.	Cinnamon	Type 2 Diabetes, cancer, inflammation, cardiovascular disease[141]
9	*Coriandrum sativum*	Coriander	Indigestion, hyperglycemia, ulcer, hypercholesterolemia[142]
10	*Curcuma longa*	Turmeric	Inflammation, arthritis, cancer, type 2 diabetes, ulcer, cardiovascular disease[143]
11	*Cymbopogon citratus*	Lemongrass	Diarrhea, inflammation, bacterial and fungal infections, bacterial and fungal infections, hypercholesterolemia, cancer, indigestion[144]
12	*Foeniculum vulgare*	Fennel	Inflammation, indigestion, type 2 diabetes, inflammation, bacterial and fungal infections[145]
13	*Glycyrrhiza glabra*	Liquorice	Inflammation, ulcer, cancer, depression, burns and wounds, sore throat, cough, respiratory distress, cardiovascular disease[146,147]
14	*Hydrocotyle sibthorpioides*	Brahmi	Memory loss, dementia, stress, anxiety, inflammation, respiratory distress, epilepsy[148,149]

TABLE 2.3 *(Continued)*

S. No.	Medicinal herb species	Common name	Medicinal purpose
15	*Laurus nobilis*	Bay leaves	Type 2 diabetes, inflammation, bacterial and fungal infections, wound[150]
16	*Mentha spicata*	Spearmint	Flatulence, inflammation, bacterial and fungal infections, respiratory distress[151–153]
17	*Moringa oleifera*	Moringa	Hypercholesterolemia, high blood pressure, nonalcoholic fatty liver disease, inflammation, type 2 diabetes, cancer[154]
18	*Murraya koenigii*	Curry leaves	Type 2 diabetes, cancer, inflammation, dyslipidemia[155,156]
19	*Nardostachys jatamansi*	Jatamansi	Insomnia, fungal infections, epilepsy, hysteria, convulsion[157,158]
20	*Ocimum tenuiflorum*	Holy basil	Cold, cough, stress, inflammation, type 2 diabetes[159,160]
21	*Phyllanthus emblica*	Indian gooseberry	Hyperglycemia, fever, hyperlipidemia, hypertension, inflammation, myocardial dysfunction, oxidative stress[161]
22	*Trigonella foenum-graecum*	Fenugreek	Cancer, hypercholesterolemia, type 2 diabetes, inflammation[162]
23	*Swertia chirata*	Chirata	Fever, anemia, asthma, hepatitis, gastritis, constipation, worms, epilepsy, ulcers, hypertension, type 2 diabetes[163]
24	*Syzygium aromaticum*	Clove	Type 2 diabetes, pain, inflammation, bacterial and fungal infections[164]
25	*Syzygium cumini*	Black plum	Hyperglycemia, inflammation, cough, dysentery, ringworm, leishmania, bacterial and fungal infections[165]
26	*Terminalia arjuna*	Arjun	Type 2 diabetes, ulcer, asthma, inflammation, atherosclerosis, anemia, allergy[166]
27	*Tinospora cordifolia*	Giloy	Type 2 diabetes, inflammation, allergy, arthritis, stress[167]
28	*Trachyspermum ammi*	Ajwain	Flatulence, diarrhea, piles, appetite loss, asthma[168]
29	*Withania somnifera*	Ashwagandha	Inflammation, arthritis, cancer, stress, memory loss[169,170]
30	*Zingiber officinale*	Ginger	Inflammation, ulcer, constipation, type 2 diabetes, indigestion, arthritis, cancer, atherosclerosis, hypertension[171]

2.2.3 *DIETARY SUPPLEMENTS*

A dietary supplement can be defined as an industrial ingestible product in the form of a pill, capsule, tablet, powder, or liquid which has been proposed to supplement one's daily diet with multiple health benefits.[172] However, they are not intended to be taken alone as a substitute for dietary foods. Dietary supplements are used worldwide and represent an expansive class of products that are distinct from conventional foods and pharmaceutical drugs. Plant-based dietary supplements claim to provide important phytonutrients which are extracted from the varieties of plants or medicinal herbs. These phytonutrients usually include vitamins, various minerals, fiber, essential fatty acids, essential amino acids, and phytochemicals, such as terpenoids, phenolic compounds, and nitrogen-containing compounds.[172] Commercially available multivitamins, multiminerals, garlic tablets, omega-3 fatty acid capsules, soy-based probiotics, weight-loss capsules, protein powder, and certain sorts of energy drinks are some common examples of plant-based dietary supplements. These dietary supplements are prepared from varieties of medicinal herbs and natural products, such as fruits, vegetables, nuts, seeds, cereals, and algae. For example, weight-loss supplements are generally prepared from apples (apple cider vinegar), green coffee beans, and *Garcinia cambogia.*

2.2.4 *FORTIFIED FOODS*

Food Safety and Standards Authority of India (FSSAI) states that "Food Fortification is a scientifically proven, cost-effective, scalable, and sustainable global intervention that addresses the issue of micronutrient deficiencies."[173] To be simply defined, fortified foods are food products with added micronutrients (vitamins and minerals) that aid the public with multiple health benefits as well as to meet the recommended daily allowance (RDA) of the micronutrients.[174] The added nutrients, however, do not provide any flavor or taste to the fortified food products. Although these nutrients are easily found in natural products, such as fruits, vegetables, nuts, cereals due to multiple factors, which include substandard dietary habits, cooking methods, processing of foods, lack of knowledge, and poverty, it often becomes challenging for the general public to get the required dietary levels of these nutrients. Micronutrient malnourishment is considered a prime contributor to the socioeconomic development of the nations which affects more than 2 billion people worldwide.[174] Fortified foods are cost-effective and they claim to prevent nutrition-related ailments, like anemia, osteoporosis, goiter, hypothyroidism, and pellagra.[174] Several nations have operationalized food fortification programs for vitamins, including folic acid, niacin, riboflavin, thiamine, vitamin A, vitamin B_6, vitamin B_{12}, vitamin D, and vitamin E and minerals, includeing calcium, fluoride, iodine, iron, selenium, and zinc.[174] Common plant-based fortified foods include breakfast cereals, which include oatmeal, cornflakes, fruit juices, bread, rice, wheat, etc. among which wheat is the most common fortified food all over the world. In 2016, FSSAI employed Food Safety and Standard Regulations, 2016 for fortifying wheat flour and rice with iron, vitamin B_{12}, and folic acid to fight the huge burden of malnourishment in India.[173]

2.3 CHALLENGES ASSOCIATED WITH PLANT-BASED NUTRACEUTICALS

Plant-based nutraceuticals are gaining attention in recent times due to the emerging awareness among the general public regarding the diverse health benefits of these products. The growing demand for nutraceuticals might boost the economic growth of many countries which are bestowed with a rich biodiversity of indigenous medicinal herbs. However, there are various challenges associated with the consumption of these plant-based nutraceuticals which should be addressed by the scientific communities as well as the marketing corporations for the holistic well-being of the consumers. The major concerns include the absence of proper scientific evidence, misidentification of the raw materials, impurity of the products, limited knowledge about the permissible limit, false advertising, mixing with illegal drugs to improve efficacy, contamination with heavy metals, and negative interaction between herbal compounds and medicinal drugs inside the body.[5]

The prime concern associated with the use of nutraceuticals is the absence of proper scientific evidence to prove the claimed health benefits. Unlike pharmaceuticals, most of the commercialized nutraceuticals are not always tested for safety and efficacy via systematic experimental procedures and most of them do not undergo randomized controlled clinical studies to establish the claimed pharmacological benefits. Besides, scientific studies related to the permissible dietary limit of nutraceuticals to attain the maximum health benefits are also insufficient to draw any conclusion. This insufficiency of accurate knowledge on the dietary reference values might result in missing the maximum health benefits offered by these products or it might also reach the toxic level inside the body, for instance, the occurrence of hypervitaminosis. Another important concern associated with the herbal products is their likely interaction with other medicinal drugs which might cause some harmful side effects to the person or the drug might lose its efficacy due to the interaction. Hence, manufacturing organizations should reveal the full composition obtained from scientific investigation for the greater well-being of consumers. Maintaining the quality of the nutraceuticals is an additional challenge. Many times, plant-based nutraceuticals are not prepared following the stringent "Good Manufacturing Practice" (GMP) conditions which might result in the accidental contamination of the nutraceuticals with microbes, pesticides, heavy metals, dust, and allergens leading to their poor quality.[5] Besides, some dietary supplements, such as those used for weight loss, bodybuilding, and libido-boosting are added with prohibited drugs to provide a faster and better result. Another important issue with herbal nutraceuticals is that many herb species have morphologically similar other varieties, and hence, misidentification of the desired herb species might result in the utilization of incorrect plant species with toxic or no biological activity. Therefore, the scientific validation, characterization, and quality assessment of medicinal herbs and their products are extremely necessary before their utilization as nutraceuticals.

Food Safety and Standards (FSS) Act, 2006 was enacted by the Government of India regarding the regulation of the manufacture, storage, distribution, sale, import, quality, safety, packaging, labeling, and safety of commercialized food including nutraceuticals.[6] The Food Safety and Standards Authority of India (FSSAI) has enacted Food Safety and Standards (health supplements, nutraceuticals, food for special dietary use, food for special medical purpose, functional food and novel food) Regulations, 2016 which came into force from January 1, 2018.[7] This act emphasizes the various criteria regarding the commercialization of nutraceuticals, including manufacturing, selling, packaging, labeling, advertisement, quality assessment, the permissible dosage of some natural products, permissible additives, and the licensing and registration of food business. Plant-based nutraceuticals should be, hence, commercialized by the manufacturing organizations only after systematic testing and rigorous investigation for quality, safety, and efficiency as per the government guidelines. Besides, awareness programs should be conducted by the government health authorities regarding the possible challenges associated with nutraceuticals and should educate the consumers to consume only those marketed herbal products which have been commercialized under proper government supervision.

CONFLICT OF INTEREST

None declared

KEYWORDS

- **nutraceutical**
- **phytochemical**
- **plant and health**

REFERENCES

1. Sachdeva, V.; Roy, A.; Bharadvaja, N. Current Prospects of Nutraceuticals: A Review. *Curr. Pharm. Biotechnol.* **2020,** *21*, 884–896. DOI: 10.2174/13892010216662001301 13441
2. Brower, V. Nutraceuticals: Poised for a Healthy Slice of the Healthcare Market? *Nat. Biotechnol.* **1998,** *16*, 728–731. DOI: 10.1038/nbt0898-728
3. Souyoul, S. A.; Saussy, K. P.; Lupo, M. P. Nutraceuticals: A Review. *Dermatol. Ther.* **2018,** *8*, 5–16. DOI: 10.1007/s13555-018-0221-x
4. Food Industry Guide to Implement GMP/GHP Requirements. Food Safety and Standards Authority of India, 2018.
5. Siddiqui, R. A.; Moghadasian, M. H. Nutraceuticals and Nutrition Supplements: Challenges and Opportunities. *Nutrients* **2020,** *12*, **1593**. DOI: 10.3390/nu12061593
6. Food Safety and Standards Act, 2006.
7. Food Safety and Standards (Health Supplements, Nutraceuticals, Food for Special Dietary Use, Food for Special Medical Purpose, Functional Food and Novel Food) Regulations, 2016.
8. Gana, W. et al. Analysis of the Impact of Selected Vitamins Deficiencies on the Risk of Disability in Older People. *Nutrients* **2021,** *13*, 3163.
9. Soetan, K.; Olaiya, C.; Oyewole, O. The Importance of Mineral Elements for Humans, Domestic Animals and Plants: A Review. *Afr. J. Food Sci.* **2009,** *4*.
10. Ryan-Harshman, M.; Aldoori, W. Health Benefits of Selected Minerals. *Can. Fam. Phys.* **2005,** *51*, 673–675.
11. Das, L.; Bhaumik, E.; Raychaudhuri, U.; Chakraborty, R. Role of Nutraceuticals in Human Health. *J. Food Sci. Technol.* **2012,** *49*, 173–183. DOI: 10.1007/s13197-011-0269-4
12. Liu, S. et al. Whole-Grain Consumption and Risk of Coronary Heart Disease: Results from the Nurses' Health Study. *Am. J. Clin. Nutr.* **1999,** *70*, 412–419. DOI: 10.1093/ajcn/70.3.412
13. Steffen, L. M. et al. Associations of Whole-Grain, Refined-Grain, and Fruit and Vegetable Consumption with Risks of All-Cause Mortality and Incident Coronary Artery Disease

and Ischemic Stroke: The Atherosclerosis Risk in Communities (ARIC) Study. *Am. J. Clin. Nutr.* **2003,** *78*, 383–390. DOI: 10.1093/ajcn/78.3.383
14. Whelton, S. P. et al. Effect of Dietary Fiber Intake on Blood Pressure: A Meta-Analysis of Randomized, Controlled Clinical Trials. *J. Hypertension* **2005,** *23*, 475–481. DOI: 10.1097/01.hjh.0000160199.51158.cf
15. Petruzziello, L.; Iacopini, F.; Bulajic, M.; Shah, S.; Costamagna, G. Review Article: Uncomplicated Diverticular Disease of the Colon. *Alimentary Pharmacol. Therap.* **2006,** *23*, 1379–1391. DOI: 10.1111/j.1365-2036.2006.02896.x
16. Brown, L.; Rosner, B.; Willett, W. W.; Sacks, F. M. Cholesterol-lowering effects of Dietary Fiber: A Meta-Analysis. *Am. J Clin. Nutr.* **1999,** *69*, 30–42. DOI: 10.1093/ajcn/69.1.30
17. Keenan, J. M.; Pins, J. J.; Frazel, C.; Moran, A.; Turnquist, L. Oat Ingestion Reduces Systolic and Diastolic Blood Pressure in Patients with Mild or Borderline Hypertension: A Pilot Trial. *J. Fam. Pract.* **2002,** *51*, 369.
18. Anderson, J. W.; Randles, K. M.; Kendall, C. W.; Jenkins, D. J. Carbohydrate and Fiber Recommendations for Individuals with Diabetes: A Quantitative Assessment and Meta-Analysis of the Evidence. *J. Am. Coll. Nutr.* **2004,** *23*, 5–17. DOI: 10.1080/07315724.2004.10719338
19. Birketvedt, G. S.; Shimshi, M.; Erling, T.; Florholmen, J. Experiences with Three Different Fiber Supplements in Weight Reduction. *Med. sci. Monitor: Int. Med. J. Exp. Clin. Res.* **2005,** *11*, Pi5–8.
20. Cummings, J. **2001,** 183–252.
21. Trumbo, P.; Schlicker, S.; Yates, A. A.; Poos, M. Dietary Reference Intakes for Energy, Carbohydrate, Fiber, Fat, Fatty Acids, Cholesterol, Protein and Amino Acids. *J. Am. Diet. Assoc.* **2002,** *102*, 1621–1630. DOI: 10.1016/s0002-8223(02)90346-9
22. Siscovick, D. S. et al. Omega-3 Polyunsaturated Fatty Acid (Fish Oil) Supplementation and the Prevention of Clinical Cardiovascular Disease: A Science Advisory From the American Heart Association. *Circulation* **2017,** *135*, e867–e884. DOI: 10.1161/cir.0000000000000482
23. Benes, L. B.; Bassi, N. S.; Davidson, M. H. Omega-3 Carboxylic Acids Monotherapy and Combination with Statins in the Management of Dyslipidemia. *Vasc. Health Risk Manage.* **2016,** *12*, 481–490. DOI: 10.2147/vhrm.s58149
24. Kastelein, J. J. et al. Omega-3 Free Fatty Acids for the Treatment of Severe Hypertriglyceridemia: The EpanoVa for Lowering Very High TriglyceridEs (EVOLVE) Trial. *J. Clin. Lipidol.* **2014,** *8*, 94–106. DOI: 10.1016/j.jacl.2013.10.003
25. Spooner, M. H.; Jump, D. B. Omega-3 Fatty Acids and Nonalcoholic Fatty Liver Disease in Adults and Children: Where Do We Stand? *Curr. Opin. Clin. Nutr. Metabol. Care* **2019,** *22*, 103–110. DOI: 10.1097/mco.0000000000000539
26. Arendt, B. M. et al. Altered Hepatic Gene Expression in Nonalcoholic Fatty Liver Disease Is Associated with Lower Hepatic n-3 and n-6 Polyunsaturated Fatty Acids. *Hepatology* **2015,** *61*, 1565–1578. DOI: 10.1002/hep.27695
27. Gioxari, A.; Kaliora, A. C.; Marantidou, F.; Panagiotakos, D. P. Intake of ω-3 Polyunsaturated Fatty Acids in Patients with Rheumatoid Arthritis: A Systematic Review and Meta-Analysis. *Nutrition* **2018,** *45*, 114–124.e114. DOI: 10.1016/j.nut.2017.06.023
28. Swan, K.; Allen, P. J. Omega-3 Fatty Acid for the Treatment and Remission of Crohn's Disease. *J. Complement. Integrat. Med.* **2013,** *10*. DOI: 10.1515/jcim-2012-0010

29. Hooper, L. et al. Omega-6 Fats for the Primary and Secondary Prevention of Cardiovascular Disease. *Cochrane Database System. Rev.* **2018,** *7*, Cd01**1094**. DOI: 10.1002/14651858.CD011094.pub3
30. Jang, H.; Park, K. Omega-3 and omega-6 Polyunsaturated Fatty Acids and Metabolic Syndrome: A Systematic Review and Meta-Analysis. *Clin. Nutr.***2020,** *39*, 765–773. DOI: 10.1016/j.clnu.2019.03.032
31. Simopoulos, A. P. The Importance of the Ratio of Omega-6/Omega-3 Essential Fatty Acids. *Biomed. Pharmacother* **2002,** *56*, 365–379. DOI: 10.1016/s0753-3322(02)00253-6
32. Johnson, E. J. The Role of Carotenoids in Human Health. *Nutr. Clin. Care* **2002,** *5*, 56–65. DOI: 10.1046/j.1523-5408.2002.00004.x
33. McLaughlin, P. J.; Weihrauch, J. L. Vitamin E Content of Foods. *J. Am. Diet. Assoc.* **1979,** *75*, 647–665
34. Booth, S. L. Vitamin K: Food Composition and Dietary Intakes. *Food Nutr. Res.* **2012,** *56*. DOI: 10.3402/fnr.v56i0.5505
35. Lebiedzińska, A.; Szefer, P. Vitamins B in Grain and Cereal–Grain Food, Soy-Products and Seeds. *Food Chem. 95*, 116–122. DOI: 10.1016/j.foodchem.2004.12.024
36. Miller, J. W.; Rucker, R. B. in *Present Knowledge in Nutrition*; Marriott, B. P.; Birt, D. F.; Stallings, V. A.; Yates, A. A., 11th edn.; Academic Press, 2020; pp 273–287.
37. Staggs, C. G.; Sealey, W. M.; McCabe, B. J.; Teague, A. M.; Mock, D. M. Determination of the Biotin Content of Select Foods Using Accurate and Sensitive HPLC/Avidin Binding. *J. Food Compost. Anal.* **2004,** *17*, 767–776. DOI: 10.1016/j.jfca.2003.09.015
38. Scott, J.; Rébeillé, F.; Fletcher, J. Folic Acid and Folates: The Feasibility for Nutritional Enhancement in Plant Foods. *J. Sci. Food Agric.* **2000,** *80*, 795–824. DOI: https://doi.org/10.1002/(SICI)1097-0010(20000515)80:7<795::AID-JSFA599>3.0.CO;2-K
39. Theobald, H. E. Dietary Calcium and Health. *Nutr. Bull.* **2005,** *30*, 237–277. DOI: https://doi.org/10.1111/j.1467-3010.2005.00514.x
40. https://ods.od.nih.gov/factsheets/Magnesium-HealthProfessional/
41. https://ods.od.nih.gov/factsheets/Phosphorus-HealthProfessional/
42. Hewlings, S.; Kalman, D. Sulfur and Human Health. **2019,** *14*, 785
43. Weaver, C. M. Potassium and Health. *Adv. Nutr.* **2013,** *4*, 368S–377S. DOI: 10.3945/an.112.003533
44. http://www.nutrition-atlas.nin.res.in/nutritionatlas/sodium.php
45. Subar, A. F.; Krebs-Smith, S. M.; Cook, A.; Kahle, L. L. Dietary Sources of Nutrients Among US Adults, 1989 to 1991. *J. Am. Diet. Assoc.* **1998,** *98*, 537–547. DOI: https://doi.org/10.1016/S0002-8223(98)00122-9
46. https://ods.od.nih.gov/factsheets/Iodine-HealthProfessional/
47. https://ods.od.nih.gov/factsheets/Iron-HealthProfessional/
48. https://ods.od.nih.gov/factsheets/Manganese-HealthProfessional/
49. https://ods.od.nih.gov/factsheets/Selenium-HealthProfessional/
50. https://ods.od.nih.gov/factsheets/Zinc-HealthProfessional/
51. Satija, A.; Hu, F. B. Cardiovascular Benefits of Dietary Fiber. *Curr. Atherosclerosis Rep.* **2012,** *14*, 505–514. DOI: 10.1007/s11883-012-0275-7
52. Saini, R. K.; Keum, Y.-S. Omega-3 and Omega-6 Polyunsaturated Fatty Acids: Dietary Sources, Metabolism, and Significance—A Review. *Life Sci.* **2018,** *203*, 255–267. DOI: https://doi.org/10.1016/j.lfs.2018.04.049

53. Yoo, S.; Kim, K.; Nam, H.; Lee, D. Discovering Health Benefits of Phytochemicals with Integrated Analysis of the Molecular Network, Chemical Properties and Ethnopharmacological Evidence. *Nutrients* **2018,** *10*, 1042. DOI: 10.3390/nu10081042
54. Arora, S. **2020,** 205–227
55. Campos-Vega, R.; Oomah, B. D. In *Handbook of Plant Food Phytochemicals*; 2013; pp 5–48.
56. Basu, H. N.; Vecchio, A. J. D.; Flider, F.; Orthoeter, F. T. Nutritional and Potential Disease Prevention Properties of Carotenoids. *J. Am. Oil Chem. 'Soc.* **2001,** *78*, 665–675. DOI: 10.1007/s11746-001-0324-x
57. Franceschi, S. et al. Tomatoes and Risk of Digestive-Tract Cancers. *Int. J. Cancer* **1994,** *59*, 181–184. DOI: 10.1002/ijc.2910590207
58. Ziegler, R. G. Vegetables, Fruits, and Carotenoids and the Risk of Cancer. *Am. J Clin. Nutr.* **1991,** *53*, 251s–259s. DOI: 10.1093/ajcn/53.1.251S
59. Bramley, P. M. Is Lycopene Beneficial to Human Health? *Phytochemistry* **2000,** *54*, 233–236. DOI: 10.1016/s0031-9422(00)00103-5
60. Linnewiel-Hermoni, K. et al. The Anti-Cancer Effects of Carotenoids and Other Phytonutrients Resides in Their Combined Activity. *Arch. Biochem. Biophys.* **2015,** *572*, 28–35. DOI: 10.1016/j.abb.2015.02.018
61. Liebler, D. C. Antioxidant Reactions of Carotenoids. *Ann. NY Acad. Sci.* **1993,** *691*, 20–31. DOI: 10.1111/j.1749-6632.1993.tb26154.x
62. Fiedor, J.; Fiedor, L.; Haessner, R.; Scheer, H. Cyclic Endoperoxides of B-carotene, Potential Pro-Oxidants, as Products of Chemical Quenching of Singlet Oxygen. *Biochim. Biophys. Acta* **2005,** *1709*, 1–4. DOI: 10.1016/j.bbabio.2005.05.008
63. Yeum, K.-J.; Aldini, G.; Russell, R. M.; Krinsky, N. I. In *Carotenoids: Volume 5: Nutrition and Health*; Britton, G.; Pfander, H.; Liaaen-Jensen, S.; Eds.; Birkhäuser Basel, 2009; pp 235–268.
64. Steinmetz, K. A.; Potter, J. D. Vegetables, Fruit, and Cancer Prevention: A Review. *J. Am. Diet. Assoc.* **1996,** *96*, 1027–1039. DOI: 10.1016/s0002-8223(96)00273-8
65. Seddon, J. M. et al. Dietary Carotenoids, Vitamins A, C, and E, and Advanced Age-Related Macular Degeneration: Eye Disease Case-Control Study Group. *JAMA* **1994,** *272*, 1413–1420
66. Rao, A.; Koratkar, R.
67. Awad, A. B.; Fink, C. S. Phytosterols as Anticancer Dietary Components: Evidence and Mechanism of Action. *J. Nutr. 2000; 130*, 2127–2130. DOI: 10.1093/jn/130.9.2127
68. Ling, W. H.; Jones, P. J. Dietary Phytosterols: A Review of Metabolism, Benefits and Side Effects. *Life Sci.* **1995,** *57*, 195–206. DOI: 10.1016/0024-3205(95)00263-6
69. Cabral, C. E.; Klein, M. R. S. T. Phytosterols in the Treatment of Hypercholesterolemia and Prevention of Cardiovascular Diseases. *Arq. Bras. Cardiol.* **2017,** *109*, 475–482. DOI: 10.5935/abc.20170158
70. El-Readi, M. Z.; Hamdan, D.; Farrag, N.; El-Shazly, A.; Wink, M. Inhibition of P-Glycoprotein Activity by Limonin and Other Secondary Metabolites from Citrus Species in Human Colon and Leukaemia Cell Lines. *Eur. J. Pharmacol.* **2010,** *626*, 139–145. DOI: 10.1016/j.ejphar.2009.09.040
71. Kim, W. et al. Dietary Curcumin and Limonin Suppress CD4+ T-Cell Proliferation and Interleukin-2 Production in Mice. *J. Nutr.* **2009,** *39*, 1042–1048. DOI: 10.3945/jn.108.102772

72. Kim, J. H. et al. Fraxinellone Inhibits Lipopolysaccharide-Induced Inducible Nitric Oxide Synthase and Cyclooxygenase-2 Expression by Negatively Regulating Nuclear Factor-Kappa B in RAW 264.7 Macrophages Cells. *Biol. Pharm. Bull.* **2009,** *32*, 1062–1068. DOI: 10.1248/bpb.32.1062
73. Sun, C.; Chen, K.; Chen, Y.; Chen, Q. Contents and Antioxidant Capacity of Limonin and Nomilin in Different Tissues of Citrus Fruit of Four Cultivars During Fruit Growth and Maturation. *Food Chem.* **2005,** *93*, 599–605. DOI: 10.1016/j.foodchem.2004.10.037
74. Yu, J. et al. Antioxidant Activity of Citrus Limonoids, Flavonoids, and Coumarins. *J. Agric. Food Chem.* **2005,** *53*, 2009–2014. DOI: 10.1021/jf0484632
75. Uddin, S. J. et al. Gedunin, a Limonoid from Xylocarpus Granatum, Inhibits the Growth of CaCo-2 Colon Cancer Cell Line In Vitro. *Phytother. Res.: PTR21* **2007,** 757–761. DOI: 10.1002/ptr.2159
76. Poulose, S. M.; Harris, E. D.; Patil, B. S. Citrus Limonoids Induce Apoptosis in Human Neuroblastoma Cells and Have Radical Scavenging Activity. *J. Nutr.* **2005,** *135*, 870–877. DOI: 10.1093/jn/135.4.870
77. Penido, C. et al. Anti-Allergic Effects of Natural Tetranortriterpenoids Isolated from Carapa Guianensis Aublet on Allergen-Induced Vascular Permeability and Hyperalgesia. *Inflamm. Res.* **2005,** *54*, 295–303. DOI: 10.1007/s00011-005-1357-6
78. Yoon, J. S.; Sung, S. H.; Kim, Y. C. Neuroprotective Limonoids of Root Bark of Dictamnus dasycarpus. *J. Nat. Prod.* **2008,** *71*, 208–211. DOI: 10.1021/np070588o
79. Yoon, J. S.; Yang, H.; Kim, S. H.; Sung, S. H.; Kim, Y. C. Limonoids from Dictamnus Dasycarpus Protect Against Glutamate-Induced Toxicity in Primary Cultured Rat Cortical Cells. *J. Mol. Neurosci.* **2010,** *42*, 9–16. DOI: 10.1007/s12031-010-9333-1
80. Lee, S. E. et al. Antimalarial Activity of Anthothecol Derived from Khaya Anthotheca (Meliaceae). *Phytomedicine* **2008,** *15*, 533–535. DOI: 10.1016/j.phymed.2007.08.001
81. Lucantoni, L. et al. Transmission Blocking Activity of a Standardized Neem (*Azadirachta indica*) Seed Extract on the Rodent Malaria Parasite *Plasmodium berghei* in Its Vector *Anopheles stephensi*. *Malaria J.* **2010,** *9*, 66. DOI: 10.1186/1475-2875-9-66
82. Barquero, A. A.; Alché, L. E.; Coto, C. E. Block of Vesicular Stomatitis Virus Endocytic and Exocytic Pathways by 1-Cinnamoyl-3,11-Dihydroxymeliacarpin, a Tetranortriterpenoid of Natural Origin. *J. Gen. Virol.* **2004,** *85*, 483-493. DOI: 10.1099/vir.0.19343-0
83. Balestrieri, E. et al. Antiviral Activity of Seed Extract from Citrus Bergamia Towards Human Retroviruses. *Bioorg. Med. Chem.* **2011,** *19*, 2084–2089. DOI: 10.1016/j.bmc.2011.01.024
84. Ribeiro, A. B. et al. Chemical Characterization of *Citrus sinensis* Grafted on *C. limonia* and the Effect of Some Isolated Compounds on the Growth of *Xylella fastidiosa*. *J. Agric. Food Chem.* **2008,** *56*, 7815–7822. DOI: 10.1021/jf801103p
85. Yang, C. S.; Chung, J. Y.; Yang, G.; Chhabra, S. K.; Lee, M. J. Tea and Tea Polyphenols in Cancer Prevention. *J. Nutr.* **2000,** *130*, 472s–478s. DOI: 10.1093/jn/130.2.472S
86. Rasouli, H.; Farzaei, M. H.; Mansouri, K.; Mohammadzadeh, S.; Khodarahmi, R. Plant Cell Cancer: May Natural Phenolic Compounds Prevent Onset and Development of Plant Cell Malignancy? A Literature Review. *Molecules* **2016,** *21*, 1104. DOI: 10.3390/molecules21091104
87. Chen, X.; Zou, L.; Liu, W.; McClements, D. J. Potential of Excipient Emulsions for Improving Quercetin Bioaccessibility and Antioxidant Activity: An In Vitro Study. *J. Agric. Food Chem.* **2016,** *64*, 3653–3660. DOI: 10.1021/acs.jafc.6b01056

88. Hisanaga, A.; Mukai, R.; Sakao, K.; Terao, J.; Hou, D. X. Anti-Inflammatory Effects and Molecular Mechanisms of 8-Prenyl Quercetin. *Mol. Nutr. Food Res.* **2016,** *60*, 1020–1032. DOI: 10.1002/mnfr.201500871
89. Gormaz, J. G.; Quintremil, S.; Rodrigo, R. Cardiovascular Disease: A Target for the Pharmacological Effects of Quercetin. *Curr. Topics Med. Chem.* **2015,** *15*, 1735–1742. DOI: 10.2174/1568026615666150427124357
90. Vaidya, B. et al. Effectiveness of Periodic Treatment of Quercetin against Influenza A Virus H1N1 through Modulation of Protein Expression. *J Agric Food Chem64*, 4416–4425. DOI: 10.1021/acs.jafc.6b00148
91. Al-Jabban, S. M. et al. Synthesis and Anti-Proliferative Effects of Quercetin Derivatives. *Nat. Prod. Commun.* **2015,** *10*, 2113–2118
92. Sharma, S.; Ali, A.; Ali, J.; Sahni, J. K.; Baboota, S. Rutin: Therapeutic Potential and Recent Advances in Drug Delivery. *Expert Opin. Investig. Drugs* **2013,** *22*, 1063–1079. DOI: 10.1517/13543784.2013.805744
93. Kessler, M.; Ubeaud, G.; Jung, L. Anti- and Pro-Oxidant Activity of Rutin and Quercetin Derivatives. *J. Pharm. Pharmacol.* **2003,** *55*, 131–142. DOI: 10.1211/002235702559
94. Arts, I. C.; Hollman, P. C.; Feskens, E. J.; Bueno de Mesquita, H. B.; Kromhout, D. Catechin Intake Might Explain the Inverse Relation Between Tea Consumption and Ischemic Heart Disease: The Zutphen Elderly Study. *Am. J Clin. Nutr.* **2001,** *74*, 227–232. DOI: 10.1093/ajcn/74.2.227
95. Friedman, M. Overview of Antibacterial, Antitoxin, Antiviral, and Antifungal Activities of Tea Flavonoids and Teas. *Mol. Nutr. Food Res.* **2007,** *51*, 116–134. DOI: 10.1002/mnfr.200600173
96. Park, K. D. et al. Antimicrobial Activity of 3-O-acyl-(-)-Epicatechin and 3-O-acyl-(+)-Catechin Derivatives. *Planta Medica* **2004,** *70*, 272–276. DOI: 10.1055/s-2004-818923
97. Chen, Y. K. et al. Effects of Green Tea Polyphenol (-)-Epigallocatechin-3-Gallate on Newly Developed High-Fat/Western-Style Diet-Induced Obesity and Metabolic Syndrome in Mice. *J. Agric. Food Chem.* **2011,** *59*, 11862–11871. DOI: 10.1021/jf2029016
98. Lin, J. K.; Lin-Shiau, S. Y. Mechanisms of Hypolipidemic and Anti-Obesity Effects of Tea and Tea Polyphenols. *Mol. Nutr. Food Res.* **2006,** *50*, 211–217. DOI: 10.1002/mnfr.200500138
99. Wolfram, S.; Wang, Y.; Thielecke, F. Anti-Obesity Effects of Green Tea: From Bedside to Bench. *Mol. Nutr. Food Res.* **2006,** *50*, 176–187. DOI: 10.1002/mnfr.200500102
100. Xu, Y. et al. The Anti-Obesity Effect of Green Tea Polysaccharides, Polyphenols and Caffeine in Rats Fed with a High-Fat Diet. *Food Funct.* **2015,** *6*, 297–304. DOI: 10.1039/c4fo00970c
101. Zhang, Y.; Chen, A. Y.; Li, M.; Chen, C.; Yao, Q. Ginkgo Biloba Extract Kaempferol Inhibits Cell Proliferation and Induces Apoptosis in Pancreatic Cancer Cells. *J. Surg. Res.* **2008,** *148*, 17–23. DOI: 10.1016/j.jss.2008.02.036
102. Song, W. et al. Kaempferol Induces Cell Cycle Arrest and Apoptosis in Renal Cell Carcinoma Through EGFR/p38 Signaling. *Oncol. Rep.* **2014,** *31*, 1350–1356. DOI: 10.3892/or.2014.2965
103. Lee, J.; Kim, J. H. Kaempferol Inhibits Pancreatic Cancer Cell Growth and Migration Through the Blockade of EGFR-Related Pathway In Vitro. *PLoS One* **2016,** *11*, e0155264. DOI: 10.1371/journal.pone.0155264

104. Tang, X. L. et al. Protective Effect of Kaempferol on LPS Plus ATP-Induced Inflammatory Response in Cardiac Fibroblasts. *Inflammation* **2015,** *38*, 94–101. DOI: 10.1007/s10753-014-0011-2
105. Zang, Y.; Zhang, L.; Igarashi, K.; Yu, C. The Anti-Obesity and Anti-Diabetic Effects of Kaempferol Glycosides from Unripe Soybean Leaves in High-Fat-Diet Mice. *Food Funct.* **2015,** *6*, 834–841. DOI: 10.1039/c4fo00844h
106. del Valle, P. et al. Antimicrobial Activity of Kaempferol and Resveratrol in Binary Combinations with Parabens or Propyl Gallate Against *Enterococcus faecalis*. *Food Control* **2016,** *61*, 213–220. DOI: https://doi.org/10.1016/j.foodcont.2015.10.001
107. Harikumar, K. B.; Aggarwal, B. B. Resveratrol: A Multitargeted Agent for Age-Associated Chronic Diseases. *Cell Cycle* **2008,** *7*, 1020–1035. DOI: 10.4161/cc.7.8.5740
108. Baur, J. A.; Sinclair, D. A. Therapeutic Potential of Resveratrol: The In Vivo Evidence. *Nat. Rev.: Drug Discov.* **2006,** *5*, 493–506. DOI: 10.1038/nrd2060
109. Aguirre, L.; Fernández-Quintela, A.; Arias, N.; Portillo, M. P. Resveratrol: Anti-Obesity Mechanisms of Action. *Molecules* **2014,** *19*, 18632–18655. DOI: 10.3390/molecules191118632
110. Olas, B. et al. Antioxidant Activity of Resveratrol in Endotoxin-Stimulated Blood Platelets. *Cell Biol. Toxicol.* **2001,** *17*, 117–125. DOI: 10.1023/a:1010962222305
111. Elmali, N.; Baysal, O.; Harma, A.; Esenkaya, I.; Mizrak, B. Effects of Resveratrol in Inflammatory Arthritis. *Inflammation* **2007,** *30*, 1–6. DOI: 10.1007/s10753-006-9012-0
112. Sun, A. Y.; Wang, Q.; Simonyi, A.; Sun, G. Y. Resveratrol as a Therapeutic Agent for Neurodegenerative Diseases. *Mol. Neurobiol.* **2010,** *41*, 375–383. DOI: 10.1007/s12035-010-8111-y
113. Shin, J. A. et al. Therapeutic Effects of Resveratrol During Acute Periods Following Experimental Ischemic Stroke. *J. Neuroimmunol.* **2010,** *227*, 93–100. DOI: 10.1016/j.jneuroim.2010.06.017
114. Bradamante, S.; Barenghi, L.; Villa, A. Cardiovascular Protective Effects of Resveratrol. *Cardiovasc. Drug Rev.* **2004,** *22*, 169–188. DOI: 10.1111/j.1527-3466.2004.tb00139.x
115. Sun, W. et al. Anti-Cancer Effect of Resveratrol Is Associated with Induction of Apoptosis via a Mitochondrial Pathway Alignment. *Adv. Exp. Med. Biol.* **2008,** *614*, 179–186. DOI: 10.1007/978-0-387-74911-2_21
116. Hussain, G. et al. Role of Plant Derived Alkaloids and Their Mechanism in Neurodegenerative Disorders. *Int. J. Biol. Sci.* **2018,** *14*, 341–357. DOI: 10.7150/ijbs.23247
117. Kainsa, S.; Kumar, P.; Rani, P. Medicinal Plants of Asian Origin Having Anticancer Potential: Short Review. *Asian J. Biomed. Pharm. Sci.* **2012,** *2*, 1–7
118. Cushnie, T. P.; Cushnie, B.; Lamb, A. J. Alkaloids: An Overview of Their Antibacterial, Antibiotic-Enhancing and Antivirulence Activities. *Int. J. Antimicrob. Agents* **2014,** *44*, 377–386. DOI: 10.1016/j.ijantimicag.2014.06.001
119. Imenshahidi, M.; Qaredashi, R.; Hashemzaei, M.; Hosseinzadeh, H. Inhibitory Effect of *Berberis vulgaris* Aqueous Extract on Acquisition and Reinstatement Effects of Morphine in Conditioned Place Preferences (CPP) in Mice. *Jundishapur J. Nat. Pharm. Prod.* **2014,** *9*, e16145. DOI: 10.17795/jjnpp-16145
120. Jin, Y.; Khadka, D. B.; Cho, W. J. Pharmacological Effects of Berberine and Its Derivatives: A Patent Update. *Expert Opin. Therap. Patents* **2016,** *26*, 229–243. DOI: 10.1517/13543776.2016.1118060

121. Jiang, W.; Li, S.; Li, X. Therapeutic Potential of Berberine Against Neurodegenerative Diseases. *Sci. China: Life Sci.* **2015,** *58*, 564–569. DOI: 10.1007/s11427-015-4829-0
122. Jiang, W.; Wei, W.; Gaertig, M. A.; Li, S.; Li, X. J. Therapeutic Effect of Berberine on Huntington's Disease Transgenic Mouse Model. *PLoS One* **2015,** *10*, e0134142. DOI: 10.1371/journal.pone.0134142
123. Mishra, A.; Punia, J. K.; Bladen, C.; Zamponi, G. W.; Goel, R. K. Anticonvulsant Mechanisms of Piperine, a Piperidine Alkaloid. *Channels* **2015,** *9*, 317–323. DOI: 10.1080/19336950.2015.1092836
124. Hritcu, L. et al. Methanolic Extract of *Piper nigrum* Fruits Improves Memory Impairment by Decreasing Brain Oxidative Stress in Amyloid Beta(1–42) Rat Model of Alzheimer's Disease. *Cell. Mol. Neurobiol.* **2014,** *34*, 437–449. DOI: 10.1007/s10571-014-0028-y
125. Al-Baghdadi, O. B.; Prater, N. I.; Van der Schyf, C. J.; Geldenhuys, W. J. Inhibition of Monoamine Oxidase by Derivatives of Piperine, an Alkaloid from the Pepper Plant *Piper nigrum*, for Possible Use in Parkinson's Disease. *Bioorga. Med. Chem. Lett.* **2012,** *22*, 7183–7188. DOI: 10.1016/j.bmcl.2012.09.056
126. Liu, Y. J.; Peng, W.; Hu, M. B.; Xu, M.; Wu, C. J. The Pharmacology, Toxicology and Potential Applications of Arecoline: A Review. *Pharm. Biol.* **2016,** *54*, 2753–2760. DOI: 10.3109/13880209.2016.1160251
127. Zhu, H. L. et al. Medicinal Compounds with Antiepileptic/Anticonvulsant Activities. *Epilepsia* **2014,** *55*, 3–16. DOI: 10.1111/epi.12463
128. Dall'Acqua, S. Plant-Derived Acetylcholinesterase Inhibitory Alkaloids for the Treatment of Alzheimer's Disease. *Botanics: Targets Ther.* **2013,** *1179–98973*, 19–28. DOI: 10.2147/BTAT.S17297
129. Dillard, C. J.; German, J. B. Phytochemicals: Nutraceuticals and Human Health. *J. Sci. Food Agric.* **2000,** *80*, 1744–1756. DOI: https://doi.org/10.1002/1097–0010(20000915)80:12<1744::AID-JSFA725>3.0.CO;2-W
130. Rasouli, H.; Farzaei, M. H.; Khodarahmi, R. Polyphenols and Their Benefits: A Review. *Int. J. Food Prop.* **2017,** *20*, 1700–1741. DOI: 10.1080/10942912.2017.1354017
131. Panche, A. N.; Diwan, A. D.; Chandra, S. R. Flavonoids: An Overview. *J. Nutr. Sci.* **2016,** *5*, e47. DOI: 10.1017/jns.2016.41
132. Koushki, M.; Amiri-Dashatan, N.; Ahmadi, N.; Abbaszadeh, H.-A.; Rezaei-Tavirani, M. Resveratrol: A Miraculous Natural Compound for Diseases Treatment. *Food Sci. Nutr.* **2018,** *6*, 2473–2490. DOI: 10.1002/fsn3.855
133. Zhang, Y.; Wang, G. J.; Song, T. T.; Murphy, P. A.; Hendrich, S. Urinary Disposition of the Soybean Isoflavones Daidzein, Genistein and Glycitein Differs Among Humans with Moderate Fecal Isoflavone Degradation Activity. *J. Nutr.* **1999,** *129*, 957–962. DOI: 10.1093/jn/129.5.957
134. Rahman, S.; Parvin, R. Therapeutic Potential of *Aegle marmelos* (L.)-An Overview. *Asian Pac. J. Trop. Dis.* **2014,** *4*, 71–77. DOI: 10.1016/S2222-1808(14)60318-2
135. Bayan, L.; Koulivand, P. H.; Gorji, A. Garlic: A Review Of Potential Therapeutic Effects. *Avicenna J. Phytomed.* **2014,** *4*, 1–14
136. Surjushe, A.; Vasani, R.; Saple, D. G. Aloe Vera: A Short Review. *Indian J. Dermatol.* **2008,** *53*, 163–166. DOI: 10.4103/0019-5154.44785
137. Okhuarobo, A. et al. Harnessing the Medicinal Properties of *Andrographis paniculata* for Diseases and Beyond: A Review of Its Phytochemistry and Pharmacology. *Asian Pac. J. Trop. Dis.* **2014,** *4*, 213–222. DOI: 10.1016/S2222-1808(14)60509–0

138. Alzohairy, M. A. Therapeutics Role of Azadirachta indica (Neem) and Their Active Constituents in Diseases Prevention and Treatment. *Evid.-Based Complement. Altern. Med.* **2016,** *2016*, 7382506–7382506. DOI: 10.1155/2016/7382506
139. Oladeji, O. S.; Adelowo, F. E.; Oluyori, A. P.; Bankole, D. T. Ethnobotanical Description and Biological Activities of Senna alata. *Evid.-Based Complement. Altern. Med.* **2020,** *2020*, 2580259–2580259. DOI: 10.1155/2020/2580259
140. Pham, H. N. T.; Vuong, Q. V.; Bowyer, M. C.; Scarlett, C. J. Phytochemicals Derived from *Catharanthus roseus* and Their Health Benefits. *Technologies* **2020,** *8*, 80
141. Kawatra, P.; Rajagopalan, R. Cinnamon: Mystic Powers of a Minute Ingredient. *Pharmacogn. Res.* **2015,** *7*, S1–S6. DOI: 10.4103/0974-8490.157990
142. Nadeem, M. et al. Nutritional and Medicinal Aspects of Coriander (*Coriandrum sativum* L.) A Review. *Br. Food J.* **2013,** *115*, 743–755. DOI: 10.1108/00070701311331526
143. Hewlings, S. J.; Kalman, D. S. Curcumin: A Review of Its Effects on Human Health. *Foods* **2017,** *6*, 92. DOI: 10.3390/foods6100092
144. Shah, G. et al. Scientific Basis for the Therapeutic Use of *Cymbopogon citratus*, Stapf (Lemon Grass). *J. Adv. Pharm. Technol. Res.* **2011,** *2*, 3–8. DOI: 10.4103/2231-4040.79796
145. Rather, M. A.; Dar, B. A.; Sofi, S. N.; Bhat, B. A.; Qurishi, M. A. Foeniculum Vulgare: A Comprehensive Review of Its Traditional Use, Phytochemistry, Pharmacology, and Safety. *Arab. J. Chem.* **2016,** *9*, S1574–S1583. DOI: https://doi.org/10.1016/j.arabjc.2012.04.011
146. Pastorino, G.; Cornara, L.; Soares, S.; Rodrigues, F.; Oliveira, M. B. P. P. Liquorice (Glycyrrhiza glabra): A Phytochemical and Pharmacological Review. *Phytother. Res.* **2018,** *32*, 2323–2339. DOI: 10.1002/ptr.6178
147. Hasan, M. K.; Ara, I.; Mondal, M. S. A.; Kabir, Y. Phytochemistry, Pharmacological Activity, and Potential Health Benefits of Gly Cyrrhiza Glabra. *Heliyon* **2021,** *7*, e07240. DOI: 10.1016/j.heliyon.2021.e07240
148. Thorat, B.; Bagkar, T.; Patil, R. "Brahmi" The Memory Booster Medicinal Herb. *J. Med. Plants Stud.* **2018,** *6*, 185–187
149. Mukhgerjee, A.; Gombar, V.; Shamsi, Y.; Gupta, M.; Sinha, S. Effectiveness of Brahmi in Various Illnesses: Review Paper. *Herb. Medicine: Open Access03*. DOI: 10.21767/2472-0151.100024
150. Batool, S.; Khera, R. A.; Hanif, M. A.; Ayub, M. A. In *Medicinal Plants of South Asia*; Hanif, M. A., Nawaz, H., Khan, M. M., Byrne, H. J., Eds.; Elsevier, 2020; pp 63–74 .
151. Kee, L.; Shori, A. B.; Hj Baba, A. S. Bioactivity and Health Effects of *Mentha spicata. Integrat. Food Nutr. Metabol.* **2017,** *5*. DOI: 10.15761/IFNM.1000203
152. Mahendran, G.; Verma, S. K.; Rahman, L. U. The Traditional Uses, Phytochemistry and Pharmacology of Spearmint (*Mentha spicata* L.): A Review. *J. Ethnopharmacol.* **2021,** *278*, 11**4266**. DOI: 10.1016/j.jep.2021.114266
153. Mahboubi, M. *Mentha spicata* L. Essential Oil, Phytochemistry and Its Effectiveness in Flatulence. *J. Tradit. Complement. Med.* **2021,** *11*, 75–81. DOI: 10.1016/j.jtcme.2017.08.011
154. Vergara-Jimenez, M.; Almatrafi, M. M.; Fernandez, M. L. Bioactive Components in Moringa Oleifera Leaves Protect Against Chronic Disease. *Antioxidants* **2017,** *6*, 91. DOI: 10.3390/antiox6040091
155. Balakrishnan, R. et al. Medicinal Profile, Phytochemistry, and Pharmacological Activities of *Murraya koenigii* and Its Primary Bioactive Compounds. *Antioxidants* **2020,** *9*. DOI: 10.3390/antiox9020101

156. Parnami, M.; Varma, K. Therapeutic Potential of Murraya Koenigii (Curry Leaves) in Dyslipidemia: A Review. 2018.
157. Ahmad, A. et al. Pharmacological Importance of *Nardostachys jatamansi* DC: A Potential Therapeutic Agent in Different Pathological Ailments. *J. Chem. Pharm. Res.* **2013,** *5*, 431–438
158. Pandey, M. M.; Katara, A.; Pandey, G.; Rastogi, S.; Rawat, A. K. An Important Indian Traditional Drug of Ayurveda Jatamansi and Its Substitute Bhootkeshi: Chemical Profiling and Antioxidant Activity. *Evid.-Based Complement. Altern. Med.* **2013,** *2013*, 142517. DOI: 10.1155/2013/142517
159. Bhateja, S.; Arora, G. Therapeutic Benefits of Holy Basil (TULSI) in General and Oral Medicine: A Review. *Int. J. Res. Ayurveda Pharm.* **2012,** *3*, 761–764. DOI: 10.7897/2277-4343.03611
160. Cohen, M. M. Tulsi—*Ocimum sanctum*: A Herb for All Reasons. *J. Ayurveda Integrat. Med.* **2014,** *5*, 251–259. DOI: 10.4103/0975-9476.146554
161. Hashem-Dabaghian, F.; Ziaee, M. A Systematic Review on the Cardiovascular Pharmacology of *Emblica officinalis* Gaertn. **2018,** *10*, 118–128. DOI: 10.15171/jcvtr.2018.20
162. Syed, Q. A. et al. Nutritional and Therapeutic Properties of Fenugreek (*Trigonella foenum-graecum*): A Review. *Int. J. Food Prop.* **2020,** *23*, 1777–1791. DOI: 10.1080/10942912.2020.1825482
163. Kumar, V.; Van Staden, J. A Review of *Swertia chirayita* (Gentianaceae) as a Traditional Medicinal Plant. *Front. Pharmacol.* **2015,** *6*, 308. DOI: 10.3389/fphar.2015.00308
164. Parle, M.; Khanna, D. Clove: A Champion Spice. *Int. J. Res. Ayurveda Pharm.* **2010,** *2*
165. Ayyanar, M.; Subash-Babu, P. *Syzygium cumini* (L.) Skeels: A Review of Its Phytochemical Constituents and Traditional Uses. *Asian Pac. J. Trop. Biomed.* **2012,** *2*, 240–246. DOI: 10.1016/s2221-1691(12)60050-1
166. Amalraj, A.; Gopi, S. Medicinal Properties of *Terminalia arjuna* (Roxb.) Wight & Arn.: A Review. *J. Tradit. Complement. Med.* **2017,** *7*, 65–78. DOI: 10.1016/j.jtcme.2016.02.003
167. Saha, S.; Ghosh, S. *Tinospora cordifolia*: One Plant, Many Roles. *Ancient Sci. Life 31*, 151–159. DOI: 10.4103/0257-7941.107344
168. Bairwa, R.; Sodha, R. S.; Rajawat, B. S. *Trachyspermum ammi. Pharmacogn. Rev.* **2012,** *6*, 56–60. DOI: 10.4103/0973-7847.95871
169. Singh, N.; Bhalla, M.; de Jager, P.; Gilca, M. An Overview on Ashwagandha: A Rasayana (Rejuvenator) of Ayurveda. *Afr. J. Tradit. Complement. Altern. Med.* **2011,** *8*, 208–213. DOI: 10.4314/ajtcam.v8i5S.9
170. Rai, M.; Jogee, P. S.; Agarkar, G.; dos Santos, C. A. Anticancer Activities of *Withania somnifera*: Current Research, Formulations, and Future Perspectives. *Pharm. Biol.* **2016,** *54*, 189–197. DOI: 10.3109/13880209.2015.1027778
171. Ali, B. H.; Blunden, G.; Tanira, M. O.; Nemmar, A. Some Phytochemical, Pharmacological and Toxicological Properties of Ginger (*Zingiber officinale Roscoe*): A Review of Recent Research. *Food Chem. Toxicol.* **2008,** *46*, 409–420. DOI: 10.1016/j.fct.2007.09.085
172. Verma, R.; Pandey, M.; Saraf, S. Nutraceuticals: New Era of Medicine and Health. *Asian J. Clin. Nutr.* **2010,** *3*.
173. https://fssai.gov.in/cms/fortified-food.php
174. Guidelines on Food Fortification with Micronutrients. World Health Organization, 2006.

PART II

Nutraceutical Formulations and Regulations

CHAPTER 3

RECENT DEVELOPMENTS IN THE FORMULATION OF NUTRACEUTICALS

SIBASHISH KITYANIA[1], DEEPA NATH[2], SUBRATA DAS[1], RAJAT NATH[1], and ANUPAM DAS TALUKDAR[1]

[1]*Department of Life Science and Bioinformatics, Assam University, Silchar, India*

[2]*Department of Botany, Guru Charan College, Silchar, Assam, India*

ABSTRACT

Nutraceutical plays vital role in the treatment of different diseases. Nutraceutical compounds have potentials to be a promising alternative to modern medicine in prevention and treatment of various diseases. Aiming to enhance the nutritional value and health benefits, incorporation of nutraceutical products in daily life has increases interest. However, the low solubility and stability of nutraceutical bio-actives are limiting the direct incorporation in food products. The novel formulation strategies of nutraceutical delivery overcome the physiochemical challenges and instability of natural bioactives. The novel techniques of nutraceutical formulation includes nanotechnology, dendrimers, liposomes, nano-particles, nano-emulsion, nanoencapsulation, nanofibers, phytosomes, microsphere, micelle-based delivery system, chitosan-based delivery systems, conjugate systems, and others to improve efficacy and bioavailability of nutraceutical compounds in the body. The novel formulation and delivery system have achieved potent efficacy than their conventional counterpart and greater commercial viability. The advance formulation strategies will direct to improve therapeutic safety and fulfill the market appeal of various nutraceutical products. Recent development in

Formulations, Regulations, and Challenges of Nutraceuticals.
Tingirikari Jagan Mohan Rao, Rajesh K. Kesharwani, Raj K. Keservani, Anil Kumar Sharma (Eds.)

formulation of nutraceutical with different techniques and prudent scientific validation will give direction of nutraceutical success in future.

3.1 INTRODUCTION

"Nutraceutical" describes as medicinally active or nutritional food supplements since the late 1980.[1] "Nutraceutical" was coined by Stephen De Felice from the different two branch of science as "nutrition" and "pharmaceutical" in 1989.[2] "Nutraceuticals", that is, a combination of "nutrient" and "ceutical" where "nutrient" refers to functional foods having medicinal value and "ceutical" refers to therapy.[3] Nutraceutical are generally possessed some pharmaceutical and anti-aging properties and also the capability of preventing or reducing symptoms of many long-term diseases. It also seen that nutraceutical can improve overall wellness along with boosting immunity.[4] Nutraceutical food products are categorizing in different sub categories like polyunsaturated fatty acid, probiotics, prebiotics, dietary fiber, and antioxidants. The industries also define any nontoxic functional foods as nutraceuticals that has vital benefits on health issues and also have some potential bioactive compounds to treat/prevent many diseases.[5] These are food products that having pharmaceutical properties. Nutraceuticals are used as supplements in different countries in the world as nutraceutical are associated with many potential bio-active phytochemicals that has various health benefits.[6] There are various nutraceutical compounds also found in some higher plants, bacteria, fungi, and algae also in some Cyanobacteria. There were many studies carried out on functional foods to discover some bioactive compounds which can be used as a nutraceutical and also to make some beverages that having great health benefits. Nutraceuticals have many medicinal importances and possess various beneficial effects in the treatment of many health problems. Nutraceuticals have beneficial effects in treating various diseases include inflammation, cardiovascular diseases, obesity, hypertension, atherosclerosis, diabetes, and cancer.[7] Recently, nutraceuticals and dietary supplements are preferred approximately 80% of the global population as nutraceutical products are claim to being safe, being a drug substitute and more effective for health benefits and also it used to treat and prevent various disease without side effects.[8] Novel drug delivery system (NDDS) introduces new strategies to enhance the efficacy of nutraceuticalproducts.[9] Recently, various developments on different strategies to increase the efficiency of nutraceuticals with novel delivery system that includes dendrimers, liposomes, nano-particles, nano-emulsion, nano-encapsulation, nanofibers, phytosomes, microsphere, micelle-based delivery system, chitosan-based systems, and conjugate systems were mostly used in formulating nutraceuticals.[9–13] For the future of nutraceutical researches and development concern associated like the efficacy and presumed safety of nutraceutical need more detail study and establishing regulations to control nutraceuticals globally.[14]

3.2 CLASSIFICATION OF NUTRACEUTICALS

Nutraceuticals classified on the basis of their novelty into non-traditional and traditional nutraceuticals.[15] It also categorizes on the basis of chemical constituents like nutrients, supplements, functional food, and medical food.[16,15] On the other hand, nutraceutical can also be classify on various categories based on, their established efficiency and safety such as potential and established nutraceuticals, their structure into herbal and phytochemicals, their other parameter like use, bioavailability, and others.[17,18] The variability depends on their complex structure that results in different category but the most common and recent classification system discussed below (Fig. 3.1).[19,16]

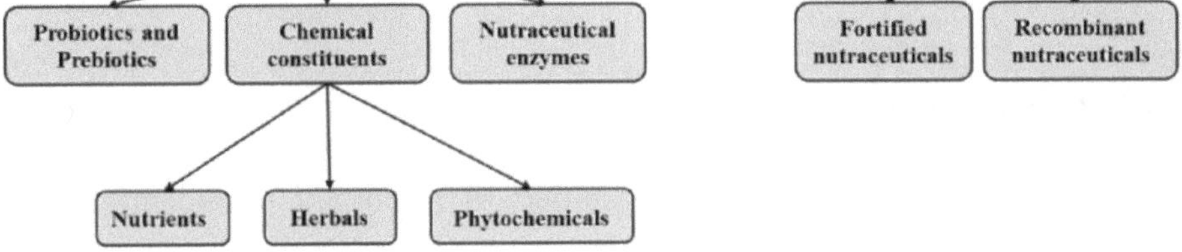

FIGURE 3.1 Outline of nutraceutical classification.

3.2.1 *TRADITIONAL NUTRACEUTICALS*

These are foods items which are not subjected to any manual changes and it includes three subcategories.[19,16] There are some examples like omega

3 fatty acid from salmon and lycopene present in tomato are traditional nutraceuticals.[15]

3.2.1.1 NUTRIENTS

The most common nutrients are vitamins and antioxidants along with some other such as nutrients amino acids, minerals, and fatty acids.[20] The soft gelatine capsule loaded with active ingredient shown significant improvement in anti-inflammatory effects.[21]

3.2.1.2 HERBALS

Herbals are mainly consisting of whole plant and/or a part of plant such as leaf, fruit, roots etc. and it also extensively used as nutraceutical for promoting health. The effects of willow bark on acne vulgaris were studied by Bassion et al. and found significant effects.[22]

3.2.1.3 PHYTOCHEMICALS

The plant secondary metabolites have specific biological effects on metabolic activity or exert biochemical reactions and also by different mechanism in the body.[19] The compound Eugenol, a phytochemical known for having anesthetic and local antiseptic effects. It is also possess antioxidant and anti-mutagenic effect in high concentration and also reported its efficacy on melanoma cell inhibition.[12,23]

3.2.2 PREBIOTICS AND PROBIOTICS

Prebiotics and probiotics also classified as important nutraceuticals that enhance health and prevents from various diseases.[24,25] Both pre- and probiotics are enhanced health by different mechanism and can prevent human from various diseases like cancer. Probiotics also have vital role in catabolizing polyphenols and also improved their absorption.[26] The use of prebiotics and probiotics for inflammatory bowel disease treatment was illustrated by Damaskos et al. and it is known as idiopathic. Insulin, lactulose, and lactosucrose are considered as most commonly used prebiotics to treat various diseases.[27]

3.2.3 NUTRACEUTICAL ENZYMES

Enzymes are mostly responsible for alleviating various health problems and regulating body function such as digestive problems, hypoglycemia, and obesity. So, the enzymes are used as nutraceutical and also to treat various diseases. A nutraceutical enzyme Nattokinase (NK), was used to treat chronic rhinosinusitis.[28]

3.2.4 FORTIFIED NUTRACEUTICALS

Foods are fortified by nutrients that include minerals, vitamins to maximize the efficacy of foods being used as a nutraceutical. It was used to treat nutrient deficiency.[29,30] The fortified food mostly effective to treat micronutrient malnutrition for many reasons such as it can be accessible to a large number of populations, socially acceptable foods and it can be introduced quickly that gives quick benefits along with its safe and cost-effective nature.[31] Folic acid[32] and Rutin,[33] a flavonol glycoside I prominently used as fortifying nutrient.

3.2.5 RECOMBINANT NUTRACEUTICALS

Recombinant nutraceutical were formulated by advance biotechnological approaches such as genetic engineering that modify the genetic makeup of the food leads to alter its properties.[34] Orcajo et al. produce milk that easily digestible through recombinant technology, generating DNA-free β-lactoglobulin gene, to minimizes the allergy effect of cow milk.[35]

3.3 NUTRACEUTICAL DELIVERY ROUTES

3.3.1 ORAL DELIVERY

Oral route of nutraceutical delivery is considered to be most preferred and acceptable method. There are various types of orally delivered nutraceuticals such as dehydroepiandrosterone, lycopene, resveratrol, glucosamine, chondroitin, lipoic acid, coenzyme-Q10, creatine, and water-soluble and fat-soluble vitamins, etc. This route of delivery system has strong preference of consumers which led scientist to focus and emphasis on improving oral delivery mechanisms.[36]

3.3.2 DERMAL DELIVERY

The dermal delivery routes result in accumulation of nutraceutical in the skin for long-term storages and repeated provision of its effects.[37] The most commonly used nutraceuticals are include N-acetylcysteine, genistein, curcumin, and fucose-rich sulfated polysaccharides for their cosmic benefits.[38,39] Nutraceuticals when applied through dermal route may become unstable due to photo degradation because it exposes to heat and light. The trans-dermal routes also used for many nutraceuticals to treat skin disease.

3.3.3 OPHTHALMIC DELIVERY

Many nutraceuticals are delivered through ophthalmic routes that can help in diseases modifying effects on pathogens. The ophthalmic route can be a safe alternative of nutraceutical delivery by simple inclusion with allopathic agents. It extents retention time of intraocular and provides synergistic clinical benefits.[40,41]

3.4 FORMULATION STRATEGIES OF NUTRACEUTICALS

Nutraceutical formulation may be hampered due to the negative effects and obstacles like the active ingredient can possess low solubility, short half-life, fast-metabolism, poor permeability, and others.[42] Novel delivery system for the development of nutraceuticals was challenging due to its complex nature, time consumption, and high price.[43] The emphasis on formulating nutraceutical is to improve the physiochemical characteristics and is more efficacious in the treatment and prevention of various diseases. To improve the efficacy and potential effects of nutraceutical, various novel strategies to formulating nutraceuticals have been incorporated for designing nutraceutical compounds to deliver on the body are discussed below (Fig. 3.2).

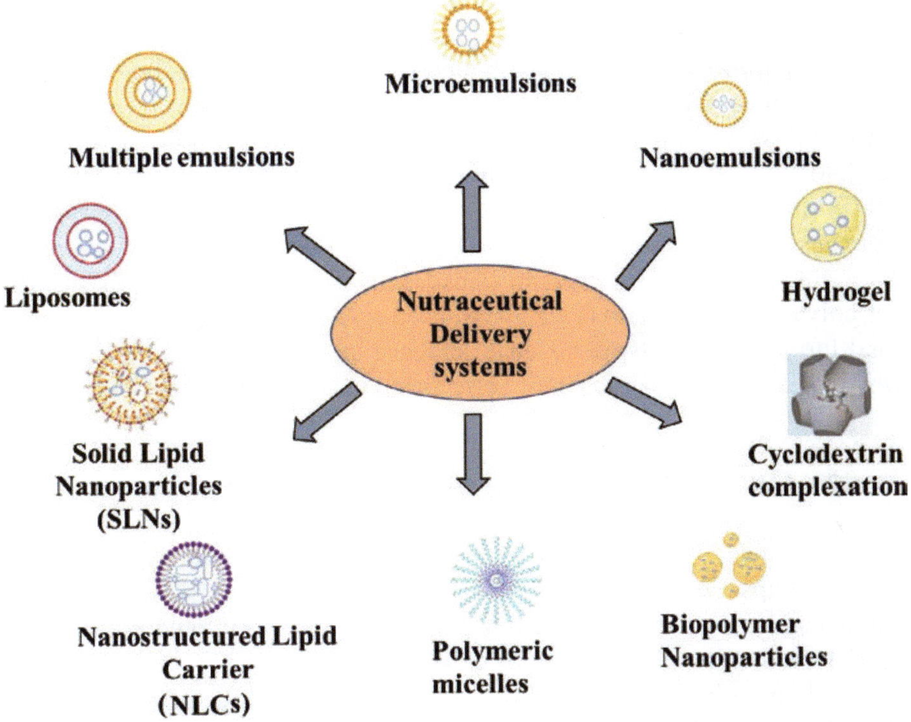

FIGURE 3.2 Different nutraceutical delivery systems.

3.4.1 LIPOSOMES

These are spherical microscopic lipid vesicle that provides an ideal delivery system for many nutraceuticals for the treatment and prevention of many skins-related problems. The delivery system uses surface charge, lipid concentration, and vesicle size as a key variable to influence the efficacy of the system. Recent study confirms the beneficial application of liposomal dermal formulation of nutraceutical delivery better than the normal dermal nutraceutical delivery system.[44] Other than the dermal formulation of liposomal technology, it has been adopted in the formulation of parenteral and oral administration of nutraceuticals in the body. Liposomal delivery system also used to improve the bioavailability and stability of resveratrol in terms of biological activity.[45] Nutraceutical like melatonin also employed liposomal systems to improve its sublingual absorption and bioavailability in comparison with conventional formulation. Some other nutraceutical formulation also used liposomal delivery system such as echinacea, milk thistle, and glucosamine.

3.4.2 ELECTROSPUN FIBER MATS

Electrospun fibers mats are formulated for the delivery of different kinds of nutraceuticals. This technology was used to deliver vitamins to skin loaded in cellulose acetate nano fiber mats.[46] The ultra-fine electrospun mats of cellulose acetate were used to deliver some popular nutraceutical including curcumin, to reduce the toxicity and improve its antioxidant, anti-inflammatory, and anti-cancerous activity.[47] The cytotoxicity of these mats is very less harmful and thus it is adopted for transdermal delivery. The nanofibrous mats of poly (ε-caprolactone) (PCL) were used to incorporate ascorbyl palmitate and further it coated with silver to increase their stability. Study also found the increase in their antibacterial effects against skin pathogen *S. aureus*.[48]

3.4.3 MICRO-SPONGES AND NANO-SPONGES

This micro and nano sponges' system of nutraceutical delivery is known as "non-collapsible" delivery system. This system of nutraceutical delivery has immense benefits due to their bioactive loading capacity and high internal surface area. This formulation of delivery system was proved to be able to protect our skin from UV-A and UV-B rays and mainly useful for reduce the allergic condition.[49] Nano-sponge formulation improves the stability of nutraceutical and it is proved to be a suitable method for buccal delivery.[50]

3.4.4 CYCLODEXTRIN COMPLEXATION

Cyclodextrin (CD) has widely used to deliver nutraceutical compounds that increase their solubility and bioavailability to the body.[51,52] This is also an effective delivery system for liver protective nutraceuticals. Silymarin complex with β-CD founds to be novel delivery system to increases its dissolution rate and efficacy compare to nutraceutical alone.[53] Garlic oil and β-CD complex also utilized to formulate novel nutraceutical that increase its antioxidant and antimicrobial properties.[54]

3.4.5 BIODEGRADABLE HYDROGELS

Biodegradable hydrogel delivery system used in both the field of pharmaceutical and nutraceutical compounds to modify and protect the compounds from degradation or denaturation. The riboflavin incorporated with soy protein cold-set hydrogel protects the vitamin from degradation in intestine.[55] It was found that milk proteins also very successful hydrogel protein carriers due to its physicochemical and ideal structure for the transport of bioactive nutraceuticals.[56] Recent studies found that the hydrogel technique of nutraceutical delivery system is very prominent in terms of controlled manner delivery and its stable response in various conditions.

3.4.6 PH RESPONSIVE FORMULATIONS

The pH responsive delivery mainly controls the nutraceutical delivery by specific pH change and it is an excellent method for site-specific delivery.[57] This system also useful in nutraceutical industry to deliver various nutraceutical compounds to get maximum bioactivity response. The bio-responsive chitosan nanoparticle loaded with EGCG that slowly release at acidic in gastric condition while rapidly at alkaline in intestinal condition shown potential health benefits.[58] Casein considered as a good candidate for targeting and delivering nutraceutical like caffeine due to its high ion binding affinity and good emulsifying properties.[59]

3.4.7 SOLID DISPERSIONS

This method is used to increase the bioavailability of nutraceutical compounds through oral delivery. The amorphous solid dispersion system that comprises of a hydrophilic polymer matrix increases the solubility and rate of dissolution of the bioactive. Resveratrol mixed with bioactive polymers to forms solid dispersion for delivery.[60]

3.4.8 SELF-EMULSIFYING DRUG DELIVERY SYSTEM

A novel strategy to improve the oral bioavailability of different nutraceutical compounds. SEDDS formulation of Vitamin E significantly improves the vitamin bioavailability via oral delivery. Recent study reported the enhancement in bioavailability of tocotrienols through SEDDS formulation.[61] Researcher formulated a self-microemulsifying DDS (SMEDDS) of plant extracts of K. Parviflora to improve their low solubility and poor oral bioavailability.[62] This strategy not only helps in increasing the bioavailability of nutraceuticals, but also helps in targeted delivery of herbal extract.

3.4.9 MICROSPHERES AND MICROCAPSULES

There are various herbal nutraceuticals that possess antioxidant and antimicrobial properties were used as traditional drinks. A novel formulation of microparticles powder of maltodextrin/pectin matrix using spray-drying method increases the efficacy of the nutraceutical.[63] Researcher formulated lycopene, using microencapsulation methods that improve the stability and antioxidant activity of lycopene in food-based inclusion.[64] Microsphere technique of nutraceutical formulation mainly used in cosmetics and protects the nutraceuticals from invasive effects. Research also reported that the successful formulation of biodegradable microsphere to treat herpes and Epstein-Bar virus containing vitamin A palmitate and acyclovir.[65]

3.4.10 NANOTECHNOLOGY-BASED FORMULATION OF NUTRACEUTICAL

Nanotechnological approaches have been used to improve nutraceutical bioavailability and solubility. This technique potentially increases the commercial value of nutraceuticals. There is various types of nano-technological approaches (Fig. 3.3) are discussed below.

3.4.10.1 NANOSUSPENSION AND NANO-EMULSION

The preparation of nanosuspension and nano-emulsion of hydrophobic nutraceuticals improves the bioavailability and enhances its efficacy. Lutein used as antioxidant, prepared by high pressure homogenization into nano size particle to produce nano-suspension.[66] Nano-emulsions have also very effective role in the delivery of micronutrient and far soluble vitamins within liposome via oral route.[67] The nano-emulsion formulation of hyaluronic acid (HA) with oil/water investigated as a carrier of lipophilic bioactive and found effective for nutraceutical delivery.[68]

3.4.10.2 SOLID LIPID NANOPARTICLES (SLNS)

It is nano-ranges spherical structure and size around 40 to 100 nm. SLNs mainly consist of surfactant and solid fat, suitable for both lipophilic and hydrophilic drug.[69] The α-lipoic acid is a useful nutraceutical for the treatment of delicate areas and it suffers from early chemical degradation. To overcome the problem, the formulation of SLNs incorporating α-lipoic acid was made. This system also useful in increasing the skin hydration and protect against UV irradiation.[70] Furthermore, it was observed that both pharmaceutical and natural ingredients can be loaded into SLNs and used in treating/preventing various diseases such as liver diseases, kidney diseases, diabetes, hyperglycemia, and others.[71]

3.4.10.3 NANOSTRUCTURED LIPID CARRIERS (NLCS)

The nano-sized lipid that comprises of lipid matrix with a special nanostructure holds potential for optimum nutraceutical delivery via oral, dermal, and tropical route. Fish oil-based NLC's was developed for the delivery of lutein to enhance dual benefits and also improve delivery mechanism of the nutraceutical compounds.[72] It is also used to deliver the herbal nutraceutical intravenously like silybin, to improve its efficacy against liver diseases. The novel formulation of silybin-NLC achieved a prolong residence time in the serum and improved its protective action.[73]

3.4.10.4 NANO MICELLES

This formulation was found to be more stable and enhance solubility that profoundly improved the anti-cancerous activity of quercetin.[74] Another group of researchers have developed nanoencapsulation of vitamin D_3 using casein micellar to improve the protection against UV degradation of the nutraceutical molecule.[75]

3.4.10.5 NANOPARTICLES, NANO-CAPSULES, AND NANOSPHERE

The nano formulation of the nutraceutical like epigallocatechin-3-gallate (EGCG) was developed to protect and enhance the delivery of nutraceuticals.[76] The milk protein like casein is being exploited to produce nano-encapsulate using β-carotene to deliver other nutraceutical molecules like vitamin D and fatty acid.[56] Nanoparticle developed by encapsulating thymol and carvacrol essential oil in zein nanoparticle possess positive antimicrobial, antifungal, and insecticidal.[77] The nano-capsules and nanosphere both are polymeric nanoparticles with a diameter of 1 μm.[78] The formulation of vitamin E-loaded nano-capsule to accelerate the stability of nutraceuticals delivery also reported.[79] Thus, the encapsulating the nutraceutical in nano-capsule improves the solubility of the nutraceutical molecule as well as preserves their activity from degradation and providing stability.

3.4.10.6 METAL NANOPARTICLES

Some metals such as silver (Ag), zinc oxide (ZnO), iron (Fe), and titanium oxide (TiO_2) is prepared in nano range to be used as a carrier for nutraceutical delivery. Metal nanoparticles have various advantages that include improved biocompatibility and catalytic activity.[80] Nutraceuticals were encapsulated into silver nanoparticles and the formulation showed potent antibacterial and antioxidant activity.[81]

3.4.10.7 COMPOSITE COLLOIDAL NANOPARTICLES

The hybrid nanoparticles that consist of an organic shell and inorganic core termed as composite nanoparticles. The formulation of nanocomposites nutraceutical to increase the efficacy of the nutraceutical compounds. This system of nutraceutical delivery increases the antioxidant, anti-cancerous, and anti-inflammatory effects of nutraceuticals. The synthesis of zein-based composite colloidal nanoparticle to encapsulated epigallocatechin-3-gallate (EGCG) has shown significant increase in its stability.[82] Curcumin also encapsulated into zein-shellac nanocomposite particles to increase its stability.[83]

3.4.10.8 SELF-ASSEMBLY NANOPARTICLES

The self-assembled nanoparticles are cost-effective and very simple to prepare.[84] The encapsulation of epigallocatechin-3-gallate into self-assembled nanoparticle that comprises of chitosan and aspartic acid has shown potential effects against atherosclerosis.[85]

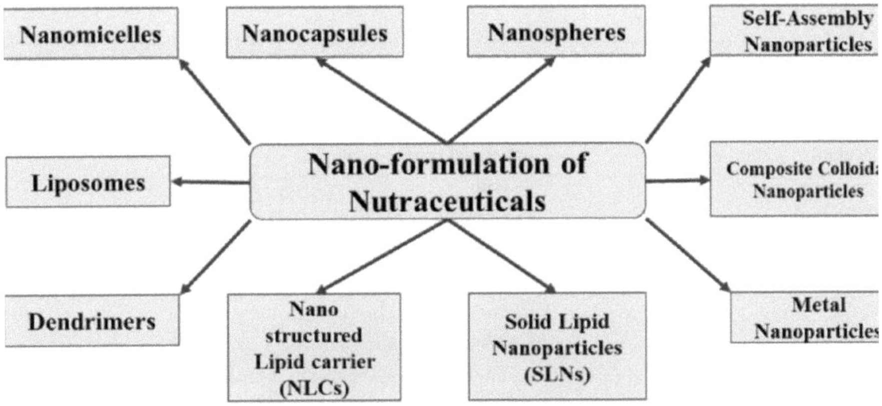

FIGURE 3.3 Overview of nano-formulation of nutraceuticals.

TABLE 3.1 Summary of Novel Formulation Strategy and Health Benefits of Nutraceutical.

Nutraceutical ingredients	Novel formulation strategy	Description	Therapeutic uses/ health benefits	References
CoenzymeQ10	Liposomal carrier systems	To improve skin penetration	Skin protection	[86]
	Self-emulsifying drug delivery system	Formulation with Self-emulsifying drug delivery system	Increase bioavailability antioxidant	[87, 88]
	Solid dispersions	Solid dispersion with solubility, stability and dissolution	Increase solubility and stability Antioxidant	[88, 89]
Vitamin E	Particle coating	To release the bioactive the contract lenses coated with Vitamin E	Increase the bioavailability to get maximum efficacy	[40]
	Nanotechnology	Nanoemulsion of hyaluronic acid with superior partitioning	Reduce the risk of cancer antioxidant	[90, 91]
		Nanocapsulation to increases stability and thermal integrity	Antioxidant	[68, 79]
	Microparticulate systems	Nylon microspheres	Antioxidant	[49]
EGCG Vitamin E	Nutraceutical-derivatives	To improve bioactivity EGCG-fatty acid ester Improve the efficacy of tocopherol-based derivatives	Anti-inflammatory, antioxidant, and antiviral, anti-cancerous	[91, 92]

TABLE 3.1 *(Continued)*

Nutraceutical ingredients	Novel formulation strategy	Description	Therapeutic uses/ health benefits	References
Vitamin E acetate	Liposomal carrier systems	Formulation with liposomal carrier to deposit in skin	Dermal protection	[44]
Melatonin	Liposomal carrier system	To improve its bioavailability via oral delivery	Antioxidant	[93]
	Nanotechnology	Solid lipid carrier formulation to achieve the sustained plasma levels	Antioxidant	[94]
Resveratrol	Liposomal carrier system	To improve sustained and stability	Anti-inflammatory, antioxidant, anti-cancer	[45]
	Solid dispersion	To improved stability	Anti-inflammatory, antioxidant, anti-cancer	[60]
	Nanotechnology	Solid lipid nanoparticle formulation for enhance efficacy and improve skin penetration	Anti-inflammatory, antioxidant, anti-cancer	[95]
	Microsponges and nanosponges	The formulation of cyclodextrin-based nanosponges to improve the solubility, stability in buccal delivery	Anti-inflammatory, antioxidant	[50]
Zinc gluconate	Liposomal carrier systems	Improved area under curve of oral formulations		[93]
Vitamin A/ α-Tocopherol	Electrospun fiber mats	Cellulose acetate nanofiber formulation to increase bioavailability		[46]
Vitamin A	Cyclodextrin complexation	Hydroxypropyl-β-cyclodextrin complex with improved solubility	Skin protection Antioxidant	[67]
Asiaticoside	Electrospun fiber mats	Cellulose acetate mats loaded with herbs for wound dressing	Wound healing	[96]
Ascorbyl palmitate	Electrospun fiber mats	To improve stability the formulation of poly (ε-caprolactone) (PCL) nanofibrous mat	Antibacterial	[48]
Melanin	Microsponges	Formulation of Melanosponge-αto protect skin from UV rays	Skin protection	[49]

TABLE 3.1 *(Continued)*

Nutraceutical ingredients	Novel formulation strategy	Description	Therapeutic uses/ health benefits	References
Acemannan	Microsponges and nanosponges	Formulation of sponges to regenerate periodontal tissues	Tooth supports	[97]
Silymarin	Cyclodextrin complexation	Formulation of β-cellulose acetate to improve sustained release	Antioxidant Anti-cancer Hepatoprotective	[53]
Garlic oil	Cyclodextrin complexation	Formulation of β-cellulose acetate to improve controlled release	Antimicrobial and antioxidant	[54]
Astaxanthin	Cyclodextrin complexation	The formulation of complex with hydroxypropyl-β-cyclodextrin (HPCD) To improved solubility and controlled release	Immuno-booster Anti-inflammatory	[52]
Curcumin	Cyclodextrin complexation	Formulation of complexes to improve	Antioxidant, anti-bacterial, anti-inflammatory Steatosis treatment	[98]
Lycopene	Microparticulate systems	Formulation of microcap-sule to improve stability		[64]
Riboflavin	Biodegradable hydrogels	Formulation of soy protein hydrogel	Gastric protection	[55. 64]
	Biodegradable hydrogels	Polyglycidyl methacrylate -g-sodium alginate hydrogel for pH sensitive delayed release	Gastric protection	[99]
Kaempferia parviflora	Self-emulsifying drug delivery system	Self-micro emulsifying drug delivery system formulation with Caco-2 cells to increase penetration and enhance oral delivery	Antioxidant Anti-cancer	[62]
	Cyclodextrin complexation	2-HP-b-CD complexes with permeation and oral bioavailability	Antioxidant Anti-cancer Cardioprotective	[62]
Ginseng	Biodegradable hydrogels	Formulation of Alginate films	Antioxidant	[100]

TABLE 3.1 *(Continued)*

Nutraceutical ingredients	Novel formulation strategy	Description	Therapeutic uses/ health benefits	References
Milk thistle	Self-emulsifying drug delivery system	Formulation of self-emulsifying drug delivery system with targeted delivery to lymphatic system	Anti-cancer Antioxidant Immunity booster	[101]
Aloe vera	Biodegradable hydrogels	Formulation of Alginate hydrogel	Wound healing dressing	[102]
Lutein	Nanotechnology	Formulation with nanocrystal suspension	Antioxidant	[66]
	Nanotechnology	Nanostructured lipid carrier formulation to increase the bioavailability	Gastric protection	[103]
Quercetin	Nanotechnology	Formulation of nano-micelles to improve stability and solubility	Anti-cancerous	[103]
β-Carotene	Nanotechnology	Formulation of Casein/dextran nanoparticles for solubility and activity	Antioxidant Preventing heart disease	[104]
Thymol	Nanotechnology	Formulate with Zein nanoparticles to improve solubility	Antioxidant and anti-microbial	[77]
Carvacrol	Nanotechnology	Formulate with Zein nanoparticles to improve solubility	Antioxidant and anti-microbial	[77]
Silybin	Nanotechnology	Nanostructured lipid carrier formulation to improve biphasic release	Antioxidant	[73]
Vitamins A, D, E, K	Nanotechnology	Formulation of nano-emulsion to increase the solubility	Improve absorption	[67]

3.5 CONCLUSION AND FUTURE PROSPECTIVE

The consumption of nutraceutical is increasing due to the increase in lifestyle diseases and adverse side-effect of allopathic medication. There are various advance formulation strategies incorporated into develop nutraceutical efficacy. Nanotechnology has a vast importance in nutraceutical formulation and its application will address several challenges currently faced by nutraceutical development. Nanotechnological approaches of nutraceutical formulation are getting attention due to its promising results regarding safety, bioavailability, stability, and targeted release of bioactive. There are lots of nano-sized nutraceuticals which are available commercially in the market as they enhance the absorption and bioavailability profile of nutraceuticals. However, there are other means of nutraceutical delivery system were adopted for novel formulation including liposomes, dendrimers, nanoemulsion, microparticles, solid dispersions, electrospun fiber mats, and others. Nutraceuticals compounds combined with multicomponent products results optimal function and these products also easily accessible and affordable. The current status of food supplement and nutraceutical in the global market is alarming. These products are claimed to have major therapeutic potential as they safe to use and mainly obtain from natural sources. The novel formulation strategies were implemented to produce nutraceutical products that increase efficacy and also sales in global market. Nutraceutical is poorly defined across the globe and no regulatory body is formed to regulate. However, clear and common regulations for nutraceuticals will be urgently needed in the near future to cope with rapidly emerging trends and demands in the global market.

KEYWORDS

- **nutraceutical**
- **novel delivery**
- **nanotechnology**
- **formulation**
- **food product**

REFERENCES

1. DeFelice, S. L. The Nutraceutical Revolution: Its Impact on Food Industry R&D. *Trends Food Sci. Technol.* **1995,** *6* (2), 59–61.
2. Maddi, V.; Aragade, P.; Digge, V.; Nitalikar, M. Short Review Importance of Nutraceuticals in Health Management. *Phcog Rev.* **2007,** *1* (2).
3. Borodina, I.; Kenny, L. C.; McCarthy, C. M.; Paramasivan, K.; Pretorius, E.; Roberts, T. J.; Kell, D. B. The Biology of Ergothioneine, an Antioxidant Nutraceutical. *Nutr. Res. Rev.* **2020,** *33* (2), 190–217.

4. Sharma, R. Nutraceuticals and Nutraceutical Supplementation Criteria in Cancer: A Literature Survey. *Open Nutraceuticals J.* **2009,** *2* (1).
5. Das, L.; Bhaumik, E.; Raychaudhuri, U.; Chakraborty, R. Role of Nutraceuticals in Human Health. *J. Food Sci. Technol.* **2012,** *49* (2), 173–183.
6. Al-Obaidi, J. R.; Alobaidi, K. H.; Al-Taie, B. S.; Wee, D. H.-S.; Hussain, H.; Jambari, N. N.; Ariffin, N. S. Uncovering Prospective Role and Applications of Existing and New Nutraceuticals from Bacterial, Fungal, Algal and Cyanobacterial, and Plant Sources. *Sustainability* **2021,** *13* (7), 3671.
7. Nasri, H.; Baradaran, A.; Shirzad, H.; Rafieian-Kopaei, M. . New Concepts in Nutraceuticals as Alternative for Pharmaceuticals. *Int. J. Prev. Med.* **2014,** *5* (12), 1487.
8. Ekor, M. The Growing Use of Herbal Medicines: Issues Relating to Adverse Reactions and Challenges in Monitoring safety. *Front. Pharmacol.* **2014,** *4,* 177.
9. Amol, K. Pratibha, P. Novel Drug Delivery System in Herbal's. *Int. J. Pharm. Chem. Sci. B.* **2014,** *4* (4).
10. Asghar, A.; Randhawa, M. A.; Masood, M. M.; Abdullah, M. Irshad, M. A. . Nutraceutical Formulation Strategies to Enhance the Bioavailability and Efficiency: An Overview. *Role Mater. Sci. Food Bioeng.* **2018,** 329–352.
11. Gonçalves, R. F.; Martins, J. T.; Duarte, C. M.; Vicente, A. A.; Pinheiro, A. C. Advances in Nutraceutical Delivery Systems: From Formulation Design for Bioavailability Enhancement to Efficacy and Safety Evaluation. *Trends Food Sci. Technol.* **2018,** *78,* 270–291.
12. Helal, N. A.; Eassa, H. A.; Amer, A. M.; Eltokhy, M. A.; Edafiogho, I.; Nounou, M. I. . Nutraceuticals' Novel Formulations: The Good, the Bad, the Unknown and Patents Involved. *Recent Pat. Drug Deliv. Formul.* **2019,** *13* (2), 105–156.
13. Saraf, S. Applications of Novel Drug Delivery System for Herbal Formulations. Fitoterapia 2010, *81* (7), 680–689.
14. Santini, A.; Novellino, E. Nutraceuticals-Shedding Light on the Grey Area Between Pharmaceuticals and Food. *Expert Rev. Clin. Pharmacol.***2018,** *11* (6), 545–547.
15. Swaroopa, G.; Srinath, D. Nutraceuticals and Their Health Benefits. *Int. J. Pure App. Biosci.* **2017,** *5* (4), 1151–1155.
16. Ruchi, S. Role of Nutraceuticals in Health Care: A Review. *Int. J. Green Pharm.* **2017,** *11* (03).
17. Chintale Ashwini, G.; Kadam Vaishali, S.; Sakhare Ram, S.; Birajdar Ganesh, O.; Nalwad Digambar, N. Role of Nutraceuticals in Various Diseases: A Comprehensive Review. *Int. J. Res. Pharm. Chem.* **2013,** *3,* 290–299.
18. Mukherjee, S. Assessment of Corticosteroid Utilization Pattern Among Dermatology Outpatients in a Tertiary Care Teaching Hospital in Eastern India. *Int. J. Green Pharm.* **2017,** *10* (04).
19. Padmavathi, D. A General Review on "Nutraceuticals": Its Golden Health Impact Over Human Community. *Int. J Food Sci. Nutri.* **2018,** *3bv* (2), 214–217.
20. Sapkale, A. P.; Thorat, M. S.; Vir, P. R.; Singh, M. C. Nutraceuticals-Global Status and Applications: A Review. *Int. J. Pharm. Chem. Sci.* **2012,** *1* (3), 1166–1181.
21. Witham, P. H.; Edward Lee, P. Formulations Containing Omega-3 Fatty Acids or Esters Thereof and Maqui Berry Extract and Therapeutic Uses Thereof. In: Google Patents, 2018.

22. Bassino, E.; Gasparri, F.; Munaron, L. Pleiotropic Effects of White Willow Bark and 1, 2-Decanediol on Human Adult Keratinocytes. *Skin Pharmacol. Physiol.* **2018,** *31* (1), 10–18.
23. Jaganathan, S. K.; Supriyanto, E. Antiproliferative and Molecular Mechanism of Eugenol-Induced Apoptosis in Cancer Cells. *Molecules* **2012,** *17* (6), 6290–6304.
24. Kerry, R. G.; Patra, J. K.; Gouda, S.; Park, Y.; Shin, H.-S.; Das, G. Benefaction of Probiotics for Human Health: A Review. *J. Food Drug Anal.* **2018,** *26* (3), 927–939.
25. Singh, J.; Ancheria, R. K.; Khinchi, M.; Nama, N.; Singh, S. A Review on Food Supplement-Nutraceuticals. *Asian J. Pharm. Res. Dev.* **2017,** 1–7.
26. Thilakarathna, W. W.; Langille, M. G.; Rupasinghe, H. V. Polyphenol-Based Prebiotics and Synbiotics: Potential for Cancer Chemoprevention. *Curr. Opin. Food Sci.* **2018,** *20,* 51–57.
27. Damaskos, D.; Kolios, G. Probiotics and Prebiotics in Inflammatory Bowel Disease: Microflora 'on the Scope'. *Br. J. Clin Pharmcol.* **2008,** *65* (4), 453–467.
28. Takabayashi, T.; Imoto, Y.; Sakashita, M.; Kato, Y.; Tokunaga, T.; Yoshida, K.; Immunology, C. Effects of Nattokinase, Profibrinolytic Enzyme, on the Patients with Chronic Rhinosinusitis with Nasal Polyp. *J. Allergy Clin. Immunol.* **2018,** *141* (2), AB165.
29. Aronson, J. K. Defining 'Nutraceuticals': Neither Nutritious Nor Pharmaceutical. *Br. J. Clin Pharmcol.* **2017,** *83* (1), 8–19.
30. Mejia, L. A. Fortification of Foods: Historical Development and Current Practices. *Food Nutr. Bull.* **1994,** *15* (4), 1–4.
31. Allen, L.; De Benoist, B.; Dary, O.; Hurrell, R. *Guidelines on Food Fortification with Micronutrients*; WHO/FAO, 2006.
32. Duthie, S. J. Folic Acid Deficiency and Cancer: Mechanisms of DNA Instability. *Br. Med Bull.* **1999,** *55* (3), 578–592.
33. Yang, J.; Guo, J.; Yuan, J. In Vitro Antioxidant Properties of Rutin. *Lebensmittel-Wissenschaft + Technologie* **2008,** *41* (6), 1060–1066.
34. Bawa, A.; Anilakumar, K. Genetically Modified Foods: Safety, Risks and Public Concerns—A Review. *J. Food Sci. Technol.* **2013,** *50* (6), 1035–1046.
35. Orcajo, J.; de Marañon, I. M.; Lavilla, M. Cow's Milk Allergen β-Lactoglobulin Immunoreactivity Affected by Pulsed Light Treatment. *Clin Transl. Allergy* **2015,** *5* (3), 1–1.
36. Braithwaite, M. C.; Tyagi, C.; Tomar, L. K.; Kumar, P.; Choonara, Y. E.; Pillay, V. Nutraceutical-Based Therapeutics and Formulation Strategies Augmenting Their Efficiency to Complement Modern Medicine: An Overview. *J. Funt. Foods* **2014,** *6,* 82–99.
37. Meinke, M. C.; Darvin, M. E.; Vollert, H.; Lademann, J. Bioavailability of Natural Carotenoids in Human Skin Compared to Blood. *Eur J. Pharm. Biopharm* **2010,** *76* (2), 269–274.
38. Schwarz, J. C.; Baisaeng, N.; Hoppel, M.; Löw, M.; Keck, C. M.; Valenta, C. Ultra-small NLC for Improved Dermal Delivery of Coenyzme Q10. *Int. J. Pharm.* **2013,** *447* (1–2), 213–217.
39. Wijesinghe, W.; Jeon, Y.-J. Biological Activities and Potential Industrial Applications of Fucose Rich Sulfated Polysaccharides and Fucoidans Isolated from Brown Seaweeds: A Review. *Carbohydrate Polym.* **2012,** *88* (1), 13–20.

40. Peng, C.-C.; Kim, J.; Chauhan, A. Extended Delivery of Hydrophilic Drugs from Silicone-Hydrogel Contact Lenses Containing Vitamin E Diffusion Barriers. *Biomaterials* **2010,** *31* (14), 4032–4047.
41. Zhang, J.; Wang, S. Topical Use of Coenzyme Q10-Loaded Liposomes Coated with Trimethyl Chitosan: Tolerance, Precorneal Retention and Anti-Cataract Effect. *Int. J. Pharm.* **2009,** *372* (1–2), 66–75.
42. Zaki, N. M. Progress and Problems in Nutraceuticals Delivery. *J. Bioequvalence Bioavailability* **2014,** *6* (3), 75.
43. Tiwari, G.; Tiwari, R.; Sriwastawa, B.; Bhati, L.; Pandey, S.; Pandey, P.; Bannerjee, S. K. Drug Delivery Systems: An Updated Review. *Int. J. Pharm.* **2012,** *2* (1), 2.
44. Padamwar, M. N.; Pokharkar, V. B. Development of Vitamin Loaded Topical Liposomal Formulation Using Factorial Design Approach: Drug Deposition and Stability. *Int. J. Pharm.* **2006,** *320* (1–2), 37–44.
45. Amri, A.; Chaumeil, J.; Sfar, S.; Charrueau, C. Administration of Resveratrol: What Formulation Solutions to Bioavailability Limitations? *J. Controlled Release* **2012,** *158* (2), 182–193.
46. Taepaiboon, P.; Rungsardthong, U.; Supaphol, P. Vitamin-Loaded Electrospun Cellulose Acetate Nanofiber Mats as Transdermal and Dermal Therapeutic Agents of Vitamin A Acid and Vitamin E. *Eur. J. Pharm. Biopharm.* **2007,** *67* (2), 387–397.
47. Suwantong, O.; Opanasopit, P.; Ruktanonchai, U.; Supaphol, P. Electrospun Cellulose Acetate Fiber Mats Containing Curcumin and Release Characteristic of the Herbal Substance. *Polymers* **2007,** *48* (26), 7546–7557.
48. Paneva, D.; Manolova, N.; Argirova, M.; Rashkov, I. Antibacterial Electrospun po(ly-Caprolactone)/Ascorbyl Palmitate Nanofibrous Materials. *Int. J. Pharm.* **2011,** *416* (1), 346–355.
49. Patravale, V.; Mandawgade, S. Novel Cosmetic Delivery Systems: An Application Update. *Int. J. Cosmetic Sic.* **2008,** *30* (1), 19–33.
50. Ansari, K. A.; Vavia, P. R.; Trotta, F.; Cavalli, R. Cyclodextrin-Based Nanosponges for Delivery of Resveratrol: In Vitro Characterisation, Stability, Cytotoxicity and Permeation Study. *AAPS PharmSciTech.* **2011,** *12* (1), 279–286.
51. Yuan, C.; Jin, Z.; Xu, X.; Zhuang, H.; Shen, W . Preparation and Stability of the Inclusion Complex of Astaxanthin with Hydroxypropyl-β-Cyclodextrin. *Food Chem.* **2008,** *109* (2), 264–268.
52. Yuan, C.; Jin, Z.; Xu, X. Inclusion Complex of Astaxanthin with Hydroxypropyl-β-Cyclodextrin: UV, FTIR, 1H NMR and Molecular Modeling Studies. *Carbohydr Polym.* **2012,** *89* (2), 492–496.
53. Ghosh, A.; Biswas, S.; Ghosh, T. Preparation and Evaluation of Silymarin β-Cyclodextrin Molecular Inclusion Complexes. *J. Young Pharm.* **2011,** *3* (3), 205–210.
54. Wang, J.; Cao, Y.; Sun, B.; Wang, C. Physicochemical and Release Characterisation of Garlic Oil-β-Cyclodextrin Inclusion Complexes. *Food Chem.* **2011,** *127* (4), 1680–1685.
55. Maltais, A.; Remondetto, G. E.; Subirade, M. Soy Protein Cold-Set Hydrogels as Controlled Delivery Devices for Nutraceutical Compounds. *Food Hydrocolloids* **2009,** *23* (7), 1647–1653.
56. Livney, Y. D. Milk Proteins as Vehicles for Bioactives. *Curr. Opin. Colloids Interface Sci.* **2010,** *15* (1–2), 73–83.
57. Zhu, Y. J.; Chen, F. pHResponsive Drug Delvery Systems. *Chem. Asian J.* **2015,** *10* (2), 284–305.

58. Khan, N.; Bharali, D. J.; Adhami, V. M.; Siddiqui, I. A.; Cui, H.; Shabana, S. M. et al. Oral Administration of Naturally Occurring Chitosan-Based Nanoformulated Green Tea Polyphenol EGCG Effectively Inhibits Prostate Cancer Cell Growth in a Xenograft Model. *Carcinogenisis* **2014,** *35* (2), 415–423.
59. Elzoghby, A. O.; El-Fotoh, W. S. A.; Elgindy, N. A. Casein-Based Formulations as Promising Controlled Release Drug Delivery Systems. *J. Controlled Release* **2011,** *153* (3), 206–216.
60. Wegiel, L. A.; Mauer, L. J.; Edgar, K. J.; Taylor, L. S. Crystallization of Amorphous Solid Dispersions of Resveratrol During Preparation and Storage—Impact of Different Polymers. *J. Pharm. Sci.* **2013,** *102* (1), 171–184.
61. Frank, J.; Chin, X. W. D.; Schrader, C.; Eckert, G. P.; Rimbach, G. Do Tocotrienols Have Potential as Neuroprotective Dietary Factors? *Ageing Res. Rev.* **2012,** *11* (1), 163–180.
62. Mekjaruskul, C.; Yang, Y.-T.; Leed, M. G.; Sadgrove, M. P.; Jay, M.; Sripanidkulchai, B. Novel Formulation Strategies for Enhancing Oral Delivery of Methoxyflavones in Kaempferia Parviflora by SMEDDS or Complexation with 2-Hydroxypropyl-β-Cyclodextrin. *Int. J. Pharm.* **2013,** *445* (1–2), 1–11.
63. Sansone, F.; Mencherini, T.; Picerno, P.; d'Amore, M.; Aquino, R. P.; Lauro, M. R. Maltodextrin/Pectin Microparticles by Spray Drying as Carrier for Nutraceutical Extracts. *J. Food Eng.* **2011,** *105* (3), 468–476.
64. Rocha, G. A.; Fávaro-Trindade, C. S.; Grosso, C. R. F. Microencapsulation of Lycopene by Spray Drying: Characterization, Stability and Application of Microcapsules. *Food Bioprod. Process.* **2012,** *90* (1), 37–42.
65. Martinez-Sancho, C.; Herrero-Vanrell, R.; Negro, S. Vitamin A Palmitate and Aciclovir Biodegradable Microspheres for Intraocular Sustained Release. *Int. J. Pharm.* **2006,** *326* (1–2), 100–106.
66. Mitri, K.; Shegokar, R.; Gohla, S.; Anselmi, C.; Müller, R. H. Lutein Nanocrystals as Antioxidant Formulation for Oral and Dermal Delivery. *Int. J. Pharm.* **2011,** *420* (1), 141–146.
67. Gonnet, M.; Lethuaut, L.; Boury, F. New Trends in Encapsulation of Liposoluble Vitamins. *J. Controlled Release* **2010,** *146* (3), 276–290.
68. Kong, M.; Chen, X. G.; Kweon, D. K.; Park, H. J. Investigations on Skin Permeation of Hyaluronic Acid Based Nanoemulsion as Transdermal Carrier. *Carbohydr. Polym.* **2011,** *86* (2), 837–843.
69. Ghasemiyeh, P.; Mohammadi-Samani, S. Solid Lipid Nanoparticles and Nanostructured Lipid Carriers as Novel Drug Delivery Systems: Applications, Advantages and Disadvantages. *Res. Pharm. Sci.* **2018,** *13* (4), 288.
70. Souto, E.; Müller, R.; Gohla, S. A Novel Approach Based on Lipid Nanoparticles (SLN®) for Topical Delivery of α-Lipoic Acid. *J. Microencapsul.* **2005,** *22* (6), 581–592.
71. Reyes, M. Use of Ellagic Acid Dihydrate in Food Products and Nutraceuticals. In: Google Patents, 2017.
72. Lacatusu, I.; Mitrea, E.; Badea, N.; Stan, R.; Oprea, O.; Meghea, A. Lipid Nanoparticles Based on Omega-3 Fatty Acids as Effective Carriers for Lutein Delivery. Preparation and In Vitro Characterization Studies. *J. Funt. Foods* **2013,** *5* (3), 1260–1269.
73. Jia, L.; Zhang, D.; Li, Z.; Duan, C.; Wang, Y.; Feng, F.Nanostructured Lipid Carriers for Parenteral Delivery of Silybin: Biodistribution and Pharmacokinetic Studies. *Colloids Surf. B Biointerfaces* **2010,** *80* (2), 213–218.

74. Tan, B.-J.; Liu, Y.; Chang, K.-L.; Lim, B. K.; Chiu, G. N. Perorally Active Nanomicellar Formulation of Quercetin in the Treatment of Lung Cancer. *Int. J. Nanomed.* **2012,** *7,* 651.
75. Semo, E.; Kesselman, E.; Danino, D.; Livney, Y. D. Casein Micelle as a Natural Nano-Capsular Vehicle for Nutraceuticals. *Food Hydrocolloids* **2007,** *21* (5–6), 936–942.
76. Shpigelman, A.; Israeli, G.; Livney, Y. D. Thermally-Induced Protein–Polyphenol Co-Assemblies: Beta Lactoglobulin-Based Nanocomplexes as Protective Nanovehicles for EGCG. *Food Hydrocolloids* **2010,** *24* (8), 735–743.
77. Wu, Y.; Luo, Y.; Wang, Q. Antioxidant and Antimicrobial Properties of Essential Oils Encapsulated in Zein Nanoparticles Prepared by Liquid–Liquid Dispersion Method. *LWT–Food Sci. Technol.* **2012,** *48* (2), 283–290.
78. Guterres, S. S.; Alves, M. P.; Pohlmann, A. R. Polymeric Nanoparticles, Nanospheres and Nanocapsules, for Cutaneous Applications. *Drug Target Insights* **2007,** *2,* 117739280700200002.
79. Khayata, N.; Abdelwahed, W.; Chehna, M. F.; Charcosset, C.; Fessi, H. Preparation of Vitamin E Loaded Nanocapsules by the Nanoprecipitation Method: From Laboratory Scale to Large Scale Using a Membrane Contactor. *Int. J. Pharm.* **2012,** *423* (2), 419–427.
80. Zahmakıran, M.; Özkar, S. Metal Nanoparticles in Liquid Phase Catalysis; From Recent Advances to Future Goals. *Nanoscale* **2011,** *3* (9), 3462–3481.
81. Otunola, G. A.; Afolayan, A. J.; Ajayi, E. O.; Odeyemi, S. W. Characterization, Antibacterial and Antioxidant Properties of Silver Nanoparticles Synthesized from Aqueous Extracts of *Allium sativum*, *Zingiber officinale*, and *Capsicum frutescens*. *Pharmacogn. Mag.* **2017,** *13* (Suppl 2), S201.
82. Donsì, F.; Voudouris, P.; Veen, S. J.; Velikov, K. P. Zein-Based Colloidal Particles for Encapsulation and Delivery of Epigallocatechin Gallate. *Food Hydrocolloids* **2017,** *63,* 508–517.
83. Sun, C.; Xu, C.; Mao, L.; Wang, D.; Yang, J.; Gao, Y. Preparation, Characterization and Stability Of Curcumin-Loaded Zein-Shellac Composite Colloidal Particles. *Food Chem.* **2017,** *228,* 656–667.
84. Xinyu, W.; Pan, D.; Min, Y. 89Zr Labelled Self-Assembled Nanoparticles for PET Imaging. In: Soc Nuclear Med, 2018.
85. Hong, Z.; Xu, Y.; Yin, J.-F.; Jin, J.; Jiang, Y.; Du, Q. Improving the Effectiveness of (−)-Epigallocatechin Gallate (EGCG) Against Rabbit Atherosclerosis by EGCG-Loaded Nanoparticles Prepared from Chitosan and Polyaspartic Acid. *J. Agric. Food Chem.* **2014,** *62* (52), 12603–12609.
86. Lee, W.-C.; Tsai, T.-H. Preparation and Characterization of Liposomal Coenzyme Q10 for In Vivo Topical Application. *Int. J. Pharm.* **2010,** *395* (1–2), 78–83.
87. Onoue, S.; Uchida, A.; Kuriyama, K.; Nakamura, T.; Seto, Y.; Kato, M.; Yamada, S. Novel Solid Self-Emulsifying Drug Delivery System of Coenzyme Q10 with Improved Photochemical and Pharmacokinetic Behaviors. *Eur J. Pharm.* **2012,** *46* (5), 492–499.
88. Zaki, N. M. Strategies for Oral Delivery and Mitochondrial Targeting of CoQ10. *Drug Deliv.* **2016,** *23* (6), 1868–1881.
89. Nepal, P. R.; Han, H.-K.; Choi, H.-K. Enhancement of Solubility and Dissolution of Coenzyme Q10 Using Solid Dispersion Formulation. *Int. J. Pharm.* **2010,** *383* (1–2), 147–153.

90. Gong, Y.; Wu, Y.; Zheng, C.; Fan, L.; Xiong, F.; Zhu, J. An Excellent Delivery System for Improving the Oral Bioavailability of Natural Vitamin E in Rats. *AAPS PharmSciTech* 2012, *13* (3), 961–966.
91. Chen, W.; Park, S. K.; Yu, W.; Xiong, A.; Sanders, B. G.; Kline, K. Synthesis and Screening of Novel Vitamin E Derivatives for Anticancer Functions. *Eur. J. Med. Chem.* **2012,** *58*, 72–83.
92. Zhong, Y.; Chiou, Y.-S.; Pan, M.-H.; Shahidi, F. Anti-Inflammatory Activity of Lipophilic Epigallocatechin Gallate (EGCG) Derivatives in LPS-Stimulated Murine Macrophages. *Food Chem.* **2012,** *134* (2), 742–748.
93. Keller, B. C. Liposomes in Nutrition. *Trends Food Sci. Technol.* **2001,** *12* (1), 25–31.
94. Kandimalla, K. K.; Babu, R.; Singh, M. Biphasic Flux Profiles of Melatonin: The Yin–Yang of Transdermal Permeation Enhancement Mediated by Fatty Alcohol Enhancers. *J. Pharm. Sci.* **2010,** *99* (1), 209–218.
95. Teskač, K.; Kristl, J. The Evidence for Solid Lipid Nanoparticles Mediated Cell Uptake of Resveratrol. *Int. J. Pharm.* **2010,** *390* (1), 61–69.
96. Suwantong, O.; Ruktanonchai, U.; Supaphol, P. Electrospun Cellulose Acetate Fiber Mats Containing Asiaticoside or Centella Asiatica Crude Extract and the Release Characteristics of Asiaticoside. *Polymer* **2008,** *49* (19), 4239–4247.
97. Chantarawaratit, P.; Sangvanich, P.; Banlunara, W.; Soontornvipart, K.; Thunyakitpisal, P. J. Acemannan Sponges Stimulate Alveolar Bone, Cementum and Periodontal Ligament Regeneration in a Canine Class II Furcation Defect Model. *J. Periodontal Res.* **2014,** *49* (2), 164–178.
98. Tønnesen, H. H.; Másson, M.; Loftsson, T. Studies of Curcumin and Curcuminoids. XXVII: Cyclodextrin Complexation: Solubility, Chemical and Photochemical Stability. *Int. J. Pharm.* **2002,** *244* (1–2), 127–135.
99. Abd El-Ghaffar, M.; Hashem, M.; El-Awady, M.; Rabie, A. pH-Sensitive Sodium Alginate Hydrogels for Riboflavin Controlled Release. *Carbohydr. Polym.* **2012,** *89* (2), 667–675.
100. Norajit, K.; Kim, K. M.; Ryu, G. H. Comparative Studies on the Characterization and Antioxidant Properties of Biodegradable Alginate Films Containing Ginseng Extract. *J. Food Eng.* **2010,** *98* (3), 377–384.
101. Iosio, T.; Voinovich, D.; Perissutti, B.; Serdoz, F.; Hasa, D.; Grabnar, I. et al. Oral Bioavailability of Silymarin Phytocomplex Formulated as Self-Emulsifying Pellets. *Polymer* **2011,** *18* (6), 505–512.
102. Pereira, R.; Mendes, A.; Bártolo, P. Alginate/Aloe Vera Hydrogel Films for Biomedical Applications. *Procedia CIRP* **2013,** *5*, 210–215.
103. Liu, C.-H.; Wu, C.-T. Optimization of Nanostructured Lipid Carriers for Lutein Delivery. *Colloids Surf. A: Physiochem. Eng. Aspects* **2010,** *353* (2–3), 149–156.
104. Pan, X.; Yao, P.; Jiang, M. Simultaneous Nanoparticle Formation and Encapsulation Driven by Hydrophobic Interaction of Casein-Graft-Dextran and β-Carotene. *J. Colloids Intersurface Sci.* **2007,** *315* (2), 456–463.

CHAPTER 4

FUTURE TRENDS IN THE FORMULATION OF NUTRACEUTICALS

NIDHI PARIHAR, TULASI NARAHARISETTI, PRATHAMESH PATIL, SRI VATHSAN, AND DEEPAK BHARADWAJ PVP

Department of Pharmacology & Toxicology, National Institute of Pharmaceutical Education and Research Guwahati, Assam, India

ABSTRACT

In the past few years, food industry development has focused on foods that provide specific nutrients to the body and prevent or cure disease by promoting healthcare benefits. Also, after the recent pandemic, consumers have played a significant role in understanding the importance of healthy or functional foods. Competition among food industries has increased because of rising concerns in consumers' minds. Therefore, nutraceuticals have evolved as biologically active cost-effective materials with known medicinal uses. This resulted in the development of food technology for health and wellness products that enhance the potential benefits to consumers. However, formulating plant-based materials into nutraceuticals has regulatory hurdles for solubility, stability, and bioavailability. Therefore, in this current chapter, we have discussed several formulations, regulatory challenges regarding the formulation, their market potential, and new technologies in the formulation of nutraceuticals. We have also discussed its scope as an alternative to pharmaceuticals for various pharmacological activities.

4.1 INTRODUCTION

Stephen De Felice first introduced the term "Nutraceutical" in 1989. Nutraceuticalsare foods or natural substances allegedly used in health and

Formulations, Regulations, and Challenges of Nutraceuticals.
Tingirikari Jagan Mohan Rao, Rajesh K. Kesharwani, Raj K. Keservani, Anil Kumar Sharma (Eds.)

medicinal benefits, including prevention and treatment of diseases. They are either consumed as dietary supplements or as functional food materials. Dietary supplements are foods or sole components of foods used to supplement the diet and are generally not used to treat/prevent diseases. Functional foods are used as foods that are used to treat infections. These foods include health-promoting benefits.1 But people often are confused with nutraceuticals and functional foods. As functional foods provide some extra benefits apart from their primary nutritional effects, nutraceuticals are defined as food substances that provide clinical usefulness of a particular component that intends to show that specific benefit. Nutraceuticals are categorized as traditional (natural) (includes nutrients, herbals, plant-derived chemicals, probiotics, and nutraceutical enzymes) and non-traditional (artificial) nutraceuticals (includes the nutraceuticals prepared by fortification and recombinant technologies).

The global industry of nutraceuticals and/or functional foods is expanding rapidly and is estimated at over 100 billion USD. Researchers have identified nutraceuticals as leading foods because natural foods have been playing a significant role for centuries in preventing and treating diseases, hence increasing life expectancy. The market potential of these foods confirms their perceived nutritional benefits and, provides evidence of their health claims. This increases product innovations, customer response toward healthy-living lifestyles through nutrition and lead to a shift from pharmaceutically derived supplements.[2]

Researchers are more curious to know which constituent is responsible for its therapeutic effect, thereby pursuing food development. From the past 20 years, food industries had a 10–20% increase in sales. Globally, the food industry has become the billion-trillion industries, and the USA is the largest market in the world which is followed by Japan which has a growth rate of 9.6%. India is growing 21% per annum in the nutraceuticals market. Foods are being consumed in various forms and becoming more attractive. The food industries develop foods by keeping in mind a person's compliance and their specific needs.[3]

4.2 CURRENT PORTFOLIO OF THE NUTRACEUTICALS

Scientists have been focusing on developing different foods or nutraceuticals for the past few years. With the change in lifestyle, stress, and consumption of junk or non-healthy food, oxidative stress results from an imbalance of the antioxidant system in our body. Due to oxidative stress, there is increased disease incidence. In the past few years, new pattern has been evolved in the healthcare systems.[4] People have started to eating good diet, understanding the role of fruits, vegetables, and nutraceuticals, and incorporating them into their daily diet for a better quality of life. Nutraceuticals have health benefits, and their regular intake (within their acceptable recommended dietary intakes) may keep diseases away contributing for an overall good health. All age groups widely accept them due to their safety, higher quality, purity, efficacy, health-promoting, and disease curing activities.[5]

4.2.1 MARKET POTENTIAL

The nutraceuticals market has gained momentum as people express their interest and acceptance in self-managing their risk of diseases. After the COVID-19 pandemic, people's mindset toward healthcare has changed significantly. With the increase in healthcare costs and geriatric population general population has leaned more toward dietary supplements and functional foods.[6] The advent of these foods' nutraceuticals innovation has gained much higher demand in the market from the past few years. Consumer's lifestyle informed and educated consumers, storage advantages, and awareness for health have increased consumption of nutritious diet, which has ultimately raised the demand for nutraceuticals available in many forms in the market. Due to this, the nutraceutical market is rapidly growing globally, estimated at over 417.66 billion USD in 2020. The expected growth rate is 8.9% in the world market from 2020–2028. In Europe and North America, nutraceuticals are coming as the daily component of the consumer's diet in these provinces. This greater prevalence of the diseases linked to lifestyle and social habits. This functional food stuff grabs the huge position in the natural products domain among the nutraceuticals, after functional health drinks and food-based supplements.[7] The nutraceutical market is saturated in Europe. The Asian nutraceutical market is also growing at rapid rate.[7,8]

New concepts such as personalized nutraceuticals have come into the picture that fulfills the body's specified need. Due to all these factors, consumers are aware of their healthcare diets. The best example of this is the immunity boosters' importance after the COVID-19 outcome. Thus, nutraceuticals that boost immunity have been introduced. Due to increased concern about food quality, the government has decided to take several measures to check the quality, purity, and efficacy of present supplements in the market in collaboration with AYUSH ministry in India. In this way, the filthy and false products from the market have been reduced and forced to develop new and advanced food delivery systems.[9]

4.2.2 CHALLENGES TO BE ADDRESSED IN NUTRACEUTICALS

Nutraceuticals offer both the therapeutic and physiological benefits besides fulfilling the fundamental nutritional requirements and include a wide coverage of peptides, phenolic, lipids, carotenoids, and vitamins, etc.[10] Though nutraceuticals are the ideal and reliable sources for bioactive components, their actual effectiveness is based on preserving their availability to body. The research on herbal medicines is not well understood or overstretched. Therefore, several challenges and factors come to the picture in developing and manufacturing nutraceuticals. Examples of nutraceuticals include vitamins, polyphenols, carotenoids, fibers, fatty acids, and essential minerals. The main challenges of incorporating nutraceuticals are their instability during food storage or within gastrointestinal tract, their poor water solubility, and bioavailability.

Encapsulation of nutraceuticals helps to protect, incorporate, mask, control release, and preserve/maximize their functional properties till they reach the target site. The safety profiles should be checked before they are manufactured or marketed. But this is the major challenge or concern related to nutraceuticals as there is no authoritative or regulatory involvement. Information about the nutraceuticals, including compound purity and quality, drug and supplement interactions, heavy metal contamination, and authentic raw material should be checked before manufacturing. The nutraceuticals consumed as dietary supplements for weight loss, bodybuilding, and sex enhancement do not follow proper GMP guidelines and may have unintentional contamination.[11] The premarket approval is not observed for herbal drugs, and people consider it safe but are not aware that plant-based medications can also cause adverse effects. For instance, Ginseng has various varieties, all are thought to have the same properties and are sold under the name of ginseng, but very few belong to the genus Panax which has the original content of ginseng. Thus, it can be harmful to consume as its toxicity profile is not fully understood.

Another example is digitalis which is mainly used for cardiovascular disorders. However, it is a plant-based medicine but causes severe adverse effects. This should be thoroughly studied. Exporting consumables requires a free sale certificate as per guidelines and nutraceuticals are not covered in it. Thus, there should be different guidelines for nutraceuticals as their demand increases, so there should be more research. Nutraceuticals has good scope in the Indian market. Indian government should take some measures or collaborations with AYUSH or several other industries to overcome the challenges followed by the nutraceuticals industry.[12]

4.2.3 CURRENT EXPORT POTENTIAL

Many countries in the world are facing a huge problem of malnutrition. So, there is a dire need to develop new nutritional food technological ideas to fight and address the malnutrition issues.[13] This help us to show a position in the world export market with the startups. India's current position in the pharmaceuticals sets up the base for the nutraceuticals market worldwide as it has established in the medications.

4.2.4 PROSPECTS OF THE NUTRACEUTICALS AND THEIR FORMULATIONS

Nutraceuticals have more chances of growth and expansion in terms of health benefits as they have multiple therapeutic benefits and provide protection against various diseases. They include anti-cancer, cardiovascular, anti-diabetic, anti-obese, anti-inflammatory agents, and immunomodulators, etc.[14] COVID-19 has raised the demand for nutraceutical products across the globe. Pandemic has changed people's perspective toward health, and people became aware of self-care and a healthy lifestyle like a nutritious diet and exercise. Vitamins and minerals are in the first place those consumed in this pandemic time to fight against the virus. The changing consumer preferences and shift toward preventive care are making the nutraceutical sector sustainable. Nutraceuticals contribute to economic growth and have a 387 billion dollar market. India's nutraceutical market is poised to be a global leader at USD 4–5 billion.[15]

4.3 TECHNOLOGY TRENDS IN THE FORMULATION OF THE NUTRACEUTICALS

The use of technology and scientific deliverables in addressing and resolving the issues related to population, environmental changes, healthcare, food supply is constantly progressing. The increasing awareness of consumers about food quality and its safety is engrossing food researchers to discover new ways to enhance the quality of food and its supply.[16] Some of the essential technology-assisted nutraceuticals systems have been discussed in the following sub-sections.

4.3.1 DRUG DELIVERY SYSTEMS

The development of the nutraceutical delivery systems needs a thorough understanding of their physicochemical properties and their production techniques. The choice of the appropriate encapsulation method is an essential step in delivering the many nutraceuticals task.[12] Several mechanisms involved in the nutraceuticals release from the drug delivery systems are diffusion-controlled release, swelling, erosion-controlled release, fragmentation-controlled release, dissolution-controlled release, and stimuli-directed release.[12,17]

4.3.1.1 IDEAL FEATURES OF NUTRACEUTICAL DELIVERY SYSTEMS

a. Food-grade status: Nutraceuticals have natural sources such as proteins, lipids, minerals, and vitamins or the constituents generally considered safe. Despite pharmaceuticals products having harmful chemicals or synthetic polymers, consumers may take in a large amounts. If nutraceuticals overcome the regulatory hurdles in developing delivery systems, they can be used as a good source for drug delivery as it constitutes substances which are generally recognized as safe (GRAS). Unlike pharmaceutical materials, the nutraceutical delivery applied in dietary products cannot be comprised using chemical polymers. As they are being incorporated into foods, consumers have no control over the intake amounts. Therefore, there is a challenge for formulation scientists and regulatory bodies for developing delivery systems of nutraceuticals. If there hurdle can be overcome, the materials obtained from plant sources can be useful as they are generally considered as safe.

b. Protection for active ingredients: Delivery systems are generally framed to protect active ingredients or drugs from several factors such as pH, light, temperature, and GI environment to protect from acid environment.[18]

c. Food matrix compatibility: Food matrix is directly related to the bioavailability of the encapsulated nutraceutical. The interaction of the food matrix with mixed micelles and the components of the partially digested food components will change the properties as absorption and digestion of the nutraceuticals encapsulated. Therefore, the delivery systems should not react with food constituents. It can be incorporated into different matrices. Delivery systems can be integrated as suspensions in aqueous solutions (e.g. beverages and drinks), entrapped inside a biopolymer (e.g. desserts and yogurts), or embedded in a solid matrix (e.g. powders and cereal products) (Yao et al., 2014). In the case of clear beverages, it is possible to use nano-emulsions because the size of particles is so small that they do not strongly disperse the light.

d. Controlled-release property: Another characteristic of the drug delivery systems is that it should have the ability of controlled release which slowly pours encapsulated material in body in specific amounts within a specific period of time.

e. Loading capacity: High loading capacity of the drug delivery systems should be high so that it can effectively retain the encapsulate materials until it reaches the targeted site. It also depends on the properties of the bioactive compound and encapsulated material used. The loading property is linked to the bioactive compound and the encapsulated material. The properties which affect the loading capacity are chemical nature, molecular weight, volatility and polarity and its interaction with the food matrix.[18,19]

f. Bioavailability: Bioavailability is another essential characteristic that affects absorption, one of the important pharmacokinetic properties. Several factors such as high instability, low solubility in GI fluids, less absorption from the intestine, first-pass metabolism, and vehicle used for encapsulation and its surrounding matrix affect the bioavailability of the material encapsulated. These are some common issues discussed that should be avoided while formulating.[20] Furthermore, the bioavailability is affected by the materials used to encapsulate and their physical state, the encapsulation vehicle and the surrounding food matrix.[21] The challenge of designing nutraceutical delivery is to have the capacity to release the nutraceutical on demand. The stimuli can be in the form of external factors such as temperature, enzymes, moisture, pH, and shear.[22,23]

4.3.2 NANOTECHNOLOGY-ASSISTED NUTRACEUTICALS

With the advent of nanotechnology and nanoencapsulation, researchers have come up with numerous ideas to add value to food production, processing, preservation, analysis, and packaging. Lots of work have been streamlined to study it with different nanocarriers than nanoencapsulation and various types of nano systems recently have been created for the delivery of various nutraceuticals.[24,25,8] Nanocarriers (<1000 nm in general) can enhance the bioavailability of encapsulated materials by providing more surface-to-volume ratio. This causes an increase in the capacity of adhesion to the different parts of the gastrointestinal mucosal layer and interacts much more with various enzymes. In addition to this, NPs may cross through cell walls and penetrate the target cells. Considering technological point, the delivery of nanoparticles can increase the solubility of the various nutraceutical hydrophobic molecules without affecting the sensory appearance of food products' stability.[24] At the same time, the some biopolymeric nanoparticles has been reported in the production of different nutraceuticals. Hydrophobic and hydrophilic nutraceutical fragments may be entrapped or otherwise adsorbed on the surface of nanoparticles, improving its stability and increasing their biological effectiveness. Proteins, carbohydrates, that is, polysaccharides, or a mixture of both are the frequent common materials used in the preparation of these nanoparticles.[26]

The nutraceuticals with established anti-cancer activity, for example, resveratrol, genistein, curcumin, quercetin, and epigallocatechin gallate, have been encapsulated into different nanoliposomes[27] and various biocompatible polymeric nanosystems.[8] The outcome of nanotechnology-assisted nutraceuticals has shown remarkable results, improving absorption & solubility, bioavailability, and anti-cancer activity compared with nutraceuticals alone. For example, folic acid (FA) is a vitamin used for food fortification. FA encapsulation with silica particles is used for commercial citrus juice, such as lemon or orange juices. The encapsulation of FA is helpful in the protection during storage and processing, resulting in more sustainable fortification. In addition to this, sustained-release after consumption can help to improve vitamin metabolism and minimize adverse effects of folic acid overdoses.[28]

4.3.3 GENETICALLY MODIFIED NUTRACEUTICALS

Genetically modified (GM) nutraceuticals thus include genetically modified foods with desired characters leading to increasing health benefits and/or disease prevention. GM nutraceuticals also include plants and animal products that could create vaccines and drugs to treat or cure diseases. The introduction of second-generation, consumer-oriented genetically modified (GM) products has enhanced functional properties (e.g., bio-fortified GM foods) compared to first-generation GM nutraceuticals. Other third-generation GM food/nutraceutical products claim to benefit in consumers' health and disease treatment. Some examples of GM nutraceuticals are bananas (used for a Hepatitis B vaccine) (American Chemical Society 2007), and a tobacco plant (used in a vaccine for the human papillomavirus [HPV] in cervical cancer) and rice (used for a cholera vaccine) (Murnaghan 2012). At present, Astellas Pharma in Japan is at the forefront of conducting clinical trials for a vaccine in rice against diarrheal diseases such as cholera.[29] The World Food Prize (2016) award for the pioneering work on genetic engineering-based-fortification technology indicates the futuristic potential of genetically modified nutraceutical products in the consumer market.[30]

4.3.4 CHEMICALLY MODIFIED NUTRACEUTICALS

Nutraceuticals are chemically modified in order to enhance their properties. Flavonoids upon polymerization or linkage to sugar molecules by ether or carbon–carbon linkages give glycosides. Different sugars can bind to flavonoids at various sites and to some portion of the flavonoids' chemical structure, which will enhance the anti-cancer activity. Thus, flavonoids along with chemically synthesized analogs are being screened as potential new drug materials for various cancer treatments like ovarian, cervical, pancreatic, breast, and prostate cancer.

Glycine max (soybean) has beneficial effects on cardiovascular system and health-protecting effects in menopausal women. The study on soybean constituents (isoflavonoids) has revealed the presence of glyceollin's (phytoalexins) that have been studied for their anti-cancer, anti-estrogenic, and anti-angiogenic activities. Recently, glyceollin I was synthesized and producing natural or synthetic stereoisomers. All these three chiral forms of glyceollin I were exhibiting growth inhibition of cancer cell lines and have shown their anti-proliferative effects on humans breast cell line, ovarian cell lines, and prostate cancer cell lines. Glyceollins also have ability to regulate growth tumor by the regulation of HIF-1α factor (hypoxia-inducible factor). Thus, chemical modification of this natural food allowed for identifying a new class of potential drugs.

Curcumin is famous medicinal spice with many therapeutic properties, many worked on curcuminoid chemical structure to develop new drugs. Reactive β-keto moieties were deleted to make series of curcumin derivative and it was responsible for the pharmacokinetic limitation of curcumin. The mono-carbonyl stability of the analogs was greatly enhanced, and their pharmacokinetic profiles were also significantly improved *in vivo*.[31]

4.4 SCOPE OF NUTRACEUTICALS FORMULATIONS AS ALTERNATIVES TO PHARMACEUTICALS

4.4.1 CARDIOVASCULAR NUTRACEUTICALS

Cardiovascular disorders such as myocardial infarction, heart failure, coronary heart disease, stroke, and hypertension are the most predominant cause of death worldwide. It is challenging to detect symptoms in hypertension, the so-called silent killer. Lifestyle, along with nutraceuticals, has become a new frontier for treating cardiovascular diseases. Foods act by different mechanisms depending on the presence of bioactive compounds. Nuts, margarine, legumes, and soy proteins act by lowering blood cholesterol. Fish acts by reducing blood triglycerides. Citrus fruits, dark chocolate and ginseng help in decreasing blood pressure whereas vegetables, whole grains and oils lower the blood homocysteine. Nutraceuticals like secondary metabolites in plant kingdom such as polyphenols, alkaloids, flavonoids, and terpenes. High level of polyphenols is presents in grape seeds and red wine extract which is beneficial for health. Olive oil from Mediterranean Basin used to reduce blood pressure by increasing plasma conc. of nitrate and decreasing serum asymmetric dimethyl arginine. The reduced levels of serum dimethyl arginine inhibit the synthesis of nitric oxide.

a. **Vitamins:** Vitamin concentration in the body also affects cardiovascular disease positively or negatively. Vitamin E, Vitamin D and Vitamin C help in preventing cardiovascular diseases. Higher production of reactive oxygen species is the crucial mechanism in hypertension. Radiation levels of plasma vitamin E are related to increasing cardiovascular diseases. Similarly, as the concentration of vitamin C increases, the associated blood pressure also decreases. Vitamin D plays an essential role in cardio-protection by modulation of the skeleton's blood pressure, vascular smooth muscle cell functions, and maintaining a healthy endothelium. Vitamin D also shows anti-mitotic activity, which is used to stabilize endothelium, which is used for cardiovascular effects. Various formulations of vitamins such as Vitamin C chewable tablets and oil capsules of Vitamin E are used daily to protect against cardiovascular diseases. This formulation is used as a supplementary food and directly used as a preventive treatment for various cardiovascular diseases.

a. **Lycopene:** They are found in red fruits such as papaya, tomato, red pepper, water lemon, and others. A conjunction of lycopene lowers the incidence of myocardial infarction in traction coronary insufficiency and angina pectoris. In addition, the plasma level of lycopene in hypertension, acute myocardial infarction stroke, and atherosclerosis is useful in reducing the risk.

c. **Polyphenols**: Diet containing polyphenol-rich food, that is, fruits or vegetables, help in lowering the blood pressure. Fruits such as grapes and berries contain many polyphenols and improve endothelial function. Studies have shown that tea reduces systolic blood pressure due to polyphenol content. Armolipid Plus is a nutraceutical product containing berberine, folic acid, policosanol, astaxanthin and coenzyme Q10. It reduces systolic blood pressure in hypertensive patients.

d. **Spirulina:** Clinical trial studies of Spirulina show antihypertensive activity. Oral supplementation of Spirulina for longer period exhibits the reduction of diastolic and systolic blood pressure.

e. **Berberine:** Berberine is used to treat heart failure and hypertension. Berberine exhibits its action by various mechanisms. It blocks adrenergic receptor and also lowering the blood pressure. The central sympatholytic activity of berberine inhibits acetylcholinesterase which lowers the blood pressure. Coronary artery flow is enhanced by berberine.

4.4.2 NEUROPROTECTIVE NUTRACEUTICALS

Neurological disorders include a set of chronic diseases having intense complicated etiology. About 10 million people annually have neurological disorders, which are expected to increase in the near days. People have been dependent upon various natural and synthetic products for treating various conditions. The development and progress in science and technology have contributed to the increased utilization of many phytochemicals with therapeutic and diagnostic properties from plant and non-plant materials. This has resulted in the creation of opportunities to advance novel dietary supplements. Neurodegenerative disorders develop due to protein misfolding. Protein Aβ mainly induces cytotoxic pathways of molecular and cellular events in the case of Down's syndrome, whereas α-synuclein is responsible for Parkinson's disease. Nutraceuticals may modify the cellular and molecular pathways by acting on misfolded proteins at all levels and aid as a supplementary therapy in the management of the neurological disorders.

Bacoside A&B (saponin derivatives) and Brahmine (alkaloid derivative) are the main nutraceuticals for curing neurological disorders. They have been used in Ayurveda for its neurocognitive-enhancing properties. Neurodegeneration occurs due to increased oxidative stress and the generation of free radicals. The neuroprotective nutraceuticals were reported to show their beneficial effects by modulating the pathways related to oxidative stress. The following are some of the important chemical constitution obtained from plants that are used for various neural disorders:

a. Withanine: It is obtained from ashwagandha (Withaniasomnifera). It is used in various formulations for memory-boosting and neurocognitive-enhancing. It also improves oxidative stress-mediated neurodegeneration due to its antioxidant property.

b. Quercetin and Kaempferol: Quercetin is a polyphenol obtained from grapes, broccoli, berries, and citrus fruits, etc. Kaempferolis a flavonoid found in tea, spinach, and beans, etc. Both act by preventing NF-kB activation and pro-inflammatory cytokines, that is, interleukins.

c. Epigallocatechin 3-Gallate (EGCG): Epigallocatechin 3-Gallate is most abundantly found in green tea, catechin. It can protect neurons from oxidative stress. EGCG inhibits neuron death caused by hydrogen peroxide. SH-SY5Y human neuroblastoma cells are protected from amyloid precursor protein (APP), 6-hydroxydopamine (6-OHDA), or 3-hydroxykynurenine toxicity EGCG. EGCG inhibits Aβ fibril formation, which results in the reduction of β-amyloid-induced toxicity in hippocampal neurons. Spinal injury caused by accidents; traumatic conditions treated using EGCG.

d. Resveratrol: It is obtained from grapes, wine, peanuts, and soy. It is a polyphenolic antioxidant. It shows *in vitro* and *in vivo* neuroprotective activity. It shows its activity by protecting hippocampal from oxygen–glucose deprivation. It reduces oxidative damage.

e. Curcumin: Curcumin decreases ROS generation, resulting in enhanced cell viability. It has neuroprotective ability against the MPTP-induced neurodegeneration in rats.

4.4.3 ANTIOXIDANT NUTRACEUTICALS

Antioxidants are substances that prevent damage or destruction of cells caused by oxidation. Generation of reactive oxygen species (ROS) such as O_2, H_2O_2, and ROO^- causes oxidative stress. Oxidative stress leads to free radical generation in cells; this may cause cells damage and cells death. Antioxidant nutraceuticals prepared using plants and animals, for example, epigallocatechin 3-gallate is obtained from green tea. These nutraceuticals are having tremendous scope in the pharmacy field. These are used for the prevention of various diseases such as cancer, cardiovascular diseases, and neurodegenerative diseases. Antioxidant acts by lowering the oxidative stress in cells and inhibits diseases. It also has free radical scavenging activity. Intake of antioxidant nutraceuticals has potential benefits. Antioxidant nutraceutical's most commonly used to prevent central nervous system diseases because they are more prone to oxidative stress. For example, Parkinson's disease may be prevented by high dietary intake of vitamin E, Huntington's disease (HD). Phenolic antioxidants also decrease oxidative cell injuries and inflammatory reactions, for example, tocopherols. They also act as modulators of pro-apoptotic and inflammatory signaling pathways,[32] for example, tocopherols. They also act as modulators of pro-apoptotic and inflammatory signaling pathways.

Flavonoids are the most abundant antioxidant compounds. They also have neuroprotective properties. Flavonoids show their effect on enzymatic systems. Studies show that the flavonoid level of industrial citrus juice is greater than in homemade juice. Flavonoids can absorb ultraviolet (UV) radiation, therefore, protecting DNA from damage. Carcinogens escape from the cell detoxification in body; these escaped carcinogens interact with flavonoids protecting DNA from damage due to carcinogens. Galangin, a flavonoid metabolic derivative used to treat lymphocytes, suppressed the induction of chromosome aberrations. Flavonoids show the anti-proliferative effect on neoplastic cell lines, such as ovarian cancer cells, myeloid leukemia cells, and gastric cancer cells. The proliferation of stellate cells causes hepatic fibrogenesis. For the treatment of this, quercetin is used. Quercetin inhibits the proliferation of stellate cells.

Some examples of food sources and their nutraceutical properties

i. Citrous fruits and tomato
   Anti-proliferative effect, cardiovascular properties
ii. Grapes, beans, olives, and peppers
   Anti-proliferative effect, inhibition of angiogenesis, anti-mitotic
iii. Onions, tea, broccoli, and maize
   Protection of DNA damage, inhibition of angiogenesis
   a. Vitamins: Vitamin C and E show antioxidant activity. Both decrease oxidative stress. Vitamin C regulates hypoxia-inducible factor (HIF)-1. The anti-oxidative property of vitamin C is used for the treatment of cardiovascular diseases. It improves vascular function and the structure of blood vessels.
   b. Lycopene: Lycopene is a carotenoid found in vegetables and fruits like tomato and papaya. It inhibits the generation of reactive oxygen species and oxidative stress in cells. Due to the antioxidant property of lycopene, it is used to treat cardiovascular disease and liver diseases.
   c. Antioxidant nutraceuticals are easily available, costly. Consumption of nutraceuticals is helpful in the prevention of various diseases. These are also taken as a dietary supplement for the mitigation of multiple symptoms.

4.4.4 HEPATOPROTECTIVE NUTRACEUTICALS

Liver is an important organ of body as its vital functions such as storage, metabolism, and secretion. The liver also detoxifies endogenous and exogenous substances. Biochemical processes like supplying energy, reproducing is happened in liver. Bile is the greenish-yellow fluid released by the liver and stored in the gallbladder. It helps in digestion. Due all these processes, liver is more prone to various abnormalities. Hepatic diseases continue to grow due to changing lifestyles, environments, etc. Hepatic diseases/liver diseases negatively affect the normal functioning of the liver. It indicates damage to the structure, cells, and tissues of liver. This damage can be caused by chemicals (pesticides and weedicides) or biological factors (virus, bacteria, and fungi). Some drugs exhibit adverse effects on the liver, so there is a need for alternative plant medicine for treatment.[33]

Studies have shown impact of different phytochemicals for better functioning of liver. For this purpose, nutraceuticals play an essential role in protection of liver from hepatic diseases. Phytochemicals extracted from yeasts, bacteria, fungi, and fruits are used as nutraceuticals. Nutraceuticals having antioxidant activity show hepatoprotective effects by reducing NAFLD factors. Increased mitochondrial activity by nutraceuticals prevents hepatic diseases. Nutraceuticals increase ATP production and β-oxidation of fatty acids, promotes respiration which helps in normal functioning of liver. Monoterpenoids, diterpenoids, triterpenoids, phenyl compounds, essential oils, alkaloids, coumarins, and steroids are plant phytoconstituents that exhibit hepatoprotective activity. Fruits, plants, resin, and polysaccharides are used as nutraceuticals.

a. Milk thistle: Scientific name for milk thistle is *Silybum marianum.* Silymarin is active constitute of it. Silymarin is used for oral treatment of toxic liver damage. Also used for acute and chronic liver diseases treatment. Silymarin increases hepatocyte protein synthesis. It inhibits lipid peroxidation. Silymarin suppresses toxin penetration into hepatic cells and prevent hepatic damage. Nopal and tuna both reduce the hepatotoxicity of organophosphorus insecticide chlorpyrifos.

Propolis: It contains flavonols, terpenes, and steroids that show hepatoprotective activity. It protects against CCl_4-induced toxicity. Propolis shows hepatoprotective effects against toxic compounds such as paracetamol and D-GalN. Wheat, mango, and carrot are used as nutraceutical agents for hepatocellular toxicity. They have shown to downregulate the CCl_4-induced hepatotoxity and decreased elevated levels of cholestatic markers. Antioxidants used for the treatment of hepatic disease. They inhibit oxidation of biomolecules thus preventing cellular damage, for example, vitamin C, superoxide dismutase, and butylated hydroxyl anisole. Opuntia robusta and Opuntia streptacantha fruits are used for treating acetaminophen-induced acute liver damage.

4.4.5 ANTI-CANCER NUTRACEUTICALS

Curcumin, capsaicin, and catechin are some nutraceuticals with anticancer activity. Among phytochemicals mostly flavonoids are used as chemo-preventive agents, for example, tea polyphenols, quercetin, and genistein. Some flavonoids were proven to be apoptotic in human lymphoid leukemic cells, carcinoma cells, lung tumor cell lines, colon cancer cells, breast cancer cells, prostate cancer cells, stomach cancer cells, brain tumor cells, head and neck squamous carcinoma, and cervical cancer cells. There is impact of dietary intake of carotenoids on reduced risk of cancer and cardiovascular diseases. Caffeic acid phenethyl ester is effective in treating leukemia and carcinoma. Curcumin induces apoptosis in colon carcinoma cells, leukemic cells, prostate cancer cells, melanoma cells, and breast cancer cells. Garlic is effective anti-cancer agent. The allyl sulfur of garlic has significant anti-proliferate activity against human cancers.[8] Plant-derived nutraceuticals with anti-cancer potential, such as resveratrol, quercetin, curcumin, genistein, and epigallocatechin gallate, have been encapsulated into nanoliposomes and biocompatible polymeric nanoparticles.[8,27] Anti-cancer nutraceuticals loaded into nanoparticles result in improved solubility, absorption, bioavailability, and anti-cancer potential compared to nutraceuticals alone.

4.4.6 ANTI-INFLAMMATORY NUTRACEUTICALS

The food supplements like protein and minerals, compounds derived from various natural sources. They have used for various functions such as delaying, preventing, and treating different chronic inflammatory diseases. They show anti-inflammatory effects by inhibiting activation of NF-κB factor, blocking the extra expression of TNF and IL-1. It also downregulates the overexpression of various cell adhesion molecules and inhibits phospholipase A2 COX-2, lipoxygenase, and inhibit ROS generating enzyme activity and increasing ability for scavenging of ROS. Their antioxidative role can reduce ROS and free radicals' generation. Peoples are showing more interest in nutraceuticals is due to various diseases. According to the global market data, China will be the huge nutraceutical market due to lifestyle. Stringent regulations and various approval process of EU affected the nutraceutical market. There are different types of laws for different nutraceutical products for different countries.

Inflammation is a reaction of the body against abnormal function or cell injury, a protective response. There are two types of inflammation, that is, acute and chronic. First response mechanism is called as acute inflammation; it is against infections, trauma, and various physical and chemical agents. If this mechanism occurs persistent occurrence of acute inflammation leads to it chronic phase. It involves swollen, redness, heat production, and abnormal function. Migration of inflammatory cells, leucocytes, and other phagocytic cells occurs at affected region. Lytic enzymes are also released from lysosomes of cells as protective response. During the inflammation process, various chemical mediators are synthesized such as pro-inflammatory cytokines and stimulate release of immunoglobulins. The increased expression of phospholipase A2 causes stimulation of arachidonic acid metabolites including prostaglandins and leukotrienes. Oxidative stress leads to the oxidation of PUFA, proteins molecules, DNA, and various sterols. The oxidative stress can be reduced by incorporating the consumption of fruits and leafy vegetables in diet. Recent studies have shown a relationship between generation of ROS and various diseases such as CVS diseases, CNS diseases, and renal disease.

a. Vitamin E: Vitamin E is an antioxidant and used for treatment of rheumatoid arthritis. Also have anti-inflammatory activity. Tocopherols and tocotrienols, these two are having nonpolar structures and they consist of lipid phase. Tocopherol is act as antioxidants.
b. Carotenoids: Classified into two classes, that is, xanthophylls and carotenes. The carotenes have hydrocarbon and xanthophylls have oxygen. Carotenoids consist of alpha-carotene, lycopene, and have antioxidant activity.
c. Polyphenols and phenolic compounds: Polyphenols are phenolic compounds containing benzene ring which is having one or more hydroxyl groups attached to it. Both are obtained from plants, vegetables, citrous fruit, and fungi. Recent studies have shown that polyphenols have antioxidant activity along with other activities such as anti-inflammatory, anti-cancer, anti-bacterial, anti-atherogenic, antiangiogenic, and antimutagenic. They also have free-radical scavenging properties. Flavonoids show anti-inflammatory activity by inhibiting the enzymes which are responsible for the generation of phospholipase A2, COX, and LOX.

4.4.7 COMBINATIONAL NUTRACEUTICALS

Foods offer a wide variety of nutrients and bioactive compounds such as plant based, vitamins, minerals, and fibers.[34] Foods offer a range of nutrients and bioactive compounds such as phytochemicals, vitamins, minerals, and fibers. These days, many nutraceuticals supplement chemopreventive agents in cancer, as evidenced by promising epidemiological, preclinical, and clinical studies. Their mechanisms of action will involve the capability to target multiple molecular pathways without any toxicity. As per proven therapeutic reports, the growing evidence suggests that nutraceuticals' health benefits are attributed to the synergism of more than one bioactive compound. Therefore, increasing the potential of combinational approaches using nutraceutical compound in cancer therapy may also be evaluated.

4.5 FUTURE MARKET AND EXPORT POTENTIALS

The nutraceutical market comprises functional foods like cereals, dairy products, and beverages like energy drinks, fortified juices, and dietary supplements like vitamins, minerals, botanicals, enzymes, etc. The upscale in the nutraceutical market is due to increase in the consumers' interest and multiple benefits like health improvement, protection against chronic diseases, delay of aging process, immune-boosting, hair, and skincare, etc. The global nutraceuticals market is expected to grow at CAGR (compound annual growth rate) of 7.5% by 2026 (during the forecast period of 2020–2026). The nutraceutical market in the USA is estimated at $104.5 billion in the year 2021. It accounts for 34.57% share in the global market. Next to that, China (World's second-largest economy) is about to reach US$52.5 billion by 2026 with 9.6% of CAGR. Japan and Canada are in queue, forecast to grow at 6.3 and 6.7%. USA, Canada, Japan, China, and Europe will contribute 5.9% CAGR in the global Dietary Supplements segment. They together account for a market size of US$41.4 billion in the year 2020, countries like Australia, India, and South Korea are forecast to reach US$7 billion by 2026.

The nutraceutical industry in India has noticed rapid growth; it will play a significant role in combating health issues, and Indians have had the habit of using natural ingredients, that is, ayurvedic or herbal medicines and health supplements, since their day-to-day lives. About 150 million people in India are actively consuming nutraceuticals in day-to-day life, which contributes to international and national turnover. Sixty-five percent of the Indian nutraceutical market consists of dietary supplements like herbal and non-herbal extracts. It dominated firms like Amway, Himalaya, Dabur, and Emami. This segment is growing at a rate of 17% and hence will drive the market's growth.

4.5.1 REGULATORY ASPECTS OF THE NUTRACEUTICAL FORMULATIONS

Nutraceuticals like functional foods, dietary supplements are procured over the counter. They need a strict regulatory check at various stages like during manufacturing, packaging, transporting, and storage to ensure the quality of nutraceutical products. Regulatory systems, laws and key definitions for functional food, herbal extracts, or drugs vary from country to country. Nutraceuticals are termed as dietary supplements in USA, as complementary medicines in Australia, Natural Health Products in Canada, Food Supplements by European Union and as Foods for special dietary use in India.

In the United States, the Food and Drug Administration (FDA) regulates nutraceuticals by a set of regulations through Dietary Supplement Health Education Act (DSHEA) 1994. In India, manufacture, storage, distribution, sale, and import of nutraceuticals are regulated by the Food safety and standards act of 2006, rules and regulations 2011. Nutraceuticals are grouped under foods by FSSAI. Sections 22(1) of FSSAI defines functional foods or health supplements as foods specially formulated as dietary requirements for particular physical or physiological conditions or specific diseases or disorders. Any product in the market labeled as "Food for special dietary uses or functional foods or nutraceuticals or health supplements or similar such foods" are usually formulated as powders, capsules, tablets, liquids, and are meant for oral administration. They include plants or botanicals or their parts in powder or extract in suitable solvents. They also have minerals or vitamins and animal-related substances like fish oils, probiotics, and whey proteins. They may not cure or mitigate disease or disorder, in fact they provide health benefits and can be used as a promotion claim if FSSAI. They don't include narcotics and psychotropic substances.[35]

4.5.2 SCIENTIFIC CHALLENGES

People consider nutraceuticals to be safe and have fewer side effects. There are many challenges in manufacturing and marketing nutraceuticals like the source of raw materials, impurities, lack of proven research/experiment on the product's benefits, false claims, quality, contamination, and interactions, etc. Some manufacturers may use adulterated raw materials or may not follow GMP due to which product gets contaminated. Though they had many beneficiary properties, factors like low solubility may hinder nutraceutical product stability. Hence, they are formulated as liposomes, phytosomes, and nano-emulsions, etc. Though quality is maintained, it's sustainability throughout the product shelf life is another challenge, which may vary due to some factors like temperature, humidity, seasonal, and geographical variations. Besides all these, the major problem is that nutraceuticals are being used with lack of scientific evidence. There are no controlled experiments and clinical trials on them like in the case of pharmaceuticals.[11]

4.5.3 ADVERSE REACTION MONITORING OF THE NUTRACEUTICAL FORMULATIONS

Nutraceuticals are sold over the counter and there is no strict follow-up on their ADR reporting. As people consume them on their own belief of claimed benefits, they won't consult a physician and may also not disclose its usage during their health check-up, due to which adverse reaction monitoring for nutraceuticals is tough. Their interactions and effects also vary from person to person, which is hard to predict. They may give adverse reactions due to adulterated constituents, interaction with food or drugs, and maybe due to the expiry of product constituents. Nutraceuticals need a strict check on their exchanges and adverse reactions. Since the common public consider that the nutraceuticals are non-toxic to the body as they are natural in nature. Here, the concept of "natural" referring that it is safe for use. A best and classical example of the same is the death of the Greek Philosopher Socrates, who was given a liquid formulation of hemlock plant extract (Conium maculatum), after being sentenced to death. The plant contains toxic piperidine alkaloids like coniine and γ- coniceinewhis is relatively more toxic. In this context, the safety of the dietary supplements or nutraceuticals has become an issue of concern for the healthcare regulatory bodies. For example, use of these sometime may leads to injurious effects, that is, green tea leaves extracts, ginseng root extract, and Chinese herbs have been reported with drug-induced liver toxicity.[36]

The studies indicated that there are bunch of published reports that have mentioned about the adverse events relating to overuse of dietary supplements, but after critical evaluation of the causality, the number is tremendously reduced. Following are some examples of the top herbal supplements that commonly occurred in adverse drug reactions. Example includes soybean (Glycine max, 19.3%, mainly allergic reactions), liquorice (Glycyrrhiza glabra, 12.2%, with hypokalemia and hypertension being the most frequent adverse events), and ginkgo (Ginkgo biloba, 8.5%, adverse reactions due coagulation difficulties).[37] Drug interaction is another important safety issue with herbal supplements. It may cause the adverse reaction with the prescribed drug.[38,39] These prescribed drugs have both interactions, that is, pharmacokinetics and pharmacodynamics.[40]

4.6 CONCLUSION

In recent years, nutraceuticals incorporation in food has provided various healthy options in food products to consumers that have helped improve consumers' health and fulfill their needs. However, the effectiveness of nutraceuticals' effectiveness depends upon several factors like bioavailability, interaction with other food materials, etc. Thus, these can be modulated by incorporating the same in the delivery systems, affecting the absorption profile, bio accessibility, etc., and provide health benefits.

Therefore, for evaluating their safety, more reliable *in vitro* digestive models should be developed to understand the rate of micro and nano-delivery systems in the body; need to develop more advanced analytical instruments and methods to assess and measure nano-delivery systems accurately. For accurate delivery of nutraceuticals, there is a need to develop a delivery system with increased functionality and stability by understanding the interactions between the two components, that is, delivery systems and the food matrix. Future studies should give more focus toward improving the delivery systems' parameters such as physicochemical characteristics by carefully selecting the delivery systems' formulation on a case-by-case basis.

KEYWORDS

- **nutraceuticals**
- **formulation challenges**
- **market potential**
- **future technologies**

REFERENCES

1. RefCencic, A.; Chingwaru, W. The Role of Functional Foods, Nutraceuticals, and Food Supplements in Intestinal Health. *Nutrients* **2010,** *2* (6), 611–625. (a) Kaur, K. Functional Nutraceuticals: Past, Present, and Future. In *Nutraceuticals*; Elsevier: London, 2016; pp 41–78.
2. Vaz Patto, M. C.; Amarowicz, R.; Aryee, A. N.; Boye, J. I.; Chung, H.-J.; Martín-Cabrejas, M. A.; Domoney, C. Achievements and Challenges in Improving the Nutritional Quality of Food Legumes. *Crit. Rev. Plant Sci.* **2015,** *34* (1–3), 105–143.
3. Smith, R. E. Food Processing: A Food Scientist's Perspective. *Food Policy* **1999,** *24* (2–3), 255–264.
4. Ozioma, E.-O. J.; Chinwe, O. A. N. Herbal Medicines in African Traditional Medicine. *Herb. Med.* **2019,** *10*, 191–214.
5. Santini, A.; Tenore, G. C.; Novellino, E. Nutraceuticals: A Paradigm of Proactive Medicine. *Eur. J. Pharma. Sci.* **2017,** *96*, 53–61.
6. Kaur, N.; Singh, D. P. Retracted: Deciphering the Consumer Behaviour Facets of Functional Foods: A Literature Review. *Appetite* **2017,** *112*, 167–187.
7. Intelligence, M. Grapes Market-Growth, Trends, and Forecast (2019–2024), 2019.
8. Reque, P. M.; Brandelli, A. Encapsulation of Probiotics and Nutraceuticals: Applications in Functional Food Industry. *Trends Food Sci. Technol.* **2021**.
9. Naim, M. The Five Wars of Globalization. *Foreign Policy* **2003,** *134*, 28.
10. Ting, Y.; Jiang, Y.; Ho, C.-T.; Huang, Q. Common Delivery Systems for Enhancing In Vivo Bioavailability and Biological Efficacy of Nutraceuticals. *J. Funct. Foods* **2014,** *7*, 112–128.
11. Siddiqui, R. A.; Moghadasian, M. H. Nutraceuticals and Nutrition Supplements: Challenges and Opportunities. *Multidisc. Digit. Pub. Inst.* **2020,** *12*, 1593.
12. Gonçalves, R. F.; Martins, J. T.; Duarte, C. M.; Vicente, A. A.; Pinheiro, A. C. Advances in Nutraceutical Delivery Systems: From Formulation Design for Bioavailability Enhancement to Efficacy and Safety Evaluation. *Trends Food Sci. Technol.* **2018,** *78*, 270–291.
13. Basu, S. K.; Thomas, J. E.; Acharya, S. N. Prospects for Growth in Global Nutraceutical and Functional Food Markets: A Canadian Perspective. *Aust. J. Basic Appl. Sci.* **2007,** *1* (4), 637–649.
14. Ruchi, S. Role of Nutraceuticals in Health Care: A Review. *Int. J. Green Pharm. (IJGP)* **2017,** *11* (03).
15. Sharma, S.; Arora, M.; Baldi, A. Probiotics in India: Current Status and Future Prospects. *Pharm. Aspire* **2013,** *1*, 1–12.
16. Shende, P.; Mallick, C. Nanonutraceuticals: A Way Towards Modern Therapeutics in Healthcare. *J. Drug Deliv. Sci. Technol.* **2020,** *58*, 101838.
17. Qian, C.; Decker, E. A.; Xiao, H.; McClements, D. J. Nanoemulsion Delivery Systems: Influence of Carrier Oil on β-Carotene Bioaccessibility. *Food Chem.* **2012,** *135* (3), 1440–1447.
18. McClements, D. J.; Decker, E. A.; Park, Y.; Weiss, J. Structural Design Principles for Delivery of Bioactive Components in Nutraceuticals and Functional Foods. *Crit. Rev. Food Sci. Nutr.* **2009,** *49* (6), 577–606.
19. Augustin, M. A.; Sanguansri, L. Challenges and Solutions to Incorporation of Nutraceuticals in Foods. *Annu. Rev. Food Sci. Technol.* **2015,** *6*, 463–477.

20. Yao, M.; McClements, D.; Zhao, F.; Craig, R.; Xiao, H. Controlling the Gastrointestinal Fate of Nutraceutical Enriched Lipid Nanoparticles: From Mixed Micelles to Chylomicrons. *FASEB J.* **2015,** *29*, 249, 246.
21. Livney, Y. D. Nanostructured Delivery Systems in Food: Latest Developments and Potential Future Directions. *Curr. Opin. Food Sci.* **2015,** *3*, 125–135.
22. McClements, D. J.; Li, F.; Xiao, H. The Nutraceutical Bioavailability Classification Scheme: Classifying Nutraceuticals According to Factors Limiting Their Oral Bioavailability. *Annu. Rev. Food Sci. Technol.* **2015,** *6*, 299–327.
23. Samal, D. Use of Nanotechnology in Food Industry: A Review. *Int. J. Environ. Agric. Biotechnol.* **2017,** *2* (4), 238902.
24. Cao, H.; Saroglu, O.; Karadag, A.; Diaconeasa, Z.; Zoccatelli, G.; Conte-Junior, C. A.; Gonzalez-Aguilar, G. A.; Ou, J.; Bai, W.; Zamarioli, C. M. Available Technologies on Improving the Stability of Polyphenols in Food Processing. *Food Front.* **2021**.
25. Feng, K.; Wei, Y.-s.; Hu, T.-g.; Linhardt, R. J.; Zong, M.-h.; Wu, H. Colon-Targeted Delivery Systems for Nutraceuticals: A Review of Current Vehicles, Evaluation Methods and Future Prospects. *Trends Food Sci. Technol.* **2020,** *102*, 203–222.
26. Hu, K.; McClements, D. J. Fabrication of Biopolymer Nanoparticles by Antisolvent Precipitation and Electrostatic Deposition: Zein-Alginate Core/Shell Nanoparticles. *Food Hydrocoll.* **2015,** *44*, 101–108.
27. Dutta, S.; Ali, K. M.; Dash, S. K.; Giri, B. Role of Nutraceuticals on Health Promotion and Disease Prevention: A Review. *J. Drug Deliv. Therap.* **2018,** *8* (4), 42–47.
28. Ruiz-Rico, M.; Pérez-Esteve, É.; Barat, J. M. Use of Nanotechnology as an Antimicrobial Tool in the Food Sector. In *Nanobiotechnology*; CRC Press, 2018; pp 413–452.
29. Gul, K.; Singh, A.; Jabeen, R. Nutraceuticals and Functional Foods: The Foods for the Future World. *Crit. Rev. Food Sci. Nutr.* **2016,** *56* (16), 2617–2627.
30. Britwum, K.; Yiannaka, A.; Kastanek, K. *Public Perceptions of Genetically Engineered Nutraceuticals*, **2018**.
31. Sut, S.; Baldan, V.; Faggian, M.; Peron, G.; DallAcqua, S. Nutraceuticals, a New Challenge for Medicinal Chemistry. *Curr. Med. Chem.* **2016,** *23* (28), 3198–3223.
32. Pisoschi, A. M.; Pop, A. The Role of Antioxidants in the Chemistry of Oxidative Stress: A Review. *Eur. J. Med. Chem.* **2015,** *97*, 55–74.
33. Ekor, M. The Growing Use of Herbal Medicines: Issues Relating to Adverse Reactions and Challenges in Monitoring Safety. *Front. Pharmacol.* **2014,** *4*, 177.
34. Septembre-Malaterre, A.; Remize, F.; Poucheret, P. Fruits and Vegetables, as a Source of Nutritional Compounds and Phytochemicals: Changes in Bioactive Compounds During Lactic Fermentation. *Food Res. Int.* **2018,** *104*, 86–99.
35. Ray, A.; Joshi, J.; Gulati, K. Regulatory Aspects of Nutraceuticals: An Indian Perspective. In *Nutraceuticals*; Elsevier, 2016; pp 941–946.
36. Navarro, G.; Martínez Pinilla, E.; Ortiz, R.; Noé, V.; Ciudad, C. J.; Franco, R. Resveratrol and Related Stilbenoids, Nutraceutical/Dietary Complements with Health Promoting Actions: Industrial Production, Safety, and the Search for Mode of Action. *Compreh. Revi. Food Sci. Food Saf.* **2018,** *17* (4), 808–826.
37. Di Lorenzo, C.; Ceschi, A.; Kupferschmidt, H.; Lüde, S.; De Souza Nascimento, E.; Dos Santos, A.; Colombo, F.; Frigerio, G.; Nørby, K.; Plumb, J. Adverse Effects of Plant Food Supplements and Botanical Preparations: A Systematic Review with Critical Evaluation of Causality. *Br. J. Clin. Pharmacol.* **2015,** *79* (4), 578–592.

38. Izzo, A. A.; Ernst, E. Interactions Between Herbal Medicines and Prescribed Drugs. *Drugs* **2009,** *69* (13), 1777–1798.
39. Al Essa, M.; Alissa, A.; Alanizi, A.; Bustami, R.; Almogbel, F.; Alzuwayed, O.; Moti, M. A.; Alsadoun, N.; Alshammari, W.; Albekairy, A. Pregnant Women's Use and Attitude Toward Herbal, Vitamin, and Mineral Supplements in an Academic Tertiary Care Center, Riyadh, Saudi Arabia. *Saudi Pharma. J.* **2019,** *27* (1), 138–144.
40. Li, G.; Simmler, C.; Chen, L.; Nikolic, D.; Chen, S.-N.; Pauli, G. F.; van Breemen, R. B. Cytochrome P450 Inhibition by Three Licorice Species and Fourteen Licorice Constituents. *Eur. J. Pharma. Sci.* **2017,** *109*, 182–190.

CHAPTER 5

MAJOR FACTORS CONTRIBUTING TO THE REGULATIONS OF NUTRACEUTICALS

TESNIM ARFI[1] and SATYAJIT SAURABH[2]

[1]*Department of Bioengineering and Biotechnology, Birla Institute of Technology, Mesra, Ranchi, Jharkhand, India*

[2]*DNA Fingerprinting Laboratory, Bihar State Seed and Organic Certification Agency, Patna, Bihar, India*

ABSTRACT

Since antiquity, a large population of the world relies on traditional functional foods as precautionary and/or medicine to resist stress and disorders. Presently, the changing lifestyle and food habits increased susceptibility to stress and disorders. The term "nutraceuticals" is gaining interest rapidly among people. With remarkable increasing demand, the global market is flooded with several types of nutraceuticals, claiming their efficacy in maintaining proper health by ameliorating stress and other disorders. This warrants the need for regulations for growing nutraceutical markets to ensure safety evaluation, labeling, and commercialization. The regulatory requirements vary across the world, as a proper distinct definition for this term has not been well established. The term nutraceuticals is being defined in several ways. Nutraceuticals can be considered as food supplements that have nutritional and pharmaceutical components. It is more than food and less than pharmaceuticals. Nutraceuticals include a wide range of substances that may contain nutrients (like carbohydrates, proteins, fatty acids, vitamins, and minerals), herbal supplements, prebiotics, and probiotics and are considered

Formulations, Regulations, and Challenges of Nutraceuticals.
Tingirikari Jagan Mohan Rao, Rajesh K. Kesharwani, Raj K. Keservani, Anil Kumar Sharma (Eds.)

as the developing group of natural products that makes line between the food and pharmaceuticals used to enhance human pathophysiological conditions. With respect to the increasing demand for nutraceuticals, several aspects like validation processes, risk assessment, and marketing should be kept in mind to ensure the health, safety, efficacy, and toxicity of nutraceuticals. This chapter provides an overview of different categories of nutraceuticals, their health applications, formulations, and factors contributing to the regulation of nutraceuticals. This chapter is an attempt to provide the basic information, classification, formulation, and factors for regulation of the nutraceuticals.

5.1 INTRODUCTION

In the past few decades, a new diet health standard has evolved which gives more emphasis on nutrition and the components promoting health. Major population is focussed on earning high standards of living and so the lifestyle and feeding habits have changed, basic foods are replaced by junk food which is the immediate cause of various diseases. Heart-related diseases and cancer are the leading causes of death in the world (mostly in developing countries), excess consumption of junk had caused obesity followed by various neurological and pulmonary disorders. The expensive pharmacy and high-tech treatments have made people more ill than the actual disease so switching to nutraceuticals is an ultimate alternative to be healthy.

Often attributed to Hippocrates (400 BC), "Let food be thy medicine and medicine be thy food," quoted by Hippocrates is proof that the use of nutrient food over medicines is far better to stay fit since history.[1] Nutraceuticals have been recognized as potentially effective over medicines and so they are scientifically supported and commercialized. The global market share of nutraceuticals is estimated to be USD120 billion in 2007, with an annual increase of ~7%.[2] Various definitions of nutraceuticals are reported in the literature (Table 5.1).

Nutraceuticals are needed for its eminent property of providing extra nutrients other than food's basic nutritional value. It can be clearer with the example of whole grains such as wheat and rice. Whole grains contain three layers viz. bran, endosperm, and germ. Generally, the bran has minerals, vitamins, and fiber; the endosperm has carbohydrates and proteins primarily; and the germ has some vitamins and beneficial phytochemicals. Despite the development of technology and advancement in the field of agriculture that improved the method of cultivation, irrigation, harvesting, and storage

TABLE 5.1 Definitions of Nutraceuticals.

Sl. No.	Definitions	References
1	Nutritional products that provide health and medical benefits, including the prevention and treatment of disease.	[3]
2	A foodstuff (as a fortified food or dietary supplement) that provides health benefits in addition to its basic nutritional.	[4]
3	A diet supplement that delivers a concentrated form of a biologically active component of food in a non-food matrix to enhance health.	[5, 6]
4	Any substance that is a food or a part of a food and is able to induce medical and health benefits, including the prevention and treatment of disease.	[7]
5	Food or part of food that provides medical or health benefits, including the prevention and/or treatment of a disease.	[8]

techniques, the processing and refinement results about 50–60% reduction in the nutritional quality and this led to the reintroduction of bioactive compounds.[9] Nutraceuticals obtained from food sources are nutrient enriched that provides perquisite health. They may function as cofactors or ligands or substrates in promoting and improving the digestion, absorption and assimilation of nutrients, and the metabolic reactions by optimizing activity of metabolic enzymes. They may also function as agents to promote the growth of beneficial gut bacteria. They could be involved in regulation of metabolism, and purifying the body by neutralizing the oxidants and removing toxins from the body.

5.2 CLASSIFICATION OF NUTRACEUTICALS

There are various nutraceuticals categorized on different bases such as their source, mode of action, chemical nature, etc. however there are two major categories of nutraceuticals-traditional nutraceuticals and non-traditional nutraceuticals (Fig. 5.1). Traditional nutraceuticals are natural products without any alteration, for example, omega-3 fatty acids in walnut and salmon, and lycopene in tomatoes whereas non-traditional nutraceuticals are the artificial foods developed from natural foods with the help of technology such as vitamin-added cereals and milk products. Various nutraceuticals from different sources and their uses are mentioned in Table 5.2.

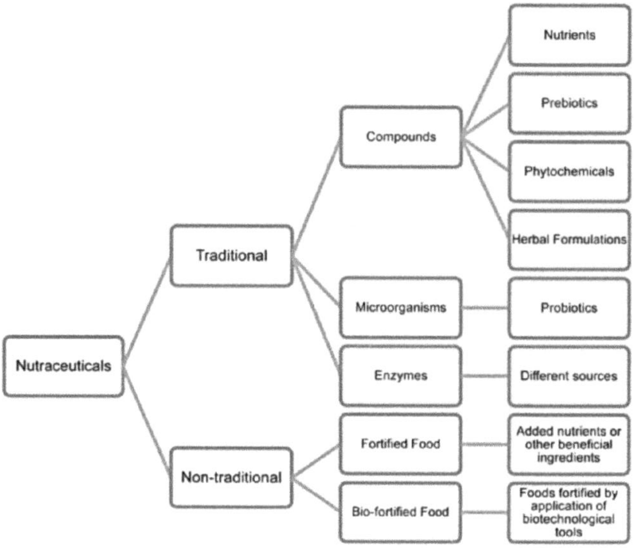

FIGURE 5.1 Classification of nutraceuticals.

5.2.1 TRADITIONAL FOODS

Traditional foods are the naturally occurring substances or food products with known health benefits. Such natural products are meant to be consumed directly by consumers, that is are not subjected to any alterations (especially chemical) during processing and packaging. It includes cereal grains, fruits, vegetables, dairy, fish, egg, meat, and tea. These nutraceuticals can be grouped as:

Herbal formulations like peppermint contain menthol, green tea contains epigallocatechin gallate, other herbal formulations contain flavonoids that are analgesic, anti-inflammatory, astringent, anti-pyretic, and anti-arthritic which can cure cold, cough, hypertension, depression, asthma, ulcers, etc.[10]

Phytochemicals are plant nutrients such as carotenoids found in vegetables and fruits that are anti-carcinogenic, boost immunity, as well as prevent eye damage from UV rays. Non-carotenoids present in legumes like chickpeas and grains lower cholesterol. The berries, fruits, and vegetables contain poly-phenolics like flavonoids. The flavonoids are potent antioxidants required for health, phytoestrogens to prevent cancers (like breast and prostate), and control diabetes. Non-flavonoids are present in raisins, berries, and turmeric and are strong anti-inflammatory, anti-oxidants, and anti-clotting agents.[11]

Prebiotics are chemicals that serve as a source of nutrients for probiotic bacteria. They are not digested in our body but support the growth of probiotic bacteria in a favorable environment and lower the germs that multiply in our digestive system. For example, Inulin is a prebiotic, commonly found in processed meals. It is a form of fiber obtained from the roots of plants such as chicory and Jerusalem artichoke.[12]

Nutraceutical enzymes are obtained from different plants, animals, and microbes such as hemicellulase, catalase, chymotrypsin, and lysozymes that helps in eliminating digestive problems, blood sugar disorders, and obesity.

Probiotics are live microorganisms that can be consumed in some amount that promote the digestion and metabolism of the host.[13] They act by adhesion to the gastrointestinal tract at specific sites and eliminate pathogens by competing for essential nutrients required for the growth of pathogens (bacteria, viruses, etc.), and produce an antitoxin effect. They can enhance the production of a specific enzyme (ß-galactosidase) that can hydrolyse the lactose into its components of sugar. The enhanced enzyme production may cure lactose intolerance.[14]

TABLE 5.2 Examples and Uses of Nutraceuticals Obtained from Different Sources.

Category	Source name	Product	Some uses	References
Herb	*Camellia sinensis*	Epigallocatechin gallate	Antioxidant	[15]
	Curcuma longa	Curcuminoids	Antioxidant	[16]
	Zingiber officinale (Ginger)	Zingiberene and gingerols	Stimulant, chronic bronchitis, hyperglycemia	[17]
	Allium sativum (Garlic)	Alliin and Allicin	Anti-inflammatory, immunomodulation, nervine tonic, anti-cancer	[18]
	Azadirachta indica (Neem)	Neem	Immune system, radiant skin, support healthy digestion, boosts liver function	[19, 20]
	Mucuna pruriens	L-Dopa	Enhance dopamine levels and protects neurons, promote brain health	[21]
	Glycyrrhiza glabra	Liquorice	Antioxidant, anti-inflammatory, and anti-allergy	[22, 23]
	Ginkgo biloba	Extract ***EGb 761***	Memory enhancement, improvement in blood circulation	[24, 25]
	Coffea arabica	Caffeine	Regulate oxidative stress	[26]
	Panax ginseng	Ginsenoside	Anti-inflammatory,	[27, 28]
		Panaxosides	stimulate immune and nervous system	

TABLE 5.2 *(Continued)*

Category	Source name	Product	Some uses	References
	Scutellaria baicalensis	Baicalein	Antioxidant, anti-inflammatory,	[29]
	Larrea tridentata	Nordihydroguaiaretic acid	antioxidant, antiviral, anti-inflammatory	[30]
	Centella asiatica	Brahmi	Nerve tonic, anti-anxiety, spasmolytic	[31]
	Aegle marmelos	Bael	Digestive, appetizer, treatment of diarrhoea and dysentery	[32]
	Valeriana officinalis	Valerenic acid and valerate	Tranquillizer, migraine, intestinal cramps, bronchial spasm	[33]
	Hydrastis Canadensis	Hydrastine and berberine	Anti-microbial, astringent, treatment of mucosal inflammation	[34]
	Withania somnifera	Ashwagandha	Aphrodisiac, liver tonic, anti-inflammatory agent, astringent, asthma	[35]
	Emblica officinalis	Amla	Anti-inflammatory, fever, anaemia	[36]
	Cassia angustifolia	Sennosides	Purgative	[37]
	Plantago ovata	Psyllium husk	Hypocholesterolemic	[38]
	Tinospora cordifolia	Giloy	Immune response, anti-inflammatory, and anti-pyretic activity	[39]
	Echinacea purpurea	alkylamide and echinacoside	Anti-inflammatory, anti-viral, and immunomodulatory	[40]
	Cissus quadrangularis	Hadjodwa	Treats bone fractures, joints related problems, arthritis, prevent osteoporosis	[41]
	Euphorbia royleana	Shilajit	Anti-oxidant, anti-ulcerogenic, anti-diabetic, cognitive, and memory enhancer	[42]
	Ferula assafoetida	Asafoetida, ferulic, and umbellic acid	Stimulant, expectorant, carminative, laxative	[43]
	Sesbania grandiflora	Agastya/Hadga	Anti-glycation potential	[44]
	Honey (formed from flower nectar by honeybees)	Honey	Anti-inflammatory, anti-viral and anti-diabetic, various protective functions	[45]
Amino acids		L-Arginine	Reduction of menopause symptoms	[46]
		Agmatine	Enhance the rate of absorption	[47]

TABLE 5.2 *(Continued)*

Category	Source name	Product	Some uses	References
		BCAA (branched chain amino acids)	Increase immunity, provide neuroprotection as well as benefit people with liver disease	[48, 49]
		Aromatic amino acids	Relief from stress, anxiety, and improve mental health	[50]
Minerals		Calcium	Strength to the bone	[51]
		Magnesium	Required for the active transport of ions like K and Ca and provide strength to the bone mass	[52]
		Glucosamine and chondroitin	Treatment of osteoarthritis	[53, 54]
Vitamins		Vitamin D and Vitamin C	Maintain calcium homeostasis, aging Immuno-modulatory effects, regulate dopamine levels	[55]
		Vitamin C and E	Free radical scavenger	[56]
		Vitamin B complex	Neuro modulatory function, bone strength	[57, 58]
Prebiotic		Raffinose, galacto-oligosaccharide, isomalto-oligosaccharides	Prevention of cholesterol gallstones	[59]
		Lactulose	Treatment of liver diseases and relief from symptoms of constipation	[60, 61]
		Fructo-oligosaccharides	Treatment of inflammatory bowel disease and relief from constipation	[62, 63]
Probiotic	*Lactobacillus*		Production of fermented dairy products	[64]
	Bifidobacterium		Part of intestinal microflora helps in digestion and absorption	[64]
	L. acidophilus and *L. bulgaris*		Treatment of antibiotic-associated diarrhoea	[64]

TABLE 5.2 *(Continued)*

Category	Source name	Product	Some uses	References
Microalgae	*Arthrospira platensis*		Anti-oxidant and antibacterial property	[65]
	Chlorella vulgaris		Health food, food supplement, feeds	[66, 67]
	Dunaliella salina	Lutein, betacarotene	Free radical scavenger and provide hepato-protection	[68, 69]
	Haematococcus pluvialis	Astaxanthin	Health food, feeds	[67]
	Schizochytrium	Docohexaenoic acid	Dietary, nutritional supplements	[70]
	Aphanizomenon flos-aquae	Mycosporine-like amino acids	UV-screening agent; sunscreen	[70]
	Crypthecodinium cohnii	DHA source	Oil for the infant formula	[71]

5.2.2 NON-TRADITIONAL NUTRACEUTICALS

These nutraceuticals are distinguished from food-derived products, food supplements, and drugs based on their appropriate target, safety assessment, mechanism of action, efficacy, possible side effects, and interactions with other products like food supplements and drugs. Some commercially available nutraceuticals and their manufacturers are listed in Table 5.3. There are also some bioactive compounds that are artificial foods engineered by recombination at the genetic level for promoting human health. The non-traditional nutraceuticals can be grouped as:

Fortified Nutraceuticals—These nutraceuticals are obtained from agricultural practices like breeding or fortification by added nutrients. As, for example, cereals are commercially available with added vitamins or minerals for nutritional security, folic acid-fortified flours, calcium-fortified orange juice, cholecalciferol-fortified milk with prebiotic and probiotic. Milk with *Bifidobacterium lactis* HN019 is used in diarrhoea, respiratory disorders, and illness in children.[72]

Recombinant Nutraceuticals—These includes the making of probiotics and the extraction of bioactive components by utilizing the biotechnological tools such as genetic engineering and fermentation. The fermentation efficiency is being enhanced by utilizing genetically modified microorganisms. Some examples for products of fermentation technology are fermented starch, bread, yoghurt, cheese, alcohol, vinegar, etc. The genetically modified cow with recombinant human lactoferrin (rhLf) is developed to address lactoferrin deficiency.[73]

TABLE 5.3 List of Commercial Nutraceuticals Products and Their Manufacturer.

Trade name/ products	Nutrient constituent	Manufacturing company
Calcirol D-3	Calcium and vitamins	Cadila Healthcare limited
Omega women	Antioxidants and vitamins	Wassn Surrey, UK
Proteinex	Prodigested proteins, vitamins, and carbohydrates	Pfizer limited India
Chyawanparash	Amla, aswagandha, and pipali	Dabur India Ltd.
Benoela EFLA943	Benoela olive leaf extract	Frutarom USA
Mushroom optimizer	Mushroom polysaccharide and folic acid	Jarrow formulas Los Angeles, CA, USA
Triphala Churna	For improved gastrointestinal health	Dabur India Ltd.
Sitopladi Churna	Suppress coughing and various lung-related problems, such as bronchitis, cold & cough, etc.	Dabur India Ltd.
Lavan Bhaskar Churna	For abdominal ailments like indigestion and anorexia	Dabur India Ltd.
Musli Pak (Laghu)	Improves sexual wellness, promotes hair health, and relieves respiratory complications	Dabur India Ltd.
Testosterone Booster	Promotes muscle growth, boosts fertility in men, and reduces stress and anxiety	Dr. Morepen

5.3 FORMULATION OF NUTRACEUTICALS

Nutraceuticals are available in pharmaceutical form like powder, capsules, tablets, etc. so that they should be stable, soluble, and bioavailable to the body. In order to make them viable following are the major factors to be kept in mind before the formulation of nutraceuticals.[74]

5.3.1 SELECTION OF BIOMATERIAL

Natural or synthetic polymers can be used as biomaterials depending upon the morphology, mechanical strength, and stability in the gastrointestinal tract. Sterilization and toxicity are also important factors to be considered.

Commonly used biomaterials for probiotics encapsulation are alginate, chitosan, carrageenan, locust bean gum, gelatin, whey proteins, and starches.[75]

5.3.2 SOLVENT TYPE

In order to maximize the dissolution and absorption rate, suitable solvents should be used so that the absorption efficacy in the GI tract is enhanced. However, the common solvents used in the extraction, purification, and formulation of nutraceuticals are usually organic solvents like methanol, ethanol, petroleum ether, ethyl acetate, acetonitrile, etc.

5.3.3 METHOD OF DELIVERY

Encapsulation method and emulsion preparation techniques are the common procedure which is being utilized for delivery of nutraceuticals. Although it needs to be optimized as it has been a challenge for a controlled release of nutraceuticals products into the internal environment of the body. For example, curcumin has low bioavailability because of its limited tissue distribution, low serum levels, rapid metabolism, and short half-life. Probiotics encapsulation technology (PET) for delivery of probiotics and Ionotropic gelation to encapsulate β-carotene using furcellaran beads shows up to 97% incorporation efficiency.[76] Cyclodextrins entrapment, liquid emulsions are also an alternative set of vectors for the delivery of foods. The alginate-chitosan microcapsules are used as vector to endure in vitro digestion with increased mechanical strength.[77] The emerging science of nanotechnology and reported studies has opened a new avenue for the stability, solubility, and/or permeability enhancement of nutraceuticals.

5.4 DEVELOPMENT OF NUTRACEUTICALS

Nutraceuticals are available in form of tablets, capsules, solutions, etc. Tablets are the most commonly used form of nutraceuticals because of its uniform size and shape, longer shelf life, easy for storage, and more economical. Anew nutraceutical product covers a long journey to come to the market which includes the process, food matrix selection for distribution and dissolution and chemical characterization and finally product processing techniques. The process of development starts with the identification of health conditions (nutritional imbalance). Fractionation of food is done to extract the required ingredients followed by biological and functional activity assay (biochemical and pharmacological behaviour). The phytochemicals or the product obtained from animal sources are identified and their metabolic profile is studied to check their bioavailability, rapid dissolution in the internal environment of the body. Identification and characterization are done using different analytical techniques like HPLC, MS, HPLC-MS, NMR, etc. These analysis may helps in revealing the mechanism of action through studies related to biochemistry. In vivo efficacy and safety of the product are assessed after clinical trials and the toxicity of overconsumption is being informed. Most importantly it must follow the regulations made by the government of a country where it is made before being available in the market for use.[78] A brief route of the development of nutraceuticals is presented in Figure 5.2.

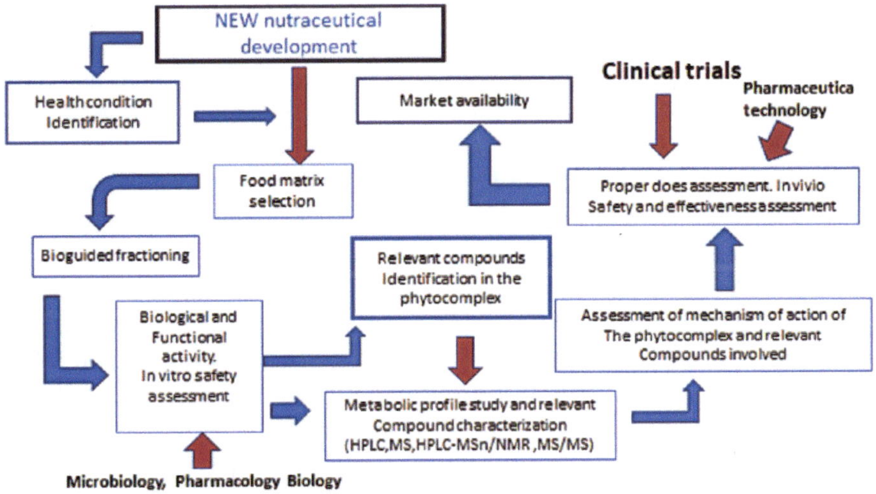

FIGURE 5.2 Route of nutraceuticals from development to market.

5.5 GLOBAL MARKET REGULATIONS FOR NUTRACEUTICALS

Nutraceuticals, functional foods, and dietary supplements requires regulatory framework for entry in the market so that there should be a balance between the consumer access and the risks associated with it. Every country has regulatory body which certifies the products such as food-like category are subjected to notification or registration based system while drug-like category tends to be subjected to a pre-market approval system. These approaches help to achieve the availability, safety, and acceptability of new products globally. There are a specific set of regulations governing dietary supplements or nutraceuticals forex in USA, EU, ASEAN, and India; for the entry of new products (nutraceutical) into the market, registration of the product is required followed by their formal evaluation. In most countries, there are separate lists of positive and negative ingredients allowed to be used as food and dietary supplements.[79]

The U.S. Food and Drug Administration (USFDA) maintains a list of ingredients that are generally recognised as safe (GRAS) to be used in foods and as ingredients of dietary supplements in the United States. In Australia, the Therapeutic Goods Administration (TGA) regulates dietary supplements as a Complementary Medicines and have a positive list of allowable ingredients which includes vitamins, minerals, and botanicals. China utilizes both a positive and negative list, positive list includes ingredients that can be added to foods, drugs and functional foods while negative list has banned ingredients. Canada employs a unique monograph system in which the ingredients that comply with the system as regarded as Natural Health Products (NHPs), while those do not comply the already established monograph require approval before reaching to the market. In the EU, there is a published list of vitamin and mineral ingredients allowable in foods and food supplements which are available as an online catalogue where a traffic light system is used to highlight whether the ingredient is novel, not novel, can only be used in food supplements, and if it is not certain then the product ingredient will undergo authorization for market entry. The FSSAI in India has proposed new regulations that includes positive lists for several categories of ingredient from botanical sources for "ASU" (Ayurveda, Siddha, and Unani) ingredients while negative lists include lack of quality standards, lack of transparency in the supply chain, and the and preparations method of botanical ingredients.

Every country has its own regulatory body which authorizes and allows the commercialization of nutraceuticals through regulations that are minimal as compared to drugs. In Europe, the European Food Safety Authority (EFSA) regulates authorization and commercialization of products claiming health benefits. In United States, the Dietary Supplement Health and Education Act (DSHEA), 1994 defined the dietary supplement as *a product that is intended to supplement the diet, contains dietary ingredients such as vitamins, minerals, amino acids, herbs, and botanicals*. They can be ingested in form of pill, capsule, tablet, liquid, or powder. Dietary supplements do not need to be authorized by the US Food and Drug Administration (FDA) before being marketed, but the production facilities are registered with FDA.[80] Japan, issued the Foods for Specified Health Use (FOSHU) in 1991 approves on the basis of a voluntary request from stake holders, which later on evolved as Health Promotion Law in 2003.

In India, the government had approved Food Safety and Standard Act in 2006 to integrate and streamline several regulations that cover nutraceuticals, food supplements, and dietary supplements which created the Food Safety and Standards Authority of India (FSSAI). In 2015, India notified the World Trade Organization (WTO)of a draft regulation for (i) defining the categories based on the use of ingredients, additives, labeling, contaminants (like allergens and toxins), claims (like health, nutritional, and pharmacological), recommended doses, and consumption levels, and (ii) determining the criteria for the manufacturing processes and trade of these defined categories. The new regulation is framed on the basis of section 22 of FSSAI as *Food Safety and Standards for Food for Health Supplements, Nutraceuticals, Food for Special Dietary Use, Food for Special Medical Purposes, Functional Food and Normal Food Regulations 2016*. The details of the Regulation can be accessed http://www.fssai.gov.in/ and http://www.fssai.gov.in/portals/0/pdf/food-act.pdf.

5.6 FACTORS CONTRIBUTING REGULATIONS OF NUTRACEUTICALS

The global use of nutraceuticals varies from country-to-country, their uses are either limited to nutrition or medicinal purposes. The registration and approval of nutraceuticals help worldwide availability of information to the individual, to research, and also help to get cheaper pharmacological treatment. To abide by the regulatory requirements and marketing authorization procedures different factors are to be considered, first is the identification of the health-promoting benefits, second is the complete and accurate assessments of the efficacy, toxicity and safety on human health accompanied with clinical trials before commercial release.

5.6.1 HEALTH BENEFITS

The nutraceuticals are obtained from plants as well as animal sources and offer several advantages in promoting health and reducing the risk of diseases. In various countries, nutraceuticals are commercialized especially to minimize the causative factors of different diseases. One study in the United States demonstrated the positive result on the use of single and combined nutraceutical supplements to treat hypertension accompanied by optimal nutrition& lifestyle modifications. Nutraceuticals provide immunity against various diseases like CVDs, Parkinson's and Alzheimer's diseases, neurological diseases.[81] Nutritional therapy is regarded as a healing system using nutraceuticals and it is based on the belief that foods can not only be sources of nutrients and energy but could also provide medicinal benefits.

5.6.2 SAFETY

The nutraceuticals were perceived as safe and having less or no side effects but the global demand for nutraceuticals have increased up to such an extent that adulteration of synthetic drugs, contamination of pesticides, microbes, toxins, etc., as well as analogue substitution, has exposed another level of health risks.[82] For instance, herbal preparations for slimming purposes are always trendy and top-selling product worldwide and the addition of sibutramine hydrochloride monohydrate (an anti-obesity drug molecule that inhibits serotonergic and noradrenergic reuptake) into these products caused trouble in some countries like Turkey, it led to abortion in pregnant women after taking this product and in Hong Kong, it was reported to show mental disorder due to neuropsychiatric side effects of adulteration. Although advanced analysis techniques are being developed to detect these illegal issues with herbal medicinal products (HMP), application of Good Agricultural and Collection Practices (GACP) and Good Manufacturing Practices (GMP) might help to diminish the threat of adulteration to herbal formulations.[83] The governmental health agencies obviously must provide safety and good quality of HMPs to avoid intended adulteration. However, consumers should be aware that adulteration or using synthetic fillers in these products may cause some serious effects.

5.6.3 EFFICACY

Nutraceuticals are efficient at their tasks and so are in great demand all over the world. For instance, they help in detoxifying the body, avoiding deficiencies of essential vitamins and minerals, and restoring healthy digestion. Herbal preparations and phytonutrients are pharmacologically effective, and appear to be disease-preventive food components. They act as cofactors of an enzyme, substrate for biochemical reactions, increase the rate of digestion, absorption, and elimination of undesirable components from the intestines. They also function as selective inhibitors of harmful intestinal bacteria and scavengers of reactive compounds. However, the risk-benefit of nutraceuticals is discussed. An efficient nutraceutical must have a natural dietary supplement, with the least side effect and increase the health value to improve the medical condition of humans and lastly, it should be easily available and economically affordable.

5.6.4 TOXICITY OF NUTRACEUTICALS

Consumption of nutraceuticals is associated with some kind of toxicity when taken in excess. Various literatures reported on overdoses of vitamins and minerals for example photosensitivity and neurotoxicity appeared at doses higher than their RDA (500 mg/d) of pyridoxine (vitamin B6) and Vitamin E (α-tocopherol) doses of 800–1200 mg/d can cause antiplatelet action & associated bleeding and if doses exceeds1200 mg/d may cause weakness, blurred vision, diarrhoea, and gonadal dysfunction.[84] Male smokers taking β-carotene supplements are at increased risk of lung cancer while alcohol users showed the incidence of prostate cancer and increased mortality rate. Other studies suggested that excess vitamin A lowers the mineral density& increases fracture risk in general, while pregnant women consuming excess vitamin A are prone to have congenital abnormalities in a foetus. Excess of iron and multimineral supplements can cause hyperchromatosis causing liver damage.[85]

5.7 THE FUTURE OF NUTRACEUTICALS

There are different definitions of nutraceuticals creating confused, not harmonized, and complex legislation globally. These compounds are considered safe and are being consumed for improvements in health-related conditions such as improved mental and/or physical health, as remedial agents, and as treatment of disease/disorder. The awareness for fitness and a healthy lifestyle, propelled the growth of nutraceuticals in the world and will dominate next millennium as the *nutraceutical revolution*. This growing sector implies that customers are looking for processed foods with additional nutritional value. This, in turn, may cause implications for the pharmaceutical and agricultural industries. The fast-growing market of these products alarmed government authorities to establish strict, simple, and harmonized legislations for operators involved in the design, testing, and marketing. The draft of legislations to ensure food safety should be based on understanding and knowledge of the product/ingredient used by scientific evaluation and assessment data from production methods/strategies, efficacy testing, interaction with other food/drugs/drink, multigenerational studies as short-term and long-term effects on consumers health. The nutraceutical industry and government agencies should respond to the lack of information about these products. The present studies utilizing advanced technologies and genome research (nutrigenomics) will play a major role and discover a new standard of food with a new function to prevent diseases, meanwhile one should be aware of the side effects and adulteration causing harm to health.

5.8 CONCLUSIONS

Nutraceuticals are bioactive compounds, present at the interface between nutrition and drugs and have the capability to prevent nutrition-related diseases. The major population is depending on nutraceuticals and dietary supplements rather than on drugs. High demand has led to the production and sale of a variety of nutraceuticals in the market and has aroused curiosity in researchers for the discovery of new nutraceuticals. On the other hand, we should be aware of the fact that the quality of nutraceuticals is becoming inappropriate due to adulteration and the addition of analogues which poses health risks so every country should implement strict regulations and authorization procedures to combat such illegal practices. Regulatory standards should ensure health, safety, and efficacy, and risk assessment to create adequate pre- and post-marketing inspections for protection. Also, the government should provide funds to agencies and academic institutions for research purposes.

KEYWORDS

- **nutraceuticals**
- **botanicals**
- **probiotics**
- **prebiotics**
- **regulations**

REFERENCES

1. Adelaja, A. O.; Schilling, B. J. Nutraceuticals: Blurring the Line Between Food and Drugs in the Twenty-First Century. *Choices***1999,** *14* (4), 35–39.
2. Ahmed, H. H.; Shousha, W. G.; Hussien, R. M.; Farrag, A. R. H. Potential Role of Some Nutraceuticals in the Regression of Alzheimer's Disease in an Experimental Animal Model. *Turk. J. Med. Sci.* **2011,** *41* (3), 455–466.
3. European Nutraceutical Association (ENA). Science Behind Nutraceuticals. In E. N. Association, Ed., Vol. 2016; Basel, Switzerland, 2016; p 594.
4. Merriam-Webster Online Dictionary. Merriam-Webster Inc.: Springfield, 2015.
5. Zeisel, S. H. Regulation of Nutraceuticals. *Science* **1999,** *285*, 1853–1855.
6. Food and Drug Administration, FDA, Dietary Supplement Health and Education Act of 1994 (DSHEA), United States.
7. Brower, V. *Nat. Biotechnol.* **1998,** *16*, 728.
8. DeFelice, S. L. The Nutraceutical Revolution: Its Impact on Food Industry R&D. *Trends Food Sci. Technol.* **1995,** *6* (2), 59–61.
9. Călinoiu, L. F.; Vodnar, D. C. Whole Grains and Phenolic Acids: A Review on Bioactivity, Functionality, Health Benefits and Bioavailability. *Nutrients* **2018,** *10* (11), 1615.
10. Chauhan, M.; Sonawane, S. K.; Arya, S. S. Nutritional and Nutraceutical Properties of Millets: A Review. *Clin. J. Nutr. Diet.* **2018,** *1* (1), 1–10.
11. Scalbert, A.; Johnson, I. T.; Saltmarsh, M. Polyphenols: antioxidants and beyond. *Am. J. Clin. Nutr.* **2005,** *81* (1), 215S–217S.
12. Parvez, S.; Malik, K. A.; Ah Kang, S.; Kim, H. Y. Probiotics and Their Fermented Food Products Are Beneficial for Health. *J. Appl. Microbiol.* **2006,** *100* (6), 1171–1185.
13. Heinrich, M.; Müller, W. E.; Galli, C. (Eds.). *Local Mediterranean Food Plants and Nutraceuticals*, Vol. 59; Karger Medical and Scientific Publishers, 2006.
14. Oak, S. J.; Jha, R. The Effects of Probiotics in Lactose Intolerance: A Systematic Review. *Crit. Rev. Food Sci. Nutr.* **2019,** *59* (11), 1675–1683.
15. Nagle, D. G.; Ferreira, D.; Zhou, Y. D. Epigallocatechin-3-Gallate (EGCG): Chemical and Biomedical Perspectives. *Phytochemistry* **2006,** *67* (17), 1849–1855.
16. Hewlings, S. J.; Kalman, D. S. Curcumin: A Review of Its Effects on Human Health. *Foods* **2017,** *6* (10), 1–11.
17. Sharma, Y. Ginger (Zingiber officinale)—An Elixir of Life a Review. *Pharm. Innov.* **2017,** *6* (10), 22–27.
18. Arreola, R.; Quintero-Fabián, S.; López-Roa, R. I.; Flores-Gutiérrez, E. O.; Reyes-Grajeda, J. P.; CarreraQuintanar, L.; Ortuño-Sahagún, D. Immunomodulation and Anti-Inflammatory Effects of Garlic Compounds. *J. Immunol. Res.* **2015,** *2015*, 1–13.
19. Bhowmik, D.; Chiranjib, Y. J.; Tripathi, K. K.; Kumar, K. S. Herbal Remedies of *Azadirachta indica* and Its Medicinal Application. *J. Chem. Pharm. Res.* 2010, *2* (1), 62–72.
20. Pingale Shirish, S. Hepatoprotection Study of Leaves Powder of *Azadirachta indica A. juss*. *Int. J. Pharm. Sci. Rev. Res.* 2010, *3* (2), 37–42.
21. Patil, R. R.; Gholave, A. R.; Jadhav, J. P.; Yadav, S. R.; Bapat, V. A. Mucunasanjappae Aitawadeet Yadav: A New Species of Mucuna with Promising Yield of Anti-Parkinson's Drug L-DOPA. *Genetic Resour. Crop Evol.* 2015, *62* (1), 155–162.

22. Kalsi, S.; Verma, S. K.; Neha, A. K.; Singh, N. J. A Review on *Glycyrrhiza glabra* (Liquorice) and Its Pharmacological Activities. *Int. J. Pharm. Sci. Drug Res.* 2016, *4* (5), 234–239.
23. Shin, Y. W.; Bae, E. A.; Lee, B.; Lee, S. H.; Kim, J. A.; Kim, Y. S.; Kim, D. H. In Vitro and In Vivo Antiallergic Effects of *Glycyrrhiza glabra* and Its Components. *Plantamedica* 2007, *73* (03), 257–261.
24. Wąsik, A.; Antkiewicz-Michaluk, L. The Mechanism of Neuroprotective Action of Natural Compounds. *Pharmacol. Rep.* 2017, *69* (5), 851–860.
25. McKeage, K.; Lyseng-Williamson, K. A. Ginkgo Biloba Extract EGb 761® in the Symptomatic Treatment of Mild-to-Moderate Dementia: A Profile of Its Use. *Drugs Ther. Perspect.* 2018, *34* (8), 358–366.
26. Martini, D.; Del, Bo. C.; Tassotti, M.; Riso, P.; Del, Rio. D.; Brighenti, F.; Porrini, M. Coffee Consumption and Oxidative Stress: A Review of Human Intervention Studies. *Molecules* 2016, *21* (8), 1–20.
27. Kim, J. H.; Yi, Y. S.; Kim, M. Y.; Cho, J. Y. Role of Ginsenosides, the Main Active Components of *Panax ginseng*, in Inflammatory Responses and Diseases. *J. Ginseng Res.* 2017, *41* (4), 435–443.
28. Kang, S.; Min, H. Ginseng, the 'Immunity Boost': The Effects of *Panax ginseng* on Immune System. *J. Ginseng Res.* 2012, *36* (4), 354.
29. Shieh, D. E.; Liu, L. T.; Lin, C. C. Antioxidant and Free Radical Scavenging Effects of Baicalein, Baicalin and Wogonin. *Anticancer Res.* 2000, *20* (5A), 2861–2865.
30. Rahman, S.; Ansari, R. A.; Rehman, H.; Parvez, S.; Raisuddin, S. Nordihydroguaiaretic Acid from Creosote Bush (Larreatridentata) Mitigates 12-O-Tetradecanoylphorbol-13-Acetate-Induced Inflammatory and Oxidative Stress Responses of Tumor Promotion Cascade in Mouse Skin. *Evid.-Based Complement. Altern. Med.* 2011, *2011*.
31. Chauhan, B.; Kumar, G.; Kalam, N.; Ansari, S. H. Current Concepts and Prospects of Herbal Nutraceutical: A Review. *J. Adv. Pharm. Technol. Res.* 2013, *4* (1), 4.
32. Neeraj, V. B.; Johar, V. Bael (Aeglemarmelos) Extraordinary Species of India: A Review. *Int. J. Curr. Microbiol. Appl. Sci.* 2017, *6* (3), 1870–1887.
33. Pilerood, S. A.; Prakash, J. Nutritional and Medicinal Properties of Valerian (Valerianaofficinalis) Herb: A Review. *Int. J. Food Sci. Nutr. Diet* 2013, *1*, 25–33.
34. Asmi, S.; Lakshmi, T. Therapeutic Aspects of Goldenseal. *IRJP* 2013, *4* (9), 41–43.
35. Sharma, V.; Sharma, S.; Pracheta, P. R. Withaniasomnifera: A Rejuvenating Ayurvedic Medicinal Herb for the Treatment of Various Human Ailments. *Int. J. Pharm. Technol. Res.* 2011, *3* (1), 187–192.
36. Jain, R.; Pandey, R.; Mahant, R.; Rathore, D. A Review on Medicinal Importance of Emblicaofficinalis. *Int. J. Pharm. Sci.* 2015, *6* (1), 72–84.
37. Ramchander, J.; Middha, A. Recent Advances on Senna as a Laxative: A Comprehensive Review. *J. Pharmacogn. Phytochem.* 2017, *6* (2), 349–353.
38. Xing, L. C.; Santhi, D.; Shar, A. G.; Saeed, M.; Arain, M. A.; Shar, A. H.; Bhutto, Z. A.; Katar, M.; Manzoor, R.; El-Hack, M. Psyllium Husk (Plantagoovata) as a Potent Hypocholesterolemic Agent in Animal, Human and Poultry. *Int. J. Pharmacol.* 2017, *13*, 690–697.
39. Saha, S.; Ghosh, S. Tinosporacordifolia: One Plant, Many Roles. *Ancient Sci. Life* 2012, *31* (4), 151.
40. Manayi, A.; Vazirian, M.; Saeidnia, S. *Echinacea purpurea*: Pharmacology, Phytochemistry and Analysis Methods. *Pharmacogn. Rev.* 2015, *9* (17), 63.

41. Mishra, G.; Srivastava, S.; Nagori, B. P. Pharmacological and Therapeutic Activity of Cissusquadrangularis: An Overview. *Int. J. Pharmtech Res.* 2010, *2* (2), 1298–1310.
42. Carrasco-Gallardo, C.; Guzmán, L.; Maccioni, R. B. Shilajit: A Natural Phytocomplex with Potential Procognitive Activity. *Int. J. Alzheimer's Dis.* 2012, *2012*, 674142. https://doi.org/10.1155/2012/674142
43. Mahendra, P.; Bisht, S. Ferula Asafoetida: Traditional Uses and Pharmacological Activity. *Pharmacogn. Rev.* 2012, *6* (12), 141.
44. Prasanna, G.; Saraswathi, N. T. The Therapeutic Role of Sesbaniagrandiflora as an Inhibitor of Advanced Glycationendproduct (AGE) Formation and the Discovery of Lead Compounds for Managing Hyperglycaemia. *PlantaMedica* 2013, *79* (13), PN84.
45. Samarghandian, S.; Farkhondeh, T.; Samini, F. Honey and Health: A Review of Recent Clinical Research. *Pharmacogn. Res.* 2017, *9* (2), 121–127. https://doi.org/10.4103/0974–8490.204647
46. Stanislavov, R.; Rohdewald, P. PACR (Pine Bark Extract, L-Arginine, L-Citrulline, Rose Hip Extract) Improves Emotional, Physical Health and Sexual Function in Peri-Menopausal Women. *J. Women's Health Care* 2014, *3* (6), 1–6.
47. Schwedhelm, E.; Maas, R.; Freese, R.; Jung, D.; Lukacs, Z.; Jambrecina, A. et al. Pharmacokinetic and Pharmacodynamic Properties of Oral L Citrulline and L Arginine: Impact on Nitric Oxide Metabolism. *Br. J. Clin. Pharmacol.* 2008, *65* (1), 51–59.
48. Shimomura, Y.; Yamamoto, Y.; Bajotto, G.; Sato, J.; Murakami, T.; Shimomura, N. et al. Nutraceutical Effects of Branched-Chain Amino Acids on Skeletal Muscle. *J. Nutr.* 2006, *136* (2), 529S–532S.
49. Tajiri, K.; Shimizu, Y. Branched-Chain Amino Acids in Liver Diseases. *World J. Gastroenterol.* 2013, *19* (43), 7620.
50. Fernstrom, J. D.; Fernstrom, M. H. Tyrosine, Phenylalanine, and Catecholamine Synthesis and Function in the Brain. *J. Nutr.* 2007, *137* (6), 1539S–1547S.
51. Reid, I. R.; Bristow, S. M.; Bolland, M. J. Calcium Supplements: Benefits and Risks. *J. Intern. Med.* 2015, *278* (4), 354–368.
52. Bhutto, A.; Mastoi, A. A.; Memon, S. A.; Qureshi, G. A.; Qureshi, A. A. Magnesium and Its Essential Role in Health. *Jlumhs* 2005, *4* (1), 33–35.
53. Huskisson, E. C. Glucosamine and Chondroitin for Osteoarthritis. *J. Int. Med. Res.* 2008, *36* (6), 1161–1179.
54. Vasiliadis, H. S.; Tsikopoulos, K. Glucosamine and Chondroitin for the Treatment of Osteoarthritis. *World J. Orthopedics* 2017, *8* (1), 1.
55. Veldurthy, V.; Wei, R.; Oz, L.; Dhawan, P.; Jeon, Y. H.; Christakos, S. Vitamin D, Calcium Homeostasis and Aging. *Bone Res.* 2016, *4* (1), 1–7.
56. Grosso, G.; Bei, R.; Mistretta, A.; Marventano, S.; Calabrese, G.; Masuelli, L. et al. Effects of Vitamin C on Health: A Review of Evidence. *Front. Biosci.* (Landmark Ed.) 2013, *18* (3), 1017–1029.
57. Kennedy, D. O. B Vitamins and the Brain: Mechanisms, Dose and Efficacy—A Review. *Nutrients* 2016, *8* (2), 68.
58. Dai, Z.; Koh, W. P. B-Vitamins and Bone Health–A Review of the Current Evidence. *Nutrients* 2015, *7* (5), 3322–3346.
59. Mitsuoka, T.; Hidaka, H.; Eida, T. Effect of Fructo Oligosaccharides on Intestinal Microflora. *Food/Nahrung* 1987, *31* (5–6), 427–436.

60. Gibson, G. R. Fibre and Effects on Probiotics (the Prebiotic Concept). *Clin. Nutr. Suppl.* 2004, *1* (2), 25–31.
61. Delzenne, N. M. Oligosaccharides: State of the Art. *Proc. Nutr. Soc.* 2003, *62* (1), 177–182.
62. Schultz, M.; Munro, K.; Tannock, G. W.; Melchner, I.; Gottl, C.; Schwietz, H.; Rath, H. C. Effects of Feeding a Probiotic Preparation (SIM) Containing Inulin on the Severity of Colitis and on the Composition of the Intestinal Microflora in HLA-B27 Transgenic Rats. *Clin. Vaccine Immunol.* 2004, *11* (3), 581–587.
63. Kelly, D.; Conway, S.; Aminov, R. Commensal Gut Bacteria: Mechanisms of Immune Modulation. *Trends Immunol.* 2005, *26* (6), 326–333.
64. Heyman, M.; Ménard, S. Probiotic Microorganisms: How They Affect Intestinal Pathophysiology. *Cell. Mol. Life Sci.* 2002, *59* (7), 1151.
65. Nuhu, A. A. Spirulina (Arthrospira): An Important Source of Nutritional and Medicinal Compounds. *J. Marine Biol.* 2013, *2013*.
66. Batista, A. P.; Gouveia, L.; Bandarra, N. M.; Franco, J. M.; Raymundo, A. Comparison of Microalgal Biomass Profiles as Novel Functional Ingredient for Food Products. *Algal Res.* 2013, *2* (2), 164–173.
67. Paniagua-Michel, J. Microalgal Nutraceuticals. In *Handbook of Marine Microalgae*; Academic Press, 2015; pp 255–267.
68. Murthy, K. C.; Vanitha, A.; Rajesha, J.; Swamy, M. M.; Sowmya, P. R.; Ravishankar, G. A. In Vivo Antioxidant Activity of Carotenoids from Dunaliellasalina—A Green Microalga. *Life Sci.* 2005, *76* (12), 1381–1390.
69. Hsu, Y. W.; Tsai, C. F.; Chang, W. H.; Ho, Y. C.; Chen, W. K.; Lu, F. J. Protective Effects of Dunaliellasalina–A Carotenoids-Rich Alga, Against Carbon Tetrachloride-Induced Hepatotoxicity in Mice. *Food Chem. Toxicol.* 2008, *46* (10), 3311–3317.
70. Chu, W. L. Biotechnological Applications of Microalgae. *IeJSME* 2012, *6* (1), S24–S37.
71. Saha, S. K.; Murray, P. Exploitation of Microalgae Species for Nutraceutical Purposes: Cultivation Aspects. *Fermentation* 2018, *4* (2), 46.
72. Sazawal, S.; Dhingra, U.; Dhingra, P.; Hiremath, G.; Sarkar, A.; Dutta, A. et al. Micronutrient Fortified Milk Improves Iron Status, Anemia and Growth Among Children 1–4 Years: A Double Masked, Randomized, Controlled Trial. *PloS One* 2010, *5* (8), e12167.
73. Hyvonen, P.; Suojala, L.; Orro, T.; Haaranen, J.; Simola, O.; Røntved, C.; Pyorala, S. Transgenic Cows That Produce Recombinant Human Lactoferrin in Milk Are Not Protected from Experimental *Escherichia coli* Intramammary Infection. *Infect. Immunity* 2006, *74* (11), 6206–6212.
74. Gonçalves, R. F.; Martins, J. T.; Duarte, C. M.; Vicente, A. A.; Pinheiro, A. C. Advances in Nutraceutical Delivery Systems: From Formulation Design for Bioavailability Enhancement to Efficacy and Safety Evaluation. *Trends Food Sci. Technol.* 2018, *78*, 270–291.
75. Chen, J.; Hu, L. Nanoscale Delivery System for Nutraceuticals: Preparation, Application, Characterization, Safety, and Future Trends. *Food Eng. Rev.* 2020, *12* (1), 14–31.
76. Gbassi, G. K.; Vandamme, T. Probiotic Encapsulation Technology: From Microencapsulation to Release into the Gut. *Pharmaceutics* 2012, *4* (1), 149–163. https://doi.org/10.3390/pharmaceutics4010149

77. Ahmed, T. A.; Aljaeid, B. M. Preparation, Characterization, and Potential Application of Chitosan, Chitosan Derivatives, and Chitosan Metal Nanoparticles in Pharmaceutical Drug Delivery. *Drug Design, Dev. Ther.* 2016, *10*, 483.
78. Yusof, Y.A.; Etti, C.; Chin, N. L.; Development of Nutraceutical Product. *Int. J. Adv. Sci. Eng. Inf. Technol.* 2015, *3*, 201–206. DOI: 10.18517/ijaseit.5.3.511
79. Bagchi, D. (Ed.). *Nutraceutical and Functional Food Regulations in the United States and Around the World*; Academic Press, 2019.
80. Radheshyam. A Review on: Nutraceuticals Challenges in Formulation and Its Regulatory Aspects. *J. Pharm. Res.* 2021, *6*, 88–94.
81. Keservani, R. K.; Sharma, A. K.; Jarouliya, U. Protein and Peptide in Drug Targeting and Its Therapeutic Approach. *Ars Pharm.* 2015, *56* (3), 165–177
82. Orhan, I. E.; Senol, F. S.; Skalicka-Wozniak, K.; Georgiev, M.; Sener, B. Adulteration and Safety Issues in Nutraceuticals and Dietary Supplements: Innocent or Risky. In *Nutraceuticals, Nanotechnology in the Agri-Food Industry*; Grumezescu, A. M., Ed., 2016; pp 153–182.
83. Skalicka-Woźniak, K.; Georgiev, M. I.; Orhan, I. E. Adulteration of Herbal Sexual Enhancers and Slimmers: The Wish for Better Sexual Well-Being and Perfect Body Can Be Risky. *Food Chem. Toxicol.* 2017, *108*, 355–364.
84. Gupta, R. C.; Srivastava, A.; Lall, R. Toxicity Potential of Nutraceuticals. *Comput. Toxicol.* 2018, 367–394.
85. Ronis, M. J.; Pedersen, K. B.; Watt, J. Adverse Effects of Nutraceuticals and Dietary Supplements. *Annu. Rev. Pharmacol. Toxicol.* 2018, *58*, 583–601.

CHAPTER 6

NUTRACEUTICAL REGULATION IN ASIAN COUNTRIES

NAVODIT KUMAR SINGH,[1] RICHA KATIYAR,[2] RANJANA KUMARI,[3] BISHWAMBHAR MISHRA,[4] SANJEEB KUMAR MANDAL,[4] and SANJAY KUMAR[2]

[1]*Department of Biochemical Engineering and Biotechnology, IIT Delhi, New Delhi, India*

[2]*Department of Chemical Engineering, IIT Delhi, New Delhi, India*

[3]*Department of Biotechnology, NIT Andhra Pradesh, Andhra Pradesh, India*

[4]*Department of Biotechnology, Chaitanya Bharathi Institute of Technology, Hyderabad, India*

ABSTRACT

In today's era of increased lifestyle standards and life expectancy, there has been a constant demand for supplements by consumers. Nutraceuticals are among the most popular supplements. Although there is a lot of opportunity for growth in the nutraceutical industry, there are no consistent regulatory standards globally. Several countries are looking to the nutraceutical industry to help them maintain their people healthy and safe by implementing certain laws and regulations. In general, industrialized countries have rules in place; nevertheless, certain countries, such as Asian countries, have not yet fine-tuned their regulations for nutraceutical products. This might be a massive opportunity for manufacturers to introduce their products into Asian markets. This chapter attempts to outline the regulatory process and regulations for

Formulations, Regulations, and Challenges of Nutraceuticals.
Tingirikari Jagan Mohan Rao, Rajesh K. Kesharwani, Raj K. Keservani, Anil Kumar Sharma (Eds.)

international business developers looking to launch new nutraceutical goods into Asian markets.

6.1 INTRODUCTION

Nutraceuticals, often known as health supplements, play a significant role in a person's overall health. In general, they are used to replace nutrients that are deficient in our everyday diet.[1,2]

Food supplements are consumed in many countries to supplement the normal diet and ensure adequate intake of specific components (e.g., vitamins and minerals).[3] Many health disorders require special nutritional requirements and the use of nutrients in unconventional forms; therefore, many physicians and customers seek out nutraceutical products.[4] Nutraceuticals are promoted in several countries to reduce the risk factors for various diseases.[3,5] Consumers have increased their use of nutraceuticals in the recent decade to the point of purposefully avoiding pharmaceuticals and considering prescription drugs as superfluous, too expensive, and occasionally unsafe.[6]

The global market for nutraceuticals, estimated at US$278.8 billion in 2020, is expected to rise to US$441.7 billion by 2026, growing at a 7.8% compound annual growth rate (CAGR) throughout the forecast period. The Asia–Pacific nutraceuticals product market is quickly expanding, accounting for US$74,648.1 Mn in 2018, and is predicted to rise at a CAGR of 7.3% between 2019 and 2027, accounting for US$140,178.6 Mn by 2027. China's nutraceutical market is presently the world's second largest behind the United States, and it is also one of the world's fastest expanding. A report titled "Nutraceuticals – Global Market Trajectory and Analytics" predicted a market size of $52.5 billion in 2026, with a CAGR of 9.6% throughout the analysis period.[7] Similarly, Japan and India, as well as the Association of Southeast Asian Nations (ASEAN) countries such as Singapore, Thailand, Malaysia, and the Philippines are expanding their nutraceutical industries in Asia.

As a result, the global nutraceutical market is quickly expanding, and nutraceutical rules are constantly evolving in the absence of a defined regulatory definition.[4] Because of this ambiguity, nutraceuticals are regulated as food supplements rather than pharmaceutical products in certain countries.[5]

The usage of nutraceuticals varies greatly from country to country; some countries restrict nutraceutical use to general health and well-being, while others allow for therapeutic use.[4] Nutraceuticals are widely used around the world, yet they are subject to less regulation when compared to traditional medications.[5] The global market is swamped with nutraceuticals claiming to be of natural origin and sold by major online retailers with therapeutic claims. Nutraceuticals are at the crossroads of nutrition and pharmaceuticals, and they have the potential to open the door to novel therapeutic options for the prevention of nutrition-related diseases. Although the possibilities for the development of nutraceutical products are endless, their safety is called into question due to a lack of regulation. Rather than intervening to defend public health, government personnel just provide surveillance.[8,9] Based on the country, different regulations govern Nutraceuticals under different names. The word even varies by area, from Nutraceutical to dietary supplements, and some governments include nutraceuticals under the food umbrella.[10] This chapter covers the nutraceutical regulation established to guarantee consumer safety in Asian countries such as China, Japan, India, and several ASEAN countries such as Singapore, Thailand, Malaysia, and the Philippines.

6.2 NUTRACEUTICAL REGULATION IN INDIA

The Food Safety and Standards Act (FSSA) of 2006, and Rules and Regulations of 2011, classify nutraceuticals as foods. In India, the nutraceuticals market is divided into three categories: functional foods, functional beverages, and dietary supplements. The nutraceutical market in India is expected to grow at a CAGR of 8.2%. However, during the coronavirus outbreak, consumption of nutritional supplements increased dramatically, causing the prediction to be revised, resulting in a 20.35% CAGR from 2020 to 2025.[11] As a result of these estimates, a comprehensive yet feasible regulatory framework is required to handle the concerns associated with contamination, intentional adulteration, spurious items, and misleading claims on the nutraceutical market. This resulted in the establishment of clear food safety guidelines, particularly for supplements, to assure food safety and consumer protection, as well as to achieve worldwide excellence in food safety.[12]

In India, the Food Safety and Standards Authority of India (FSSAI), was established in 2008 under the FSSA of 2006. Before the establishment of the FSSAI, food legislations in India were highly complicated and governed by various regulatory authorities in different ministries through several laws, regulations, and orders, namely, the Prevention of Food Adulteration

Act, 1954; Fruit Products Order, 1955; Meat Food Products Order, 1973; Vegetable Oil Products (Control) Order, 1947; Edible Oils Packaging (Regulation) Order, 1988; Solvent extracted oil, De-oiled meal and Edible flour (Control) Order, 1967; Milk and Milk Products Order, 1992. All the regulations, orders, and acts related to food and food-associated products were repealed by FSSAI. FSSAI releases various regulations and guidelines to control the manufacturing, distribution, storage, sale, and import of food articles such as dietary supplements and nutraceuticals so that safety could be ensured before consumption by the consumer. The evolution of laws related to food laws since the 1990s is summarized in Figure 6.1.

FIGURE 6.1 Chronology of regulation of food laws in India.

The regulatory guidelines for the approval of nutraceuticals in the Indian market are defined in FSSA, 2006. Section 22(1) of FSSA, defines "foods for special dietary uses or functional foods or Nutraceutical or health supplements" as:

1. Processed or formulated foods which fulfill the need of the particular dietary requirements and it will be needed due to a specific physiological or physical condition or particular diseases and disorders. These are presented without any change, wherein the ingredient composition of these food items notably must be different from the ordinary food's composition of similar nature, if such ordinary foods prevail, and may have more than one of the following ingredients, namely: (a) Botanicals or Plants or their portions in the form of powder, concentrate or extract in water, ethanol or hydroalcoholic extract, single or in the amalgamation; (b) Vitamins or minerals or metals or proteins or their compounds or amino acids (in amounts not surpassing the Recommended Daily Allowance for Indians) or enzymes (within permitted limits); (c) Materials from animal source; (d) A dietary material for the use by human beings to supplement the diet by increasing the total dietary intake.
2. (a) a product that is labeled as a "Food for special dietary uses or functional foods or health supplements or Nutraceuticals or similar such foods" which is not described for the use as a conventional food and whereby these products may be formulated in the form of tablets, powders, capsules, granules, jelly, liquids, and other dosage forms but not parenteral, and these are intended for oral administration; (b) These products do not consider a drug substance as per the Drugs and Cosmetics Act, 1940 [clause (b) and Ayurvedic, Siddha, and Unani drugs mentioned in Section 3; clauses (a) and (h)]; (c) These foods should not be claimed to cure or mitigate any particular disorder, disease, or condition (except some promotional claims or health benefits) as may be allowed by the regulations made under FSSA; (d) These food does not consist of a psychotropic substance or a narcotic drug (as per Narcotic Drugs and Psychotropic Substances Act, 1985) and substances included in Drugs and Cosmetics Rules, 1945 [Schedules E and E(I)].[10]

The FSS regulations (2011) were published in the official gazette and the governed areas are summarized in Figure 6.2. FSSAI also provided precise labeling advice and developed a system to authorize health claims if sufficient scientific proof was provided.[13] Table 6.1 summarizes the regulatory requirements for nutraceuticals in India[14] and Figure 6.3 depicts the steps involved in the registration process of nutraceuticals.

FIGURE 6.2 Various areas that are governed through Food Safety and Standards (FSS) regulations 2011.

TABLE 6.1 Regulatory Requirements in India for Nutraceuticals.

S.No.	Regulatory requirements	Description
1.	1. Product evaluation	Here, each active ingredient and additive is examined through the following steps: Initially, all the relevant documents related to the product are extracted, then the sample is collected in the presence of witnesses. Afterward, the sample needed to be dispatched to the concerned authority where sample analysis was to be done. If the analysis is not done within the specified period, then further action may be planned by the designated officer through adjudication proceedings including hearing, appeal procedure, holding inquiry, etc.
2.	Licenses	To register a product in India, several licenses for manufacturing, marketing, and import will be required along with several state and national level clearances/ licenses.
3.	Health and label claims	It means any illustration that states or implies or suggests that a relationship is there between health and food or its constituent.

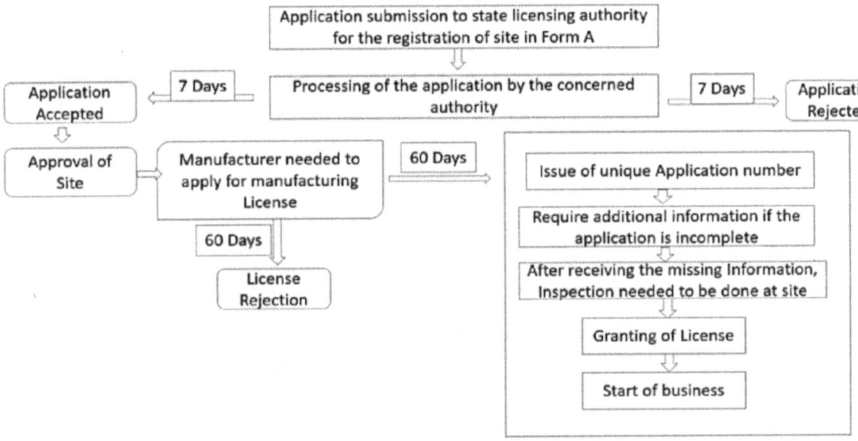

FIGURE 6.3 Registration process of nutraceuticals in India.

6.3 NUTRACEUTICAL REGULATION IN CHINA

There are some specific categories of foods viz. herbal products, innovative foods, dietary supplements, isolated nutrients, and processed food ingredients which are considered as nutraceuticals in China. Nutraceuticals have grown into a multibillion-dollar industry on a global scale. Traditional Chinese herbal medicines have a long history of usage in the treatment of various diseases and the promotion of healthcare; therefore, these are considered as important raw materials for nutraceuticals. Globally, China comes in third place in the market of nutraceuticals.[15] The Concept of yin–yang balance is one of the key factors for the growth of nutraceuticals in China, as this concept is considered as central to traditional medicine, culture, and diet in China.[16]

Despite such a huge rate of growth of nutraceuticals per year, its expansion is hampered by issues associated with its definition and management. In addition to that, it is challenging to provide a real estimate on the nutraceutical market of China due to the complicated regulations and ambiguities surrounding nutraceuticals. Nutraceutical products outside of China are being supplied and distributed through a channel named cross-border e-commerce (CBEC). However, the government of China enacted CBEC policies – 2016, which affected import tax on CBEC products and made them eligible for various tariffs, and taxes such as value-added tax, consumption taxes, etc. In view of this, recent advancements in the regulations of health foods and nutraceuticals, and the evolution of concerned regulatory authorities in China are being discussed.

The evolution of laws related to food health and their concerned regulatory authorities in China are described in Figure 6.4 whereas the registration process of (functional) food in China is summarized in Figure 6.5.

FIGURE 6.4 Evolution of laws related to food health.

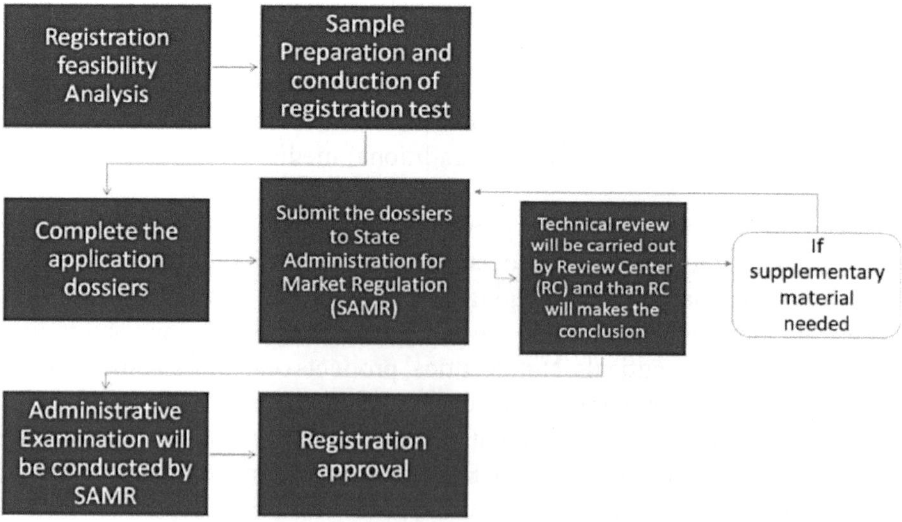

FIGURE 6.5 Registration process of (functional) food in China.

6.4 NUTRACEUTICAL REGULATION IN RUSSIA

This topic mainly focuses on regulatory aspects of functional foods, food supplements, and foods for specific dietary purposes in the territory of the Russian Federation. Various essential documents (Fig. 6.6) have been released in the recent past to secure a guarantee of state for the foods and promote public health.[17] It has been proven clinically that combining the supplements of minerals, multivitamins, prebiotics, and probiotics into the feed of hospital patients increases the rate of recovery and decreases the dosages of medication. In addition to that supplementation of minerals or multivitamins in the diet decreases the occurrences of chronic illnesses and respiratory diseases.[18,19] Hence, the inclination toward consumption patterns of functional foods, food supplements, and foods for specific dietary was observed to be high in recent years. Therefore, it was realized to regulate the process of registration, control the process of manufacturing, ensure the safety and quality norms, compile the comprehensive methods for evaluation, and regulate the claims labeling related to health and nutrition.

FIGURE 6.6 Essential documents that secure guarantee of state for the foods and promote the public health.

In the early 1990s, the Russian Federation entered into the nutraceutical market. These foods gained significant attention for the reduction in deficiencies of nutrients, as well as significant improvement in the nutrition level and quality of lifestyle. Russian Federation signed Decree No. 873-r in 2010, which approved the policy framework of the Russian Federation on Healthy

Public Nutrition up to 2020. One of the key objectives of this was to boost the production of such foods which are supplemented with functional foods, essential nutrients, foods for special dietary uses, specific baby food products, and biologically active ingredients. Moreover, a Strategy for the Improvement of the Quality of Food Products in the Russian Federation until 2030 was approved through a government order No. 1364-p in 2016, which ensures the quality of food. Furthermore, legislations related to food quality and safety have been revised on a larger scale due to the formation of the Eurasian Economic Union (A economic block of the Russian Federation, the Republic of Armenia, the Republic of Kazakhstan, the Republic of Belarus, and the Kyrgyz Republic) and membership of Russia in the World Trade Organization. It is also noteworthy that Federation of Russia is refining its national standards for the regulation of foods used in the treatment of various disorders associated with nutrient deficiency. Therefore, a committee on Technical Standardization of Functional Foods was created in 2008 which was later on reorganized into a Committee for Technical Standardization of Foods for Special Dietary Uses in 2014. Biologically active food supplements (BAFSs) are being regulated in the territory of Russia through various federal laws depicted in Figure 6.7.

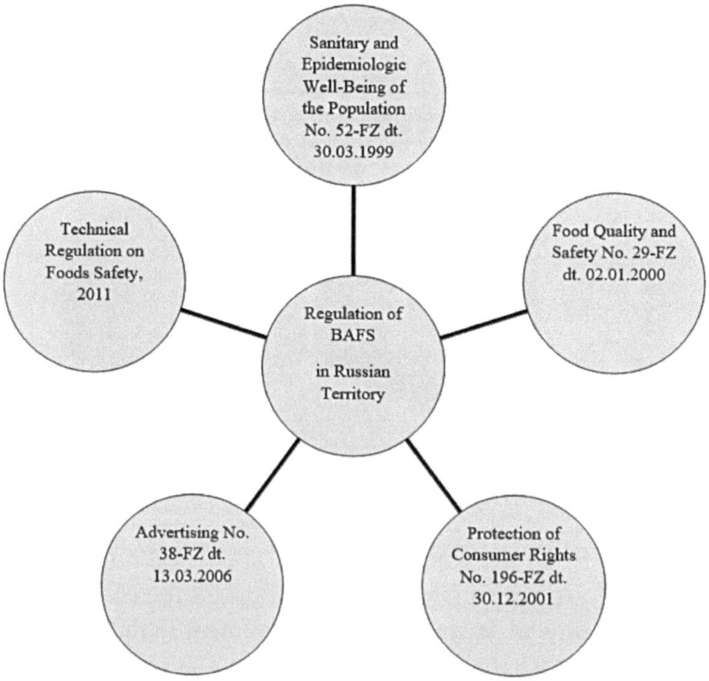

FIGURE 6.7 Various federal laws for the regulation of BAFSs in the territory of Russia.

The objectives of all the currently available regulatory documents are described in the state-released standard 58,040–2017 which regulates the complexes made up of minerals and vitamins, controls the quantities and types of vitamins and minerals, and determines the tolerated quantities for a population of different age groups.[17] The standard applies to ready-to-use complexes of vitamin and mineral complexes, direct consumption of BAFSs in the form of premixes, components used in meals of therapeutic nutrition and dietary prophylactic for the population of different age groups, food items enriched for consumption by the mass population in the form of specialized foods, functional food, and dietary supplements. In addition to that, various aspects like quality, circulation, efficacy, and safety of nutraceuticals are governed by some of the other federal documents such as Sanitary Regulations and Norms (SanPiN), and Methodical Guidelines and Recommendations. Furthermore, various ministries, authorities, and agencies which are directly or indirectly associated with BAFSs release their guidelines, orders, decrees, and other industry-related documents for the regulation of BAFSs.

Compositional regulation of BAFSs in the Russian system is more stringent as compared to the United States and Europe. Russia maintains a negative list of compounds, substances, specific tissues, species of microorganisms, and animals that are prohibited from dietary supplements. A positive list of components under nutraceuticals in Russia includes amino acids, vitamins, fatty acids, minerals, biologically active substances and enzymes, and probiotic microorganisms. Furthermore, these compounds include information about the maximum safe intakes and recommended intake.[20]

Production and distribution, and registration of BAFSs are overseen by Rospotrebnadzor. BAFSs is regulated through sanitary laws and norms. Quality and safety of the product along with the adherence to the legislative criteria in the Customs Union and the Russian Federation is ensured through the registration process of the state by Rospotrebnadzor. All the registered products are stored at State Register. Furthermore, Rospotrebnadzor regulates the manufacturing, import, transportation, storage, and sale of BAFSs in the Federation of Russian territory through SanPiN No. 2.3.2.1290-2003. In addition to that Rospotrebnadzor also controls the quality and safety of BAFSs postapproval. The hygienic and epidemiological necessities for food articles and uniform sanitary for BAFSs in the territory of the Customs Union are governed by Sanitary and Epidemiological Supervision (Control) Law. This law was amended latest by 10 May 2018. The present system of state registration of BAFSs in the Federation of Russia is aligned with the guidelines issued by Codex Alimentarius Commission, and the relevant law prevails in countries such as Great Britain, Germany, Canada, and the United States.[21]

As per the SanPiN No. 2.3.2.1290-2003, the consumer information (Ingredients) displayed on the product label in the "Indications" column is subjected to expert review during the state registration process of BAFSs. The Statement used for the description of BAFSs efficacy and other standards for product labeling such as Trade name of food supplements, Formulation of ingredients, etc., should also follow the norms of SanPinnumber (2.3.2.1290-2003). The Date of manufacturing, expiry, and conditions of storage must be labeled. The registration number of the food supplement and the date of registration should also be listed on the label. Moreover, the product label should also have the name and address of the manufacturer along with the authorized phone number. All the information mentioned on the label should be readable to consumers.[22]

The state registration process of BAFSs is controlled by Rospotrebnadzor. For the registration of BAFSs, initially, a dossier of registration is prepared which provides information on studies related to the safety of samples. The dossier includes documents related to the physicochemical composition of the product, radiological and organoleptic studies, microbiological purity, hygienic assessment, toxicological studies, etc. Dossier also includes the documents related to administration from the manufacturer/organization. Thereafter, the registration dossier is submitted before the Rospotrebnadzor. Rospotrebnadzor specialists perform an investigation of the documents supplied, following which a decision on a certificate issuing of state registration or rejection of state registration is made. The registration certificate is issued for the entire term of Russia's industrial production and import supplies.

6.5 NUTRACEUTICAL REGULATION IN JAPAN

Japan's growing geriatric population and rising healthcare costs are cited as the primary drivers of Japan's functional food industry's development. In addition, the use of nutraceuticals and dietary supplements can help to prevent and improve health issues. Dietary supplements and functional foods grew by an average of 15% per year in the Japanese nutraceutical industry. With a projected value of $13.3 billion in 2019, Japan's health food market was the world's third largest.[24]

Japanese health foods include a variety of foods that have health benefits. Currently, there are only two major food categories defined by law: those for special dietary purposes and those that have health claims attached to them. Three types of foods fall under this category: specific health food, nutrient-function food, and function food. The regulatory system for health claims allows these foods to make claims about their functionality or health benefits. The health claims system's goal is to distinguish between products that lack scientific evidence of their safety or efficacy and those that do, to help people, live healthier lives. Consumers, on the other hand, become confused by the wide variety of names given to foods that claim to have health benefits. As a way to ensure that foods with health claims (FHC) can be used safely and effectively by the public, it is necessary to provide scientifically based information, collect data on side effects, and facilitate communication among all parties involved.[25] For the Food for Specified Health Uses (FOSHU), the Council of Pharmaceutical Affairs and Food Hygiene under the Ministry of Health, Labor, and Welfare (MHLW) scientifically evaluates food products for their efficacy and safety. After implementing a functional food system, many clinically proven FOSHU products with health benefits were developed and released into the market. Including FOSHU products, functional food sales totaled $8 billion in 2018. The terms "nutraceuticals" and "dietary food supplements" are not well known in Japan, compared to most other countries, and are instead regulated as FHC.

FHC (Fig. 6.8) is regulated by the MHLW in Japan, whereas Consumer Affairs Agency (CAA) regulates the advertising and sale of food with health claims in the country. Producing foodstuffs is regulated by local health agencies.

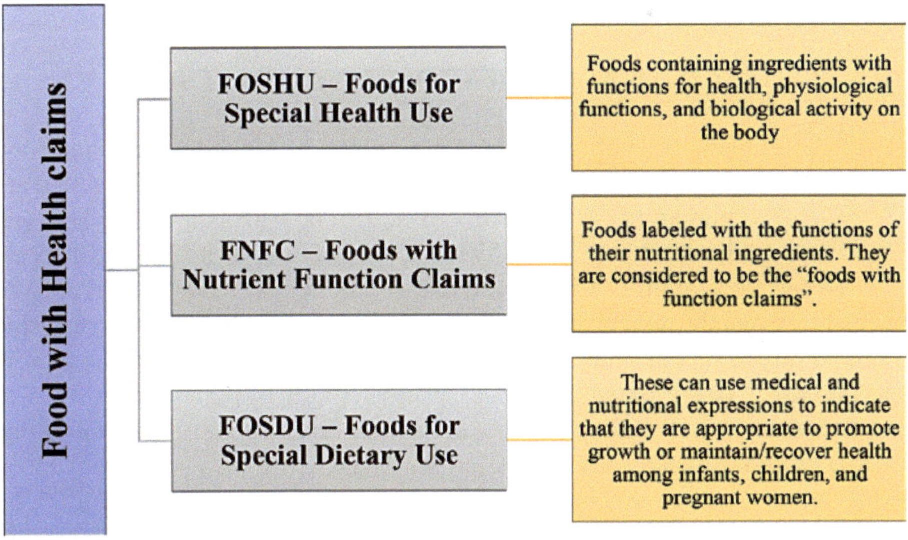

FIGURE 6.8 Foods with health claims in Japan.

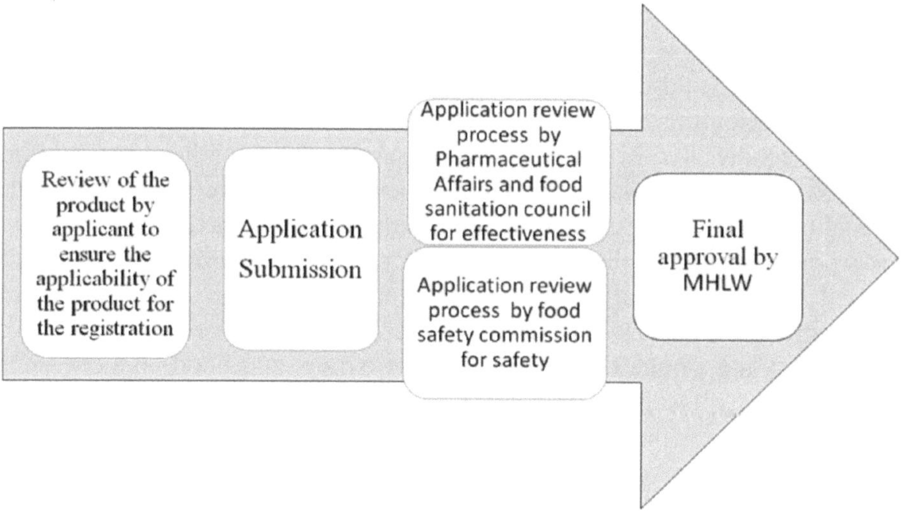

FIGURE 6.9 Registration approval process flow of nutraceuticals in Japan.

In Japan, the regulatory process is unique. Nutraceuticals are classified as either Health Foods or Supplements, and each of these classifications is subject to its own set of regulations. The sponsor must obtain an import license from the MHLW quarantine stations' food sanitation inspection section to bring the approved products into Japan after they have been approved. The approval process of nutraceuticals in Japan is shown in Figure 6.9.[24]

6.6 NUTRACEUTICAL REGULATION IN SOUTH KOREA

Dietary supplements in South Korea are known as Health Functional Food (HFF) because they are made or processed with functional materials that are beneficial to human health. In terms of nutrition, HFF is a bridge between food and medicine. Many people want to improve their health and prevent illness by consuming healthy/functional foods, which have been proven to improve eyesight, counter the effects of natural aging, and boost immune system strength. The market for these functional foods has grown rapidly in the last 5 years, rising from 2.23 trillion South Korean won in 2015 to approximately 4.18 trillion South Korean won in 2020.26 As a result of this market's rapid expansion, legislation was enacted to safeguard consumers and guarantee their right to accurate information about the functionality of food. Therefore, to better manage HFF safety in Korea, the Health/

Functional Food Act (HFFA) went into effect in 2004 (HFFA, 2004). As the Ministry of Food and Drug Safety (MFDS) became more and more aware of subordinate laws and regulations, it grew the law hierarchy. HFFA 3 and its supporting regulations, such as the Food Code 5 and the Regulation on Functional Ingredients Certification and Specification 6, as well as the HFF labeling standard, are the primary laws and requirements governing HFF. In terms of legislation, the HFF Code 5 is the most commonly referred to. In Korea, there are detailed guidelines on how to produce, test, import, store, and sell HFF. To make functional food, one of the following methods can be used. According to the Health/Functional Food Act (HFFA), in 2004, functional foods can be made either in tablets, capsules, pills, granules, liquids, powders, flakes, paste, syrups, gels, bars and films.27

HFF ingredients and raw materials can include functional materials (such as vitamins or fiber) as well as general food ingredients and specific additives, as stated in Article 2 of the HFF Code 5. Functional materials can be categorized in addition to the notified and individual components. MFDS has notified 28 functional ingredients and 68 functional raw materials. Before making or selling HFF containing unnotified ingredients, it is necessary to obtain individual recognition of the material's efficacy and safety. To date, the MFDS has published information on more than 300 different types of recognized individual ingredients. Only the applicant, however, has access to the relevant, individually recognized ingredients in their application. Manufacturers and importers must comply with the Regulations on Certification of Functional Ingredients and Specification6 to get their ingredients approved as individually recognized ingredients. They must also submit the following four documents: (1) an application letter (including the name of the company, the principal business, product information, and any other necessary information about the company); (2) dossiers (including a summary of the dossier and information about the origin of the ingredient); and (3) test results. After consulting with the HFF Deliberation Committee, MFDS will conduct the review (HFF Committee). As soon as MFDS is satisfied with the individual recognized ingredient's safety and efficacy, it will approve its use. When additional data is required, MFDS can ask the applicant to provide it. The "individually recognized ingredient" application will be rejected if the applicant does not submit additional documents. See Figure 6.1 for an illustration of this concept. It costs 1,900,000 KRW to certify a "individually recognized ingredient" (about 1710 USD). Depending on the circumstances, the evaluation period can last anywhere from 60 to 120 days.

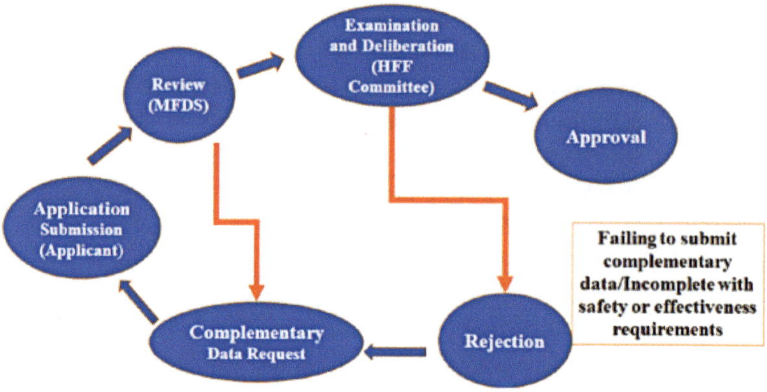

FIGURE 6.10 Evaluation procedure

FIGURE 6.11 Health functional food mark.

Source: Reprinted from chemlinked. https://food.chemlinked.com/foodpedia/south-korea-health-functional-food-dietary-supplement

Figure 6.10 illustrates the HFF certification application process for the product, which is identical to that of the new ingredient. After passing the official evaluation, the product in question could display the label "HFF" (Fig. 6.11) or advertise its efficacy.

The HFF labeling items include the HFF mark, Product name, Business entity and address, Sell-by date and storage conditions, and Net quantity. In addition to providing information about the product's nutritional value, effectiveness, dosages recommended for use, and cautionary statements about its use, the product's raw material names and contents are also included. Precautions for safe use and other specifics are specified by the labeling standard of HFF 7, but other items, such as food additives, packaging materials, and dietary recommendations, are not. The labeling standard of HFF 12 should be followed for Korean reference intakes, for example.[28]

Expanding the definition of HFFs in 2008 to include processed foods that contain functional ingredients was a major step forward.[27] The food import process in South Korea is managed by a single framework, and businesses can export Health Function Food to South Korea by following the procedure 1.[29]

6.7 NUTRACEUTICAL REGULATIONS IN NEPAL

In Nepal, doctors routinely prescribe nutraceuticals without consulting with patients to collect kickbacks from manufacturers, and street vendors peddle illegal herbal remedies and nutraceuticals.[30] Nutraceutical/dietary supplement production, market and use in Nepal must be regulated by an appropriate law. At the Ministry of Agriculture and Livestock Development, the Department of Food Technology and Quality Control released the Dietary Supplement Guideline (2016) which categorizes the dietary supplements in the nutraceutical and functional food market of Nepal (Fig. 6.12). Controlling and ensuring the composition and purity of products, as well as defining quality standards for nutritional supplements, were the primary goals for these guidelines.[31,32]

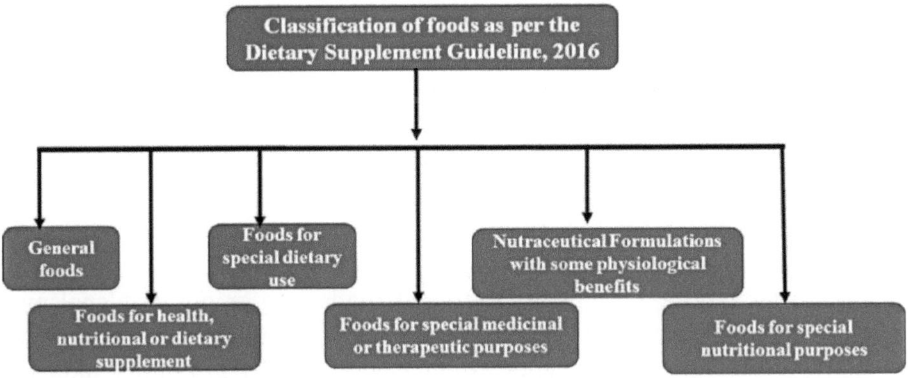

FIGURE 6.12 Classification of dietary supplements and functional foods in Nepal.

Two types of claims can be made by anyone who produces or markets nutritional foods in Nepal, according to this guideline. The first of these is a health-related one. Any statement, suggestion, or implication that connects a food component or nutrient addition to human health falls under this category. It also includes claims about the added nutrients having a functional effect on the body's physical growth and general functions. Statements such as "reduced risk of disease" or "slowed disease progression" fall under this umbrella as well as any other health-related claim. A nutritional claim is the second type of claim. There are a variety of ways in which this can be said to refer to the nutrients found in a food: as a fact or as an assumption. Because functional foods have been divided into six categories, their production, sale, export, and import can all be strictly regulated.

Nutraceutical producers, importers, exporters, and marketers in Nepal should keep the following considerations in mind[31]: Functional foods must meet the legal requirements of the Guideline and be prepared in accordance with food and nutrition principles that have been sufficiently scientifically validated to be taken seriously as food and nutrition. Aside from that, producers should make certain that: (1) The preparation of the formulation is based on food and nutrition principles, which can be verified by sufficient scientific sources; (2) There can be evidence that no hormones, steroids, or psychotropic ingredients have been added; (3) At least 15% of the recommended dietary allowance (RDA) of a specific vitamin, mineral, or nutritional ingredient is added. It's possible to consult guidelines issued by the Nepal government, FAO, or ICMR when determining the RDA for the RDA; this can be done for the RDA In this case, the food product has undergone special processing and is not identical to its natural form. To claim that a food product is a "special purpose medicine/dietary supplement," the product must be able to be used for a specific purpose without any special processing; (4) The formed formulation must adhere to Nepalese or Codex Alimentarius standards; (5) The label must include the composition, instructions for use, and the expiration date in a form that allows the target customer to read and understand the information. Even if the department has been assured of the above statements, it can still examine the claims of a producer regarding the special nutritional features they claim.

Importers and exporters of functional foods into or out of Nepal must follow the Department's detailed instructions. Importing nutraceuticals requires a recommendation letter from the Ministry of Agriculture and Livestock Development and the submission of all necessary documents to the Department. Nepal prohibits the import and sale of any product that has been banned in any other country for any reason. Only authorized dealers of the manufacturer are allowed to import products that aren't banned elsewhere and meet the aforementioned requirements. The Export–Import Inspection and Quality Certification System in Nepal (2063 BS) under the Ministry of Agriculture and Livestock Development should be followed when exporting or importing a product from or into Nepal.[31,33] Nutraceuticals can be imported into Nepal in the manner depicted in Figure 6.13. It is up to the importer to provide a sample of the product they wish to import to be evaluated by the Department and, if necessary, by a laboratory in Nepal and a foreign country. That evaluation should be paid for by the importer of the goods. Additional paperwork could be requested by the Department if it thinks it's necessary.[31] The manufacturing company must also provide a number of other documents, such as a GMP certificate, scientific evidence of the efficacy of the formulation, information on the active ingredient(s) and the method(s) used to test them, a flow chart of the production process, packaging design, and other documents, if necessary, to produce nutraceuticals in Nepal.

In addition, the Department has the following labeling requirements for the product: The product's name and the functional food category it belongs to are provided below. How much the manufacturer claims the product contains in the way of vitamins and nutrients A statement such as, "This product is not for medical treatment of disease" or something similar must appear on the label. In order for the intended recipient to be able to read and comprehend the dosage and administration method, any necessary cautionary measures have been taken (if any), a claim about nutrition(s), the product's expiration date.[34]

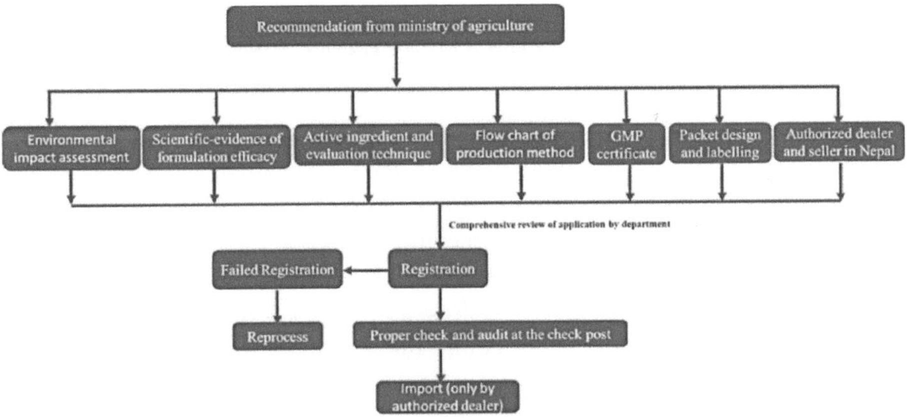

FIGURE 6.13 Registration process for import of nutraceuticals in Nepal.

6.8 NUTRACEUTICAL REGULATION IN SAUDI ARAB

Saudi nationals have undergone a dramatic shift in their lifestyles and eating habits over the years. Obesity, hypertension, and other related health issues have prompted a major shift in diet and exercise habits. Taking vitamins and dietary supplements, as well as functional foods and beverages, will be the most important part of controlling lifestyle diseases in the coming years. The Saudi Arabian Nutraceutical market is expected to grow at a five-year CAGR between 2018 and 2023. The local manufacturers are increasing their production of nutraceuticals in the FMCG sector. For the most part, functional foods rule Saudi Arabia's nutraceuticals and food markets. The functional food market is dominated by baby foods and milk formulas. It is the responsibility of the Saudi Food and Drug Authority (SFDA) to establish regulations and compliance standards for food, raw, and processed animal feed products.

When it comes to preventing and treating lifestyle diseases, vitamins, and dietary supplements along with functional foods have become the primary focus. As a natural and healthy alternative to conventional medicine, the nutraceutical industry has established itself as a leader in this field. All food businesses and operators in Saudi Arabia are to be registered with the Executive Department for Local Markets Control Electronic Services (EMCES) (KSA). New account creation, food establishment registration, and food item registration are all available as online services. End-to-end regulatory support, including HALAL certifications, SFDA food licenses, and SFDA food compliance, is provided by Freyr to manufacturers.[35]

6.9 NUTRACEUTICAL REGULATION IN ASEAN COUNTRIES

Nutraceutical products are regulated in most developed countries, but there are some countries, such as those in the Asia–Pacific region or ASEAN countries, that have yet to implement regulations. Singapore, Thailand, Malaysia, and the Philippines are just a few of the ASEAN countries with thriving commercial markets. The ASEAN region's nutraceuticals market is expected to grow at an annual rate of 8.4%. Nutraceuticals are consumed by approximately 40% of the population on a daily basis.[36] Research and safety evaluation of Nutraceuticals are necessary for better health management, as these substances are neither a drug nor a food, and thus pose a potential toxicity and safety risk. It's therefore the goal of this research project to examine the regulatory status of these products in ASEAN nations.[37] Dietary supplements will soon be regulated uniformly across the ASEAN region. As a manufacturer, this could be one of your best chances to enter the ASEAN market. Product Working Group (TMHA PWG) has a special unit to help manufacturers understand the regulatory process in these countries. Manufacturers can adhere to this working group's standards and harmonized guidelines even though countries have their unique requirements.[36] In terms of food, health status, nutrition requirements, and language, the ASEAN region consists of 10 countries. There is still a lot of work to be done by regulatory agencies in the region to make it easier for the registration and regulation of nutraceuticals.[38] Regulators have been given a variety of tools to help them carry out their duties more effectively.

In addition to the ASEAN Economic Committee (AEC), each country's regional or state body handles all regulatory issues.[39,40] To comply with the agreement on health supplements, each member state's regulatory authority must first approve the products for marketing. As of now, nations have not decided whether this approval is required.[41] The TMHSPWG was established in 2004 under the auspices of the AEC. Amendments have been made over the years, the last of which stated that the association had agreed on the guidelines for health supplements in 2018. The "ASEAN agreement on health supplements 2018 (Fig. 6.14)" is the official name of the agreement.

FIGURE 6.14 Years of progress in health supplement guidelines. Good manufacturing practices (GMP) and traditional medicines and health supplements product working group (TMHSPWG).

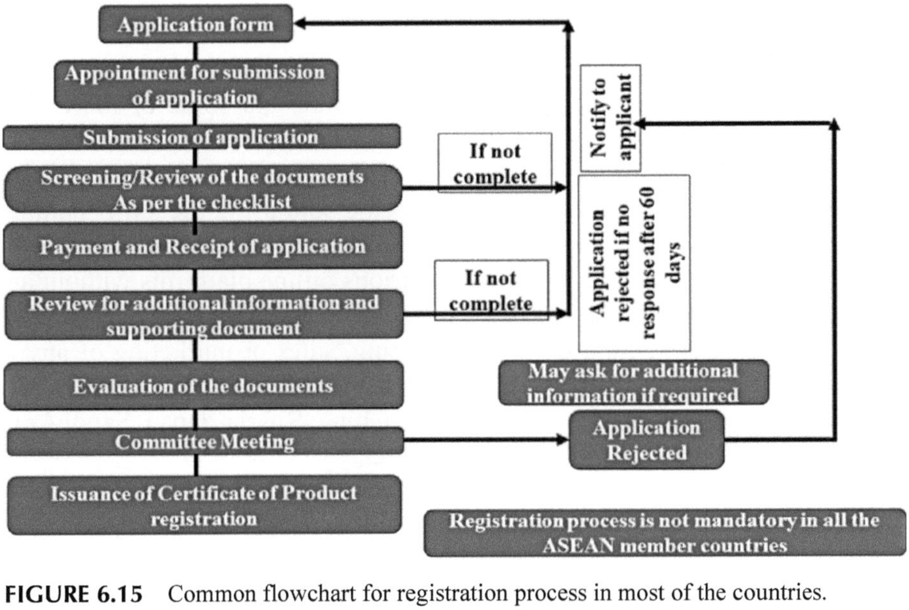

FIGURE 6.15 Common flowchart for registration process in most of the countries.

FIGURE 6.16 According to the countries' administrative data requirements. Certificates of free sales/certificates of pharmaceutical product. GMP, good manufacturing practices.

These countries' administrative data requirements (Fig. 6.16) that are mandatory to register (Fig. 6.15) are discussed here Ref. [42,43].

6.10 CONCLUSION

Asian countries contribute maximum in the world population. With the increasing trend of population in Asian region, the nutraceutical markets are also growing rapidly as the nutraceuticals are looked by the people as an alternative to food supplements. Moreover, the COVID-19 Pandemic also increased the demand of nutraceuticals in the population. The present status of nutraceuticals in the Asian countries can be seen as alarming because various countries such as Pakistan, Afghanistan, and some of the central Asian countries are not even having their own regulation system for the manufacturing, distribution, categorization, markets, and claims of nutraceuticals. Many of food alternatives are being sold using the term "natural" and claiming as a safe substitute to various medicines. Furthermore, these products can be bought easily through various online platforms without any control or regulation. Hence, there is no uniform, consistent or standardized regulation governing the manufacturing, sales or marketing of any of these products. This chapter focuses on the existing regulations available in some of the Asian countries such as India, China, Russia, Japan, Saudi Arab, Nepal, South Korea, and ASEAN countries.

KEYWORDS

- **nutraceutical**
- **regulation**
- **registration**
- **Asian countries**
- **ASEAN countries**

REFERENCES

1. Tutelyan, V. A.; Petrenko, A. S.; Smirnova, E. A.; Sukhanov, B. P.; Kutyshenko, A. V. Regulatory Aspects of Nutraceuticals: Russian Perspective. In *Nutraceuticals: Efficacy, Safety and Toxicity*; Academic Press, 2016; pp 959–970.
2. Tripathi, C.; Girme, A.; Champaneri, S.; Patel, R. J.; Hingorani, L. Nutraceutical regulations: An opportunity in ASEAN countries. *Nutrition* **2020,** *74*, 110729.
3. Domínguez Díaz, L.; Fernández-Ruiz, V.; Cámara, M. The Frontier between Nutrition and Pharma: The International Regulatory Framework of Functional Foods, Food Supplements and Nutraceuticals. *Crit. Rev. Food Sci. Nutr.* **2020,** *60* (10), 1738–1746.
4. Thakkar, S.; Anklam, E.; Xu, A.; Ulberth, F.; Li, J.; Li, B.;Tong, W. Regulatory Landscape of Dietary Supplements and Herbal Medicines from a Global Perspective. *Regul. Toxicol. Pharmacol.* **2020,** *114*, 104647.
5. Helal, N. A.; Eassa, H. A.; Amer, A. M.; Eltokhy, M. A.; Edafiogho, I.; Nounou, M. I. Nutraceuticals' Novel Formulations: The Good, the Bad, the Unknown and Patents Involved. *Recent Pat. Drug Deliv. Formul.* **2019,** *13* (2), 105–156.
6. Turner-Ravana, N. Nutraceutical and Functional Food Regulations in the United States and Around the World. *J. Nutr. Educ. Behav.* **2020,** *52* (10), 989.
7. Global Industry Analysts, Inc. Nutraceuticals - Global Market Trajectory & Analytics [Online]. https://www.researchandmarkets.com/reports/338402/ (accessed Dec 15, 2021).
8. Blaze, J. A Comparison of Current Regulatory Frameworks for Nutraceuticals in Australia, Canada, Japan, and the United States. *Innov. Pharm.* **2021,** *12* (2), 8.
9. Papetti, A. Nutraceuticals: Health Benefits and Government Regulations. *Curr. Res. Nutr. Food Sci.* **2019,** *7* (1), 01–04.
10. Ganesh, G. N. K.; Ramachandran, A.; Suresh, K. R.; Senthil, V.; Baviya Priyadharshini, R. Nutraceuticals—A Regulatory Review. *Int. J. Drug Regul. Aff.* **2015,** *3* (2), 22–29.
11. Makreo Research and Consulting. India Nutraceuticals & Immunity Boosting Supplements Market Outlook to 2025 [Online]. https://www.marketresearch.com/Makreo-Research-and-Consulting-v4231/India-Nutraceuticals-Immunity-Boosting-Supplements-30377812/(accessed Dec 15, 2021).
12. Bhupathiraju, K.; Krishnaraju, A. V.; Sengupta, K.; Golakoti, T.; Akolkar, S. K.; Datla, P. Regulations on Nutraceuticals, Functional Foods, and Dietary Supplements in India.

In *Nutraceutical and Functional Food Regulations in the United States and Around the World*; Academic Press, 2019; pp 445–464.

13. Krishnaraju, A. V; Bhupathiraju, K.; Sengupta, K.; Golakoti, T. Regulations on Nutraceuticals, Functional Foods and Dietary Supplements in India. In *Nutraceutical and Functional Food Regulations in the United States and Around the World*; Academic Press, 2014; pp 343–362.
14. Verma, B.; Popli, H. Regulations of Nutraceuticals in India & us. *Pharm. Innov.* **2018,** *7* (7), 811–816.
15. Yang, M. S. Regulatory Aspects of Nutraceuticals: Chinese Perspective. In *Nutraceuticals: Efficacy, Safety and Toxicity*, 2nd ed.; Academic Press, 2021; pp 947–957.
16. Yates M. Making Sense of Nutraceuticals in China, Feb 10, 2017 [Online]. https://www.reutersevents.com/pharma/column/making-sense-nutraceuticals-china (accessed Dec 15, 2021).
17. Tutelyan, V. A.; Petrenko, A. S.; Smirnova, E. A.; Sukhanov, B. P.; Kutyshenko, A. V. Regulatory Aspects of Nutraceuticals: Russian Perspective. In *Nutraceuticals: Efficacy, Safety and Toxicity*; Academic Press, 2016; pp 959–970.
18. Tutelyan V.; Lashneva N. Biologically Active Substances of Plant Origin. Flavones: Food Sources, Bioavailability, the Effect on the Xenobiotic Metabolism Enzymes. *Vopr. Pitan.* **2011,** *80* (5), 4–23.
19. Spirichev, V. B. Food Supplements as a Supplementary Source of Vitamins in the Nourishment of the Healthy and Ill People. *Vopr. Pitan.* **2006,** *75* (3), 50–58.
20. Tutelyan, V. A.; Sukhanov, B. Biologically Active Food Supplements: Modern Approaches to Quality and Safety. *Vopr. Pitan.* **2008,** *77* (4), 4–15.
21. Sukhanov B.; Kerimova M.; Elizarova E.; Chigireva E. Sanitary and Epidemiological Examination of Foods in the Frame of State Sanitary and Epidemiological Surveillance on the Customs Border and Customs Territory of the Customs Union. *Vopr. Pitan.* **2011,** *80* (4), 25–31.
22. Tutelyan, V. A.; Sukhanov, B. P.; Kochetkova, A. A.; Sheveleva, S. A.; Smirnova, E. A. Russian Regulations on Nutraceuticals, Functional Foods, and Foods for Special Dietary Uses. In *Nutraceutical and Functional Food Regulations in the United States and around the World*; Academic Press, 2019; pp 399–416.
23. Russian Contract Research Organisation. Registration of Biologically Active Food Supplements [Online]. http://rusclinic.ru/eng/services/registration/bafs (accessed Dec 15, 2021).
24. CREDEVO. Japan Nutraceutical Approval Process [Online]. https://credevo.com/articles/2020/11/15/japan-nutraceutical-approval-process/ (accessed Dec 15, 2021).
25. Umegaki, K. Regulatory Aspects of Nutraceuticals: Japanese Perspective. In *Nutraceuticals: Efficacy, Safety and Toxicity*, 2nd ed.; Academic Press, 2021; pp 1299–1307.
26. Statista. Size of the Health Functional Food Market in South Korea from 2015 to 2020 [Online]. https://www.statista.com/statistics/780696/south-korea-health-functional-food-market-size/ (accessed Dec 15, 2021).
27. Kim, J. Y.; Kim, S. J.; Jeong, S. Regulations on Health/Functional Foods in Korea. In *Nutraceutical and Functional Food Regulations in the United States and around the World*; Academic Press, 2019; pp 497–507.

28. ChemLinked. South Korea Health Functional Food Regulation [Online]. https://food.chemlinked.com/foodpedia/south-korea-health-functional-food-dietary-supplement (accessed Dec 15, 2021).
29. Ministry of Food and Drug Safety. Food Labeling System [Online]. https://www.mfds.go.kr/eng/wpge/m_14/de0110051001.do (accessed Dec 15, 2021).
30. Shrestha, R.; Shrestha, S.; KC, B.; Shrestha, S. Evaluation of Nutritional Supplements Prescribed, its Associated Cost and Patients Knowledge, Attitude and Practice Towards Nutraceuticals: A Hospital based Cross-sectional Study in Kavrepalanchok, Nepal. *PloS One* **2021,** *16* (6), e0252538.
31. *Dietary Supplement Guideline;* Department of Food Technology and Quality Control, Ministry of Agriculture and Livestock Development, Government of Nepal. http://dftqc.gov.np/downloadfile/supplement_1575441578.pdf (accessed Dec 15, 2021).
32. *Food Act, 2023 (1967)*; Department of Food Technology and Quality Control, Ministry of Agriculture and Livestock Development, Government of Nepal. http://nnfsp.gov.np/PublicationFiles/fa11d36f-b393-49f8-b449-5f7dfbaa774d.pdf (accessed Dec 15, 2021).
33. *Directives on Export-Import Inspection and Quality Certification System in Nepal*; Department of Food Technology and Quality Control, Ministry of Agriculture and Livestock Development, Government of Nepal. http://www.spsenquiry.gov.np/downloadfile/PDMO_Text_of_Notification_NNPL5_Exp_Imp_Dir_for_Food_(1)_1559802882_1589278835.pdf (accessed Dec 15, 2021).
34. Lamichhane, G.; Pandey, P. R. Regulatory Aspects of Nutraceuticals and Functional Foods in Nepal. *Int. J. Nut. Funct. Foods Nov. Foods* **2020.**
35. Ken Research. Saudi Arabia Nutraceuticals Market Outlook to 2023 - By Vitamins and Dietary Supplements, Functional Foods (Dairy, Baby Food, Breakfast Cereals, Edible Oil, and Confectionary) and Functional Beverages (Energy Drinks, Concentrates, Fruit/Vegetable Juices, and Sports Drinks) [Online]. https://www.kenresearch.com/healthcare/pharmaceuticals/saudi-arabia-nutraceuticals-market-outlook/276726-91.html (accessed Dec 15, 2021).
36. Bhavani, G. S.; Venkatesh, M. P.; Kumar, T. M. Nutraceutical Regulations in ASEAN Countries. *Drug Invent. Today* **2020,** *13* (7).
37. Athalye, M.; Vaghela, S.; Bhavsar, N. The Study of the Registration Guideline of Nutraceutical Products in ASEAN Countries. *Appl. Clin. Res. Clin. Trials Regul. Aff.* **2020,** *7* (2), 104–116.
38. Pettman, S. Standards Harmonisation in ASEAN: Progress, Challenges and Moving beyond 2015. ERIA Discussion Paper Series, 2013; p 30.
39. Hartono, D.; Hardiwinoto, S. Legal Perspective on Asean Economic Community. *Diponegoro Law Rev.* **2018,** *3* (2), 199–222.
40. ASEAN Agreement on Regulatory Framework for Health Supplements [Online]. http://food.fda.moph.go.th/data/news/2560/601212_TrainFood/601212_25TFRF_Annex%205.pdf (accessed Dec 15, 2021).
41. Scattergood G. Thailand's Top Three Supplement and Functional Food Trends as Beauty Bubble Begins to Burst: Infographic, 2017 [Online]. https://www.nutraingredients-asia.com/Article/2017/03/06/Thailand-s-top-three-supplement-and-functional-food-trends-as-beauty-bubble-begins-to-burst-Infographic (accessed Dec 15, 2021).
42. Profile of Definition, Terminology, and Technical Requirement of Traditional Medicines and Health Supplements among ASEAN Member Countries [Online]. https://www.asean.org/storage/images/archive/20258.pdf (accessed Dec 15, 2021).

43. Chia, S. Y.; Plummer, M. G. In *ASEAN Economic Cooperation and Integration: Progress, Challenges and Future Directions*; Cambridge University Press, 2015; p 8.

CHAPTER 7

NUTRACEUTICAL REGULATION IN EUROPEAN COUNTRIES

SATYAJIT SAURABH,[1] MANALI SINGH,[2] SHIVANI BHUTANI,[2] SNEHA SRIVASTAV,[2] and SHILPI KIRAN[3]

[1]*DNA Fingerprinting Laboratory, Bihar State Seed and Organic Certification Agency, Patna, Bihar, India*

[2]*Department of Biotechnology, Invertis University, Bareilly, Uttar Pradesh, India*

[3]*Department of Botany, Patna Women's College, Patna University, Patna, Bihar, India*

ABSTRACT

An ingredient or mixture of ingredients that may be used or given to humans for restoring, repairing, or modifying the functions of the body by performing a pharmacy, immunological, or metabolic activity, or making a medical diagnosis may be a medical product. The changing attitude toward food supplements or diet to meet a recommended dietary supplement is also creating the market for nutraceuticals. People today prefer natural and organic products, active drinks, and natural additives. A growing percentage of consumers are concerned about rising health care prices. People, who are dissatisfied with health care products, depends on neutraceuticals to improve their health and avoid serious illnesses. Nutraceuticals administered exerts pharmacological effect with medical claims and administered as dietary supplements for vitamins, minerals, and other nutrients are substantially different. This substantial difference in use of nutraceuticals has been considered differently for regulatory legislations in the European Union. This chapter is an

Formulations, Regulations, and Challenges of Nutraceuticals.
Tingirikari Jagan Mohan Rao, Rajesh K. Kesharwani, Raj K. Keservani, Anil Kumar Sharma (Eds.)

attempt to discuss the role of EU Framework and European legislations on nutraceuticals.

7.1 INTRODUCTION

The term "nutraceuticals" is derived from the combination of two words, "nutrition" and "pharmaceutical," for the food or components of food having physiological effects (Fig. 7.1). DeFelice[2] defined this term as "Food or parts of food that provide medical or health benefits, including the prevention and treatment of disease.[1,2]" This might be the possible reason that nutraceuticals/ bioceuticals often have place in a gray area between the food and medicine and are not considered for legal legislations.[3] These products are being considered for health promotion by making the diet balanced and complete as "optimal nutrition" to improve both mental and physical health status and reducing the risk factors.[4]

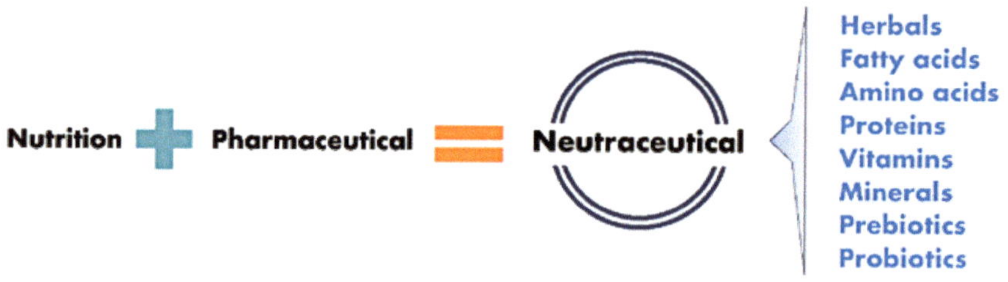

FIGURE 7.1 The nutraceuticals are considered in between the gray area of nutrition and pharmaceutical.

Nowadays, several continents have changed food habits from the traditional foods in terms of preparations and varieties. Also, due to technological advancements, the dynamic lifestyle has been changed to comfortable and sedentary lifestyle. Such changes in lifestyle and foods being consumed results in several incidences of imbalanced hormone and metabolic disorders. The most common disorders are obesity, blood sugar, blood pressure, thyroid, glucotoxicity, and lipotoxicity.[5,6] Progressing with science and technology is appreciable but understanding the past is also necessary at the same time. Ancient documents of the Ayurveda has mentioned role of foods needed for health and to cure a disease. Similarly, other ancient documents/scriptures are available in different continents/subcontinents as per the geographical availability of food source.

Sugar was the only processed food product available in drug stores, around 200 years ago.[7] These days, most of the food and/or beverages consumed are milled, processed, and packaged with food preservatives. The nutritive value of most of the foods is being compromised due to milling and processing. To compensate these nutritive losses, nutritional supplements and functional foods are becoming necessary for a healthy life.[8–12] The nutritional or dietary supplements are intended to ensure sufficient nutrition as per the recommended dietary intake, whereas the functional foods are foods fortified with additional ingredient(s) as food additives. In general, these products are being considered for health promotions. At the same time, there is a need to make strategies for ensuring food safety by strengthening the national food control legislations.

7.1.1 HEALTH BEHAVIORS AND FOOD MARKETS

For health promotions, the nutraceuticals are used in various forms as an additive or as a supplement or formulation for treating various diseases.[13] It helps to keep up with the essential requirements of minerals and vitamins and boosts up the immune system. Research on nutritious foods and dietary supplements can make a huge difference in the field of health care with your facilities and your needs and the need for society as a whole. In some medical schools, alternative medicine has been taught and recognized by many medical doctors as it provided sufficient evidence of medical recommendations (e.g., medical herbalism and enzyme therapy). More recently, a growing number of questions indicate a growing public interest in the health sector that has changed a feature that tells about healthy behavior despite having increased focus on the benefits of nutraceuticals. Consumers switch to foods that reduce the risk of chronic diseases. People have begun to use food products as a diet that will prevent disease and increase physical activity rather than take it to meet a recommended diet. People today prefer natural and organic products, active drinks, and natural additives. In addition, the media has played a very important role in attracting people's attention to scientific advances that have benefited the market for nutritious products. From the first phase to rapid market development, the growing public interest now requires new confidence and a more precise control environment. Several regulatory guidelines are being implemented globally by different regulatory authorities established in different countries. As for example, the Food Safety and Standards Authority of India (FSSAI) in India, the European Food Safety Authority (EFSA) in European Union, the United

States Food and Drug Administration (USFDA) in United States, the Food for Specified Health Uses (FOSHU) in Japan, and the Therapeutic Goods Administration (TGA) in Australia. The regulations are primarily based on registration, evaluation, safety assessment, labeling, sale, and suspension of products (Fig. 7.2). Presently, the international recognition of nutraceuticals has been considered for effective and beneficial effects on health.[14]

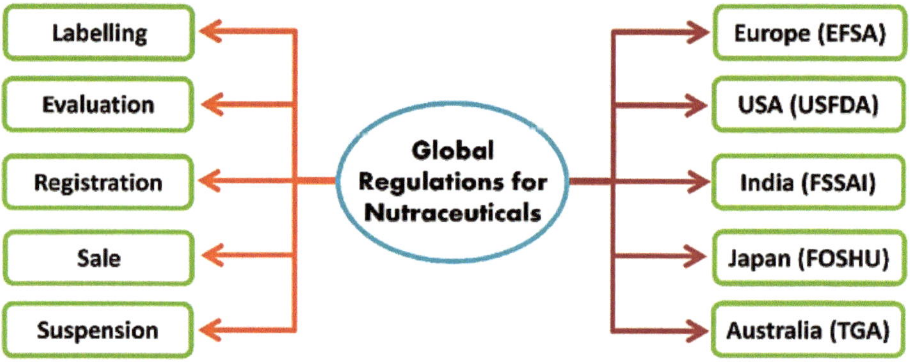

FIGURE 7.2 The global regulations for nutraceuticals are enforced in several countries by different regulatory authorities to ensure food safety.

Any action taken by a person that reflects a person's health beliefs to sustain, achieve, or reinstate good health and to prevent disorders is referred as health behavior. A better understanding of metabolism, physiology, and molecular machinery of body through functional genomics and metabolomics helps in establishing the relationships between the nutrition and lifestyle on health. As per the Centre for Disease Control and Prevention, four health risk behaviors featured for morbidity and mortality are sedentary lifestyle, malnutrition, tobacco, and alcohol. Studies based on survey reports and scientific evaluation suggests that life expectancy can be increased by changing the health behavior by appropriate interventions.[15–18]

7.1.2 MARKETING ENTRY AND COMMERCIALIZATION

It is a market where all herbs, food additives, and active foods are being sold in different places besides pharmacies (e.g., supermarkets, grocery stores, drug stores, etc.), nutraceuticals sold in pharmacies, and para-pharmacies in the form of drug claims drug properties. In addition, one large, attractive, and sometimes dangerous market is the internet. Here, some companies that want to make more money can create products that can be controlled with substandard or ineffective ingredients. Modern social treatments and positive effects that are suspected to have low side effects of such products have also led to increased use of these products. Additionally, the expensive prescription drugs and the uncertainty of some insurance companies and national health plans to address the cost of medications help nutraceuticals strengthen their presence in the global medical market and medical providers. The information service of the EC for scientific activities is provided by CORDIS. According to CORDIS, nutraceuticals in 2008 earned a profit of ~1.4 billion euros with an increase of 11% in comparison to the previous year. Several pharmaceutical and biotechnology companies, found great business opportunities and invested profoundly in the food industry to create a market of 5% of global food sales.[19]

The integration of EU food and food products for 40 years has been ongoing. In 1985, a new law has been recommended as White Paper from the European Commission. In 2000, European Commission published five horizontal framework guidelines that identified public health and safety requirements, consumer information, and general food safety standards. Since then, more than 90% of White Paper proposals have been executed. The European Food Safety Authority (EFSA) regulates plant safety inspections and "plant preparations" intended for use as dietary supplements as defined by the European Food Safety Authority. The rules of the general diet regulation apply to all foods. In other words, there are many regulations and, as such, there is no regulatory framework for "active food" or "nutrition" in EU food law. However, for all the good points mentioned above, nutraceuticals still need extensive scientific support. Studies prove *their effects reduced side effects*. Collected information about a healthy diet needs to be verified and this need represents a major challenge for many professionals for example scientists, biochemists, nutritionists, doctors, and dieticians.

By using the science-based evidence that certain nutrients affect physiological functions in a positive manner, the Functional Food Science in Europe (FUFOSE) Concerted Action was established as defined principles for scientific concepts of functional foods in Europe to reach consensus.[20] The scientific evaluations of the functional foods are being evident from human trials based on markers to claim the function. The markers are categorized as Type A for enhanced function claims and Type B for reduced risk of disease claims. This established an initiative directed for assessing scientific support for food claims, namely, the Process for the Assessment of Scientific Support for claims (PASSCLAIM).

With growing market of the nutraceuticals, the need of strong legislations should be strengthened to ensure the safety by risk assessment and monitoring in entire food/feed chain. The strengthening of legislations could be achieved by establishing some components related with food safety. These components are:

i. Establishing a framework of strong, modern, and harmonized policy and regulation.
ii. Establishing the institutional framework for setting, developing, and implementing a fit-for-purpose standards and guidelines in agreement with those of the Codex Alimentarius of FAO/WHO.
iii. Institutional framework should have competent authorities and have scientific capacity to access risk, data collection, data generation, communication, coordination, and capable of establishing emergency response plans.
iv. Reviewing performance of staff for competency and conducting training for continuous improvements of staffs.
v. Establishing strong compliance and enforcement for effective risk management by surveillance and monitoring programs.
vi. Establishing shared responsibility, coordination, communication, and collaboration among stakeholders.

7.2 EUROPEAN REGULATIONS ON NEUTRACEUTICALS

The European Regulation (EU) 2015/2283 has approved norms for identification of New Food products in the EU market in two ways based on the category or type of food product, only after undergoing the registration process for the novel food products and the traditional food from other (i.e., non-EU) country. Such traditional food go through the notification process. Whereas the New Food product requires the entire complete dossier submission and these methods are considered to be significant. Ingredients are considered foods according to EU General Food Law Regulation (EC) No 178/2002. The General Food Law regulation applies to *any substance or product, whether processed, partially processed or unprocessed, intended to be, or reasonably expected to be ingested by humans*. The food/foodstuffs include any food or drinks that undergo the process of preparation, production, treatment, and/or fortification. Even water is included, if placed in market as processed, treated, and packaged drinking water. An integrated list of vitamins and minerals used in the manufacture of food ingredients and labeling requirements for these products is established by EU legal guidelines in the area of food supplements namely, Directive 2002/46/EC, EU-based assessments. The European Commission is supported by EFSA scientific ideas. Some established legislations are tabulated in Table 7.1.

TABLE 7.1 List of Most Recent Legislation Established Under General Food Law.

Nutraceuticals	Legislations	Description
Food supplements	Directive 2002/46	Directive 2002/46/EC of the European Parliament and of the Council of June 10, 2002 on the approximation of the laws of the Member States relating to food supplements. June 12, 2002. Official Journal of the European Union: L136/85.
Dietetic foods	Directive 2009/39	Directive 2009/39/EC of the European Parliament and of the Council of May 6, 2009 on foodstuffs intended for particular nutritional uses. May 20, 2009. Official Journal of the European Union: L124/21.
Nutrition and health claims	Regulation 1924/2006	Corrigendum to Regulation (EC) No 1924/2006 of the European Parliament and of the Council of December 20, 2006 on nutrition and health claims made on foods. January 18, 2007. Official Journal of the European Union: L12/3.
Fortified foods	Regulation 1925/2006	Regulation (EC) No 1925/2006 of the European Parliament and of the Council of December 20, 2006 on the addition of vitamins and minerals and of certain other substances to foods. December 30, 2006. Official Journal of the European Union: L404/26.

7.2.1 EFFICACY AND SAFETY OF FOOD SUPPLEMENTS

In 1962, the Codex Alimentarius Commission (Codex) was established by the WHO and the FAO to harmonize the food standards, guidelines, and recommendations. The Codex Alimentarius or the food code has more than 200 standards, and become a global reference for regulatory authorities, producers, processors, traders, and consumers.[21] The appropriate balance for consumer access and benefits without any risk needs a regulation for the consumer products under the policy. This principle has been applied to regulate every commodity. Such regulations have the regulatory frameworks to ensure high safety concerns associated with the products. Mostly, the nutraceuticals are also regulated under the category of food (as conventional food). But, there are well-established regulations with specific set of guidelines in some countries and regions like Argentina, Australia, Brazil, Canada, China, India, Japan, Mexico, Russia, United States (US), Association of South East Asian Nations (ASEAN), and European Union (EU). A list of ingredients allowed in foods and nutraceuticals has been published by the EC. The ingredients acceptable across EU have not been harmonized and the member states have their different list of approved ingredients.

The EU legislature will decide on a course of action, based on a risk analysis model, in which the European Food Safety Authority (EFSA) will conduct a scientific risk assessment and the European Commission will conduct risk management, Member States, and the European Parliament. Monitoring policy and other regulatory factors should determine the best way to address the problem in the case of a risk management phase. An example of a phytosterol-containing fat distribution product that went through a novel diet process is an example of how the process of having an "active diet" approved can be an expensive, tedious, and time-consuming task. But the outcome of the market success also proves it worthy. Claims for nutrition and health and the addition of vitamins and minerals and other nutrients to the diet are two proposals currently in the legal process, which will be a key factor in the future promotion of "active food" in Europe. Ultimately, the EU legislature will decide on a course of action, based on a risk analysis model, in which the European Food Safety Authority will conduct a scientific risk assessment and the European Commission will conduct risk management, Member States, and the European Parliament. Monitoring policy and other regulatory factors should determine the best way to address the problem in the case of a risk management phase.

7.2.2 EU FRAMEWORK AND EFSA

The European Food Safety Authority (EFSA) was established legally by the European Union (EU) under the General Food Law Regulation (EC) No 178/2002 for a source of scientific opinion on risks related with the food chain. This European agency is funded by EU to operate independently with a mandate to keep food safe across Europe. It also provides scientific assessment work portal to access tools, resources, and consultations on variety of topics. The remarkable milestones achieved by EFSA are tabulated in Table 7.2.

TABLE 7.2 Milestones of EFSA.

Year	Description
2012–2021	EFSA carries out the safety assessments of six substances used as ingredients in food supplements, for which safety concerns were raised either by the EC or Member States.
2018	The mandate of the EFSA NDA Panel is broadened to include the assessment of *nutrient* sources and requests for safety assessment under Art. 8 of Regulation (EC) No 1925/2006. The EFSA Panel on Food Additives and Nutrient Sources Added to Food (ANS) adopted guidance on the evaluation of sources of nutrients and *bioavailability* of *nutrient* from the sources.
2017	EFSA published the Summary report on Dietary Reference Values for nutrients. Earlier in 2017, EFSA published the Overview on Tolerable Upper Intake Levels as derived by the Scientific Committee on Food (SCF) and the EFSA Panel on Nutrition, Dietetic Products, and Allergies (NDA).
2010	The EFSA NDA Panel publishes an opinion laying down the general principles for establishing Dietary Reference Values (*DRV*) and started the review of DRVs for macro and micronutrients established by the Scientific Committee on Food in 1993.
2005–2009	EFSA carried out a comprehensive assessment of substances that are permitted to be used as sources of vitamins and minerals in food supplements in the EU. The assessment included both the evaluation of the safety of a *nutrient* source at the *intake* levels suggested by the applicant, and the *bioavailability* of the nutrient from the source, that is, the effectiveness with which the *mineral* or *vitamin* is released into the body.
2006	EFSA and the Scientific Committee on Food (SCF) *compile all opinions identifying possible adverse health effects of individual micronutrients* at intakes in excess of dietary requirements and, where possible, establish tolerable upper *intake* levels (UL) for different *population* groups.
2002	EFSA was established legally by the European Union

The food supplements are considered as foodstuffs and regulated according to the EU General Food Law Regulation (EC) No 178/2002. The EU legislation establishes a reference EU legislation to harmonize the regulation and provisions in the area of food supplements for the substances/ingredients required in production of food supplements and the labeling requirements for these products. The EFSA provides science-based opinions for the evaluation of food supplements. The evaluations are carried by the European Commission with directives as rules and legislations. Directive 2002/46/EC is defined for ingredients/substances used and labeling of nutraceuticals.

As in EU the food supplements are considered as food, the safety has to be ensured responsibly by the operators (producers to retailers) involved in the supply chain of that food supplement. The Member States may also monitor by requesting notifications for placing a food supplement in the market of their territory. The product placed in market of the territory may be monitored by the competent authority of the respective Member Sate. Based on the information provided, the competent authorities of the Member States within Article 4(6) of Directive 2002/46/EC on food supplements has been prepared by the Health and Food Safety Directorate-General Unit E1—Food information and composition (Table 7.3).

TABLE 7.3 List of Competent Authorities of the EU Member States.

Member state	Competent authority
AUSTRIA	BundesministeriumfürGesundheit Federal Ministry for Health Department II/B/7 Radetzkystraβe. 2 A-1031 Vienna
BELGIUM	SPF Santé publique, Sécurité de la Chaînealimentaire et Environnement Service Denréesalimentaires, Aliments pour Animaux et AutresProduits de Consommation Place Victor Horta, 40 Boîte 10 Bloc II - 7° étage BE-1060 Bruxelles
BULGARIA	Food Control and Border Control Directorate Bulgarian Food Safety Agency 15 PenchoSlaveikov Blvd. BG-Sofia - 1606
CROATIA	Ministry of Health Ksaver 200a HR-10000 Zagreb
CYPRUS	Medical and Public Health Services, Ministry of Health 18 John Kennedy, Pallouriotissa CY-1449 Nicosia
CZECH REPUBLIC	Ministry of Health Department of Public Health Protection Palackéhonám. 4 CZ–128 01 Praha 2
DENMARK	Danish Veterinary and Food Administration Chemistry and Food Quality Division Stationsparken 31-33 DK-2600 Glostrup

TABLE 7.3 *(Continued)*

Member state	Competent authority
ESTONIA	Office for Retail, Organic Farming and Food of Non-animal Origin, Veterinary and Food Board Väike-Paala 3 EE-11415 Tallinn
FINLAND	ElintarviketurvallisuusvirastoMustialankatu 3 FI-00790 Helsinki
FRANCE	Direction Générale de la Concurrence, Consommation et Répression des Fraudes (DGCCRF) Bureau D3 - Secteur Nutrition 59 bd Vincent Auriol FR-75703 PARIS Cedex 13 - télédoc 251
GERMANY	BundesamtfürVerbraucherschutz und LebensmittelsicherheitMauerstrasse 39-42 D-10117 Berlin
GREECE	Particular Nutrition Food, Food Supplement and Biocides Assessment Section, National Organization of Medicines 284 Mesogeion Av. 15562 Xolargos GR-Athens
HUNGARY	OrszágosÉlelmezés- ésTáplálkozástudományiIntézet (OETI) (National Institute for Food and Nutrition Science) Gyáli u. 3/a HU-1097 Budapest
ICELAND	Environmental and Food Agency of Iceland Sudurlandsbraut 24 IS-108 Reykjavik
IRELAND	Food Safety Authority of Ireland Abbey Court Lr. Abbey Street IR-Dublin 1
ITALY	Ministerodella Salute DSVET DGSAN Viale G. Ribotta 5 IT-00144 Roma
LATVIA	Ministry of Agriculture Food Center of the Food and Veterinary Service Pelduiela 30 LV-1050 Rīga
LITHUANIA	National Nutrition Centre at the Ministry of Health Kalvariju str. 153 LT- 08221 Vilniusia

TABLE 7.3 *(Continued)*

Member state	Competent authority
LUXEMBOURG	Ministère de la Santé Allée Marconi - Villa Louvigny LU-2120 Luxembourg
MALTA	Malta Standards Authority Regulatory Affairs Directorate Foodstuffs, Chemicals, Pesticides and Cosmetics Unit 2nd floor, Evans Building, Merchants Street, MT-Valletta VLT 1179
NETHERLANDS	Ministerie van Volksgezondheid, Welzijnen Sport DirectieVoeding, GezondheidsbeschermingenPreventie Postbus 20350, NL-2500 EJ Den Haag
NORWAY	Norwegian Food Safety Authority Head office P.O. Box 383 NO-2381 Brumunddal
POLAND	Department of Health Promoting Foods
PORTUGAL	DGAV – DireçãoGeral de Alimentação e Veterinária Campo Grande 50 PT-1700-162 Lisboa
ROMANIA	Ministry of Health (MH) Institute of Public Health Timisoara Bd. Dr. V. Babes Nr. 16 RO-300 226 Timisoara
SLOVAK REPUBLIC	Department of Food Safety and Improvement of Nutrition Public Health Authority SR Trnavská 52 SK–826 45 Bratislava
SLOVENIA	Ministry of Health of Republic of Slovenia Sector for Food Safety Štefanova 5 SI-1000 Ljubljana
SPAIN	Ministerio de Sanidad y Política Social Agencia Española de SeguridadAlimentaria y Nutrición (AESAN) DirecciónEjecutiva C/ Alcalá, 56 ES-28071 Madrid
SWEDEN	National Food Administration Box 622 SE-751 26 UPPSALA

7.2.3 INGREDIENTS AUTHORIZED FOR USE IN FOOD SUPPLEMENTS

The botanicals have no consistency in legal framework among the EU countries. In broader sense, the botanicals are considered as nutraceuticals.

The large segments of nutraceuticals are represented by the botanicals, which includes novel foods, functional foods, fortified foods, and food supplements. These food/dietary supplements are mostly consumed in unit dose form as tablets capsules syrup and drops; whereas the fortified and functional foods are consumed as their traditional counterparts. Being more complex in nature and composition, it is difficult to be defined in the food law in European Union (EU). Although these botanicals are defined in European Medicines law as *whole, fragmented or cut plants, algae, fungi, lichens, and botanical preparations from these materials involving extraction, distillation, expression, fractionation, purification, concentration, and fermentation. Botanical substances are defined by the botanical name of the plant according to the binomial system (genus, species, variety, and author) and the part used (e.g., leaf, root, and fruit)*.

There are challenges for botanicals (conventional and processed) in the EU such as the identification absorption and metabolism of active ingredients. These challenges could be related with the product quality, safety, claims, and other issues. The stake holders express diverse views making the nature of issue complex with no common opinion and no establishment of a harmonized directives and regulations. Presently, the EU has no definition for maximum and minimum levels of ingredients in food supplements. Directive 2002/46/EC predicts the maximum and minimum levels, as defined by consulting the European Commission, Members States and interested stakeholders.

The EFSA is involved in ascertaining the upper levels (ULs) for intake of individual micronutrients that support the European Commission in establishing maximum limits. On the basis of information provided by the Member States and assessment by EFSA, the European Commission may take decisions on a given ingredient for its use in food is prohibited, or restricted, or under scrutiny and listed in Annex III defined in article 8 of Regulation (EC) No 1925/2006. Article 8 of the Regulation is most important element that deals with ingredients and amount for consumption to ensure the safety. There are two possibilities to deal with such risk:

i. The available information by EFSA for ingredients is assessed for its harmful effect on health with scientific certainty, and are considered to be prohibited or restricted use.
ii. The available information by EFSA for ingredients is assessed for its harmful effect on health with scientific uncertainty, the

scientific evaluation of such substances/ingredients ensuring safety and explaining the purpose of use have to be submitted to EFSA by the food business operators or any other parties.

Also, the guidelines are defined in Regulation (EC) No. 609/2013 on foods for specific groups such as infants and young children, special medical purposes, and diet replacement for weight control. Regulation (EC) No 2015/2283 is applicable on the novel food, substances that has no history of safe use in the EU before 1997. The list of vitamins (e.g., vitamin C) and minerals (e.g., Calcium) that may be added for nutrition in food supplements are listed in Annex I of Directive 2002/46/EC. A list of authorized substances as the source of vitamins (e.g., sodium-L-ascorbate for vitamin C) and minerals (e.g., calcium lactate for calcium) are listed Annex II of Directive 2002/46/EC. Ingredients to be used in food supplements, other than vitamins and minerals, may also be authorized under the specific legislations according to the nature of substance. The ingredients/substances, other than novel foods, functional foods, and fortified foods, are subject to regulations and provisions defined in specific Member States legislations. Additionally, the nutritional food supplements should have food additives (like sweeteners, color, preservatives, and tablet coating agents) from the list of permitted food additives in Regulation (EC) No. 1333/2008.

7.2.4 FOODS FOR PARTICULAR NUTRITIONAL USES (DIETETIC FOODS/PARNUTS)

The foods for particular nutritional uses are considered as a category of food products regulated under the PARNUTS framework of Directive 39/2009/EC (previously Directive 89/398/EEC). According to the Directive 39/2009/EC, the dietetic foods should have clearly distinguishable composition and/or process involved in production when compared with normal food composition. The products destined for normal consumption is not covered in this legislation. PARNUTS only includes the foods defined for particular nutritional uses. The framework is divided under five subcategories, as mentioned in Table 7.4. The revision in legislation of dietetic foods are being revised and updated regularly.

TABLE 7.4 The List of Dietetic Foods (PARNUTS) and the Relevant Legislations.

PARNUTS	Legislations	For further readings
Infant formulae and follow-on formulae	Directive 2006/141/EC (previously Directive 91/321/EEC)	Commission Directive 2006/141/EC of December 22, 2006 on infant formulae and follow-on formulae and amending Directive 1999/21/EC. December 30, 2006. Official Journal of the European Union: L401/1.
Processed cereal-based foods and baby foods for infants and young children	Directive 2006/125/EC (previously Directive 96/5/EC)	Commission Directive 2006/125/EC of December 5, 2006 on processed cereal-based foods and baby foods for infants and young children. December 06, 2006. Official Journal of the European Union: L339/16.
Foods intended for use in energy-restricted diets for weight reduction	Directive 96/8/EC	Commission Directive 96/8/EC of February 26, 1996 on foods intended for use in energy-restricted diets for weight reduction. March 06, 1996. Official Journal of the European Union: L55/22.
Foods for special medical purposes	Directive 1999/21/EC	Commission Directive 1999/21/EC of March 25, 1999 on dietary foods for special medical purposes. April 07, 1999. Official Journal of the European Union: L91/29.

7.2.5 LABELING REQUIREMENTS FOR FOOD SUPPLEMENTS

Following good agriculture practice and good manufacturing practice, the botanicals must be produced in standard procedures to maintain quality and assure safety. The labeling of products is highly appreciable by consumers. So, the beneficial nutritional attributes and health claims should also be displayed on package with factual statements like net carb, glycemic index, 0 gmtrans fat, goodness of whole grain, etc. The consumers are always in search of labels that claims safety, quality, environment friendly, no animal harm, GMO free, and organic produce; with logo of certifying body and/or regulatory organization.

The food supplements must not be presented and/or advertised by referring its properties as preventing, treating, or curing a disease. The specific requirements for labeling of food supplements have been defined in Directive 2002/46/EC. The directive of labeling states a set of specific information required on the labels. These include:

i. *the names of the categories of nutrients or substances that characterize the product or an indication of the nature of those nutrients or substances;*

ii. *the portion of the product recommended for daily consumption;*
iii. *a warning not to exceed the stated recommended daily dose;*
iv. *a statement to the effect that food supplements should not be used as a substitute for a varied diet;*
v. *a statement to the effect that the products should be stored out of the reach of young children.*

7.3 CONCLUSION

The nutraceuticals encompasses a wide range of issues with status of food or ingredients. There is no common harmonized legislation that can really determine the type and limit of biologically active ingredients. There are many rules and directives to follow, but it depends on the type of food, defined limits, EFSA recommendation, and the Member States. The principles of the General Food Law Regulation, which include the obligation for food safety, traceability, recall, and awareness, are definitely applicable to all diets. If the "new" or "active" food has been consumed to a critical level in the EU before May 15, 1997, then only the Novel Food Act will apply. Claims for the nutrition/health and the addition of vitamins/minerals/other nutrients to the diet are two proposals currently in the legal process, which will be a key factor in the future promotion of "active food" in Europe. The use of traditional botanicals is also being covered by new legislation to ensure safety. The food operators face tough situation in placing their product in the right category with appropriate claims and labels, depending upon scientific evidence available. The stakeholders should access the regulatory status of products in each Member State before market launch. These difficulties associated with the regulations to categorize the food category most appropriately should addressed by the draft of future harmonized legislation depending on the fundamental nature of food product, the biochemistry of active components, and detailed scientific assessment.

KEYWORDS

- **neutraceuticals**
- **dietary supplement**
- **pharmacological significance**
- **legislation**
- **regulation**

REFERENCES

1. Kalra, E. K. Nutraceutical-Definition and Introduction. *AAPS PharmSci.* 2003, *5* (3), E25.
2. Defelice, S. L. The Nutraceutical Revolution, its Impact on Food Industry Research and Development. *Trends Food Sci. Technol.* **1995,** *6*, 59–61.
3. Kauhanen, J.; Myllykangas, M.; Salonen, J. T.; Nissinen, A. Kansanterveystiede, 2nd ed.; WSOY: Porvoo, 1998.
4. Gulati, O. P.; Berry Ottaway, P. Legislation Relating to Nutraceuticals in the European Union with a Particular Focus on Botanical-Sourced Products. *Toxicology* **2006,** *221* (1), 75–87.
5. Fitch, K.; Pyenson, B.; Iwasaki, K. Metabolic Syndrome and Employer Sponsored Medical Benefits: An Actuarial Analysis. *Value Health* **2007,** *10* (Suppl. 1), S21–S28.
6. Poitout, V.; Hagman, D.; Stein, R.; Artner, I.; Robertson, R. P.; Harmon, J. S. Regulation of the Insulin Gene by Glucose and Fatty Acids. *J. Nutr.* **2006,** *136*, 873–876.
7. Brillat, S. A. In *Physiologie du gout*; Revel, J. F.; Paris, 1965.
8. Tovey, F. I.; Hobsley, M. Milling of Wheat, Maize and Rice: Effects on Fibre and Lipid Content and Health. *World J. Gastroenterol.* **2004,** *10*, 1695–1696.
9. Burdock, G. A.; Carabin, I. G.; Griffiths, J. C. The Importance of GRAS to the Functional Food and Nutraceutical Industries. *Toxicology* **2006,** *221*, 17–27.
10. Choi, Y. M.; Bae, S. H.; Kang, D. H.; Suh, H. J. Hypolipidemic Effect of Lactobacillus Ferment as a Functional Food Supplement. *Phytother. Res.* **2006,** *20*, 1056–1060.
11. Olmedilla-Alonso, B.; Granado-Lorencio, F.; Herrero-Barbudo, C.; Blanco-Navarro, I. Nutritional Approach for Designing Meat-based Functional Food Products with Nuts. *Crit. Rev. Food Sci. Nutr.* **2006,** *46*, 537–542.
12. Sieber, C. C. Functional Food in Elderly Persons. *Ther. Umsch.* **2007,** *64*, 141–146.
13. Richardson, D. P. Functional Foods—Shades of Grey: An Industry Perspective. *Nutr. Rev.* **1996,** *54*, S174–S180.
14. Mandel, S.; Packer, L.; Youdim, M. B. H.; Weinreb, O. Reviews: Current Topics-Proceeding from the "Third International Conference on Mechanism of Action of Nutraceuticals". *J. Nutrition. Biochem.* **2005,** *16*, 513–520.
15. Ford, E. S.; Zhao, G.; Tsai, J.; Li, C. Low-Risk Lifestyle Behaviors and All-Cause Mortality: Findings from the National Health and Nutrition Examination Survey III Mortality Study. *Am. J. Public Health* **2011,** *101* (10), 1922–1929.
16. Kinzie, M. B. Instructional Design Strategies for Health Behavior Change. *Patient Educ. Couns.* **2005,** *56*, 3–15.
17. Colella, C.; Laver, J. Setting the Stage for Changing Health Behavior. *Nurse Pract.* **2005,** *30*, 68–70.
18. Nieuwenhuijsen, E. R.; Zemper, E.; Miner, K. R.; Epstein, M. Health Behavior Change Models and Theories: Contributions to Rehabilitation. *Disabil. Rehabil.* **2006,** *28*, 245–256.
19. Mandala, F.; Guida, R.; Guida G. F.: Nutraceuticals : What They are and How They Work, An Article from the E-Journal of the ESC Council for Cardiology Practice. *E-J. Cardiol. Pract.* **2010,** *9* (11), 03 Dec 2010.
20. Contor, L. Functional Food Science in Europe. *Nutr. Metab. Cardiovasc. Dis.* **2001,** *11* (4 Suppl), 20–23..

21. Das, A.; Sen, C. K. Nutritional Supplements and Functional Foods: Functional Significance and Global Regulations; In: *Food Science and Technology, Nutraceutical and Functional Food Regulations in the United States and Around the World*, Second Ed.; Bagchi, D., Ed.; Academic Press, 2014; pp 13–39. doi.org/10.1016/B978-0-12-405870-5.00002-22

CHAPTER 8

REGULATORY GUIDELINES FOR THE DEVELOPMENT OF PLANT/PHYTOCHEMICAL-BASED NUTRACEUTICALS

GARIMA SHARMA

Department of Botany, Shrimant Madhavrao Scindia Govt. Model Science College Gwalior, Gwalior, Madhya Pradesh, India

ABSTRACT

Nutraceuticals are natural food compounds that have health-promoting and disease-preventing properties. This category of food and medicine consists of functional foods, novel foods, herbal supplements, and many more. The plant-based nutraceuticals have active phytoingredients that are known to have health-promoting effects. The regulation of the production, manufacturing, packaging, storage, and transportation of these compounds is necessary and challenging to control their quality, efficacy, and safety. In different parts of the world, these products are defined differently, and accordingly, their regulatory mechanisms also vary. In some of the countries, the production, manufacture, packaging, and marketing of these nutraceuticals are as simple as food products, whereas in some other countries including India, there are explicit set of regulations. Bringing nutraceutical products to market usually follows a registration-based approach, premarket surveillance, or robust postmarket surveillance. This chapter provides an overview of the global regulatory guidelines on the regulation of the production of plant-based nutraceuticals.

Formulations, Regulations, and Challenges of Nutraceuticals.
Tingirikari Jagan Mohan Rao, Rajesh K. Kesharwani, Raj K. Keservani, Anil Kumar Sharma (Eds.)

8.1 INTRODUCTION OF NUTRACEUTICALS

Nutraceuticals are food or foodborne compounds that impart medical and health benefits in various diseases, or in general, improve the quality of life of an individual. Nutraceuticals are continuously gaining importance in global markets due to their vast benefits, safety in promoting health, prevention of diseases, and reducing healthcare costs. Around 2000 years ago, Hippocrates highlighted the importance of nutrients in food as "*Let food be your medicine and medicine be your food.*" However, it took around 20 centuries to realize the importance of nutrients in food, and in 1989, Stephen DeFelice coined the term "Nutraceuticals"1 which are considered as functional compounds at the midpoint of food and drugs. Some of the common sources of nutraceuticals include plant materials, plant extracts, microbial products, animal and marine products, and other sources that have ingredients other than vitamins and minerals and provide health benefits. Out of these sources, plant-derived nutraceuticals have received a considerable attention because they are considered safe and without side effects, and their potential role in disease prevention and progression is documented for decades. The plant-based nutraceuticals/phytonutraceuticals may be of various types as shown in Table 8.1.2

TABLE 8.1 Types of Plant-Based Nutraceuticals.

S N	Nutraceutical type
1	**Dietary supplements** Concentrated sources of nutrients like vitamins, minerals, herbals, amino acids, or other substances with a beneficial effect. These supplements the diet of an individual with an aim to improve some physiological condition; however, they are not taken alone as food or medicine.[3,4]
2	**Functional foods** Consumed as a part of normal diet to supplement it to promote growth and developmental processes.
3	**Nutrition foods** Various plant foods that provide carbohydrates, proteins, vitamins, minerals, amino acids, fatty acids, and antioxidants.
4	**Novel foods** The foods that are new and were not consumed by humans significantly until the beginning of the 21st century. They are developed by new technologies and processes like biotechnology and nanotechnology. They are a good source of proteins, vitamins, and minerals. Examples are plant-based proteins like leghemoglobin,[5,6] mung bean proteins, exotic fruits, fermented soybean, etc.

TABLE 8.1 *(Continued)*

S N	Nutraceutical type
5	**Herbs**
	Nonnutritive phytoconstituents that play an important role in health enhancement.[7,8] For example: *Agave*, aloe vera, gingko, *Withania*, and many more.
6	**Natural plant products/phytochemicals**
	Natural plant-based products to provide them color, flavor, smell, etc. Some of them have physiological activities, prevent some diseases, and improve body functioning.[9,10] Some of the phytochemicals are flaxseeds; β carotene from yellow, red, and green leafy vegetables; lycopene in tomatoes; catechins in tea.

8.2 NUTRACEUTICALS MARKET IN INDIA AND GLOBE

The global market for nutraceuticals was assessed at USD 417.66 billion in 2020, and it is expected to grow annually at a rate of 8.9% from 2020 to 2028. In India, the growth has been estimated from USD 4 billion to USD 18 billion from year 2017 to 2025.[11,12] This increasing growth in the market of nutraceuticals is due to certain key factors which include:

1. Increased health awareness among people worldwide, especially due to chronic health diseases and pandemic situation, and also increased economic growth.
2. The scope of nutraceuticals is continuously growing in area of lifestyle-related diseases like treatment of cardiovascular disorders (CVDs), diabetes, malnutrition, cancer, etc.
3. The increasing elderly population across the world is another driving factor in the demand for nutraceuticals.
4. A rapid economic growth has led the consumer to use more dietary supplements (DS) and fortified foods.
5. New technologies of production of nutraceuticals are continuously adding to reliability and positive attitude toward the use of these products.

India has an abundant supply of plant-based medicines and herbal flora, which have been known for centuries and are used as ayurvedic medicines. The nutraceuticals market in India mainly comprises DS, which constitute over 65% of the total Indian nutraceuticals market and include herbal and nonherbal extracts. Some of the main driving forces behind the growth of the Indian market for nutraceuticals and DS are government initiatives to reduce the malnourishment among children, improved Indian economy, ideal manufacturing hub for plant-based products due to their local availability, low labor costs, better market coverage due to increased internet access, etc.

8.3 CHALLENGES IN MANUFACTURING NUTRACEUTICALS FROM PLANTS

Due to the increasing demand of nutraceuticals, it is now even more necessary that the production of nutraceuticals should be as strict as modern medicines around the globe.[13] The production of plant-based nutraceuticals requires stringent quality control and regulation throughout the whole process, which essentially comprises the following steps:

- Research and development in nutraceuticals
- Collection of plant materials
- Acquisition of other necessary chemicals
- Primary processing
- Packaging and retail.

The manufacturing of phytonutraceuticals has always been a challenging task, as the raw material has diversity in nutritional contents due to geographical region, season, age, and climatic conditions. The plant's associated microbial flora is another key factor that demands a very stringent manufacturing process so as not to introduce any disease or pathological conditions in the consumer. Some of the challenges associated with the entire process are discussed as follows:

8.3.1 REPRODUCIBILITY OF SOURCE OF PLANT MATERIAL

Various factors like location of plants, climatic conditions, time of harvesting, variety and age of plant, extraction methods, etc. affect the metabolic contents of plant. This is the biggest factor that affects the reproducibility of source material, and ultimately, the manufactured product. The standardization of biochemical content through well-known techniques, along with carefully selected controls, is used to make the manufacturing cost-effective without compromising quality.

8.3.2 THE PROCESSING OF MATERIAL AND DETERMINATION OF ACTIVE INGREDIENT

Usually, herbal products are a mixture of several constituents that are responsible for the activity. The determination of the active ingredient and its concentration are very crucial steps. Generally, the different sizes, shapes, and textures of source materials, like bark, leaves, seeds, stem, roots, etc., make the manufacturing operations difficult. For instance, powders of barks and roots require milling to generate fine particles, as they inhibit mixing during processing. The varying moisture, oil, and wax contents also cause problems during processing.[14] The presence of minerals in the formulation imparts poor taste and causes problems by reacting with other components. Therefore, plant-based preparations have specific processing-related challenges to increase their bioavailability, tolerance, and taste without compromising their efficacy.

8.3.3 MICROBIAL FLORA ASSOCIATED WITH PLANT MATERIAL

The raw plant material that is used for the production of nutraceuticals generally contains microbial flora associated with it. The microbial population should be removed thoroughly to prevent cross-contamination and control the quality of the final product.

8.3.4 THE STORAGE OF FINAL PRODUCT AND ITS STABILITY

Many phytochemicals are unstable at high temperatures; hence, they are needed to be stored at low temperatures throughout the process. Sometimes, the biochemical activity of the plant extracts stored at low temperature is reduced significantly. All these factors contribute to the challenges encountered during the manufacturing of nutraceuticals from plant products.

8.4 REQUIREMENT OF REGULATORY GUIDELINES FOR THE PRODUCTION OF PHYTONUTRACEUTICALS

The global market of plant-based nutraceuticals has increased very significantly in the last few decades. It becomes imperative to substantiate the scientific data on the safe and effective use of these compounds as well as their bioavailability. The regulatory guidelines for their production are highly needed for the following reasons[15]:

- Identification of target on which the drug works;
- Evaluating the mechanism of action of the drug;
- Safety assessment;
- Efficacy assessment validated by clinical studies;
- Evaluation of possible unwanted effects if any;
- Evaluation of possible interactions with other products like food, drugs, etc.

For plant-based nutraceuticals, the formulations of the final product and their quality and safety may be highly variable. Moreover, the phytonutraceuticals have several active ingredients; therefore, their analysis is also difficult. Due to all these factors, there is a need to continuously monitor and regulate the quality of nutraceuticals.[16]

8.5 REGULATORY GUIDELINES FOR PLANT-BASED NUTRACEUTICALS IN VARIOUS PARTS OF THE WORLD

Plant-based nutraceuticals are regulated in the same way as food through the FDA. The main objectives of regulation are to ensure safety of ingredients and make substantiated and non-misleading claims about efficacy and bioavailability in target tissues. For many years, various global governments have tried to regulate the production and consumption of food and drug items by making new guidelines and policies and amending old ones related to Good Manufacturing Practices (GMP) and Generally Regarded as Safe (GRAS). In the given section, important guidelines of regulation of plant-based nutraceuticals in different parts of the world are summarized.

8.5.1 REGULATORY GUIDELINES FOR PLANT-BASED NUTRACEUTICALS IN INDIA

In India, the manufacturing, licensing, registration, packing, and labeling of all nutraceuticals, including plant-based, are regulated by FSSAI (Food Safety and Standards Authority of India), which has defined regulatory guidelines for the approval of nutraceuticals in the Indian market through **Food Safety and Standards Act of 2006**. In Section 4 of the act, science-based standards for food items that are used as "Nutraceuticals" or "Food for special medical purpose" or "Food for special dietary use" or "Functional food" or "Novel food" are defined. Some of the salient features of the FSSAI Act, in context to the manufacturing, labeling, packaging, transport, and storage of plant-based nutraceuticals, are as follows:

- Decentralization of licensing for food products authority to state for issuing license to food products.
- Effective, transparent, and answerable regulatory framework for production of food-based items.
- Surveillance and monitoring of production, packaging, and transportation mechanism of food-based items.
- Regulation on food import within the country.
- Developing a large network of food testing laboratories.
- Consistency between national and global standards of food safety measures without compromising consumer health and interest.
- Training and awareness program on food safety for stakeholders.

As per FSSAI regulations, there is no differentiation in functional foods, nutraceuticals, and DS, and they all are indicated as food for a special dietary application. A regulation was drafted for the World Trade Organization by India in 2015 about nutraceuticals and foods for special diets and medical purposes. These food materials were defined based on their ingredients, labeling, additives, contaminants, and health and nutritional claims. This draft regulation also determined the criteria for the manufacturing and sale of these materials as foods and their recommended doses. The full text of the regulation can be accessed from the FSSAI website (http://www.fssai.gov.in).

The FSSAI has initially drafted six principal regulations (which came into force on August 5, 2011), which are further reviewed, and five new regulations, notified in 2017, considering the latest developments and technologies in food science and food additives and to become at par with international standards in food product technology. The regulations of FSSAI are:

1. Regulation of Food Safety and Standards, 2010 (Authority of India Business Transaction Regulation)
2. Regulation of Food Safety and Standards, 2010 (Licensing and Registration of Food Business Regulation)
3. Regulation of Food Safety and Standards, 2011 (Food Products Standards and Additives Regulation)
4. Regulation of Food Safety and Standards, 2011 (Prohibition and Restriction on Sale Regulation)

5. Regulation of Food Safety and Standards, 2011 (Packaging and Labeling Regulation)
6. Regulation of Food Safety and Standards, 2011 (Laboratory and Sampling Analysis Regulation)
7. Regulation of Food Safety and Standards, 2016 (Food or Health Supplements, Nutraceuticals, Foods for Special Dietary Uses, Foods for Special Medical Purpose, Functional Foods, and Novel Foods)
8. Regulation of Food Safety and Standards, 2017 (Food Recall Procedure)
9. Regulation of Food Safety and Standards, 2017 (Import)
10. Regulation of Food Safety and Standards, 2017 (Approval for Non-Specified Food and Food Ingredients)
11. Regulation of Food Safety and Standards, 2017 (Organic Food)

In FSSAI, plant-based nutraceuticals are referred to as "Specialty food containing plant or botanical ingredients" and defined as "food article which is shown to be containing plant or botanical ingredient with the history of safe usage." Some of the salient features of these regulations for plant-based nutraceuticals are as follows (Notified in Gazette of India, December 26, 2016):

i. Schedule IV of this regulation specifies the plants, their parts, and recommended doses that may be used for the manufacturing of specialty food products in which these plants or botanical ingredients are used. In this list, about 400 entries are included.
ii. The plants or botanical ingredients other than those specified in these regulations in Schedule IV, whose safety has been established in India or in any other country, can only be manufactured or sold in India after the prior approval of the Food Authority.
iii. The approval application of any plant or botanical ingredient to the Food Authority should also have attached the relevant documents related to the previous use of the botanical ingredient for at least the last 15 years in India or 30 years in the country of origin.
iv. When a compound used as a nutraceutical is manufactured from plant-based components, it may contain ingredients that are specified in Schedule VI of the FSSAI. These ingredients can either be used in individual formulations or in various combinations of plant-based ingredients. The ingredients may be used either in their raw or unprocessed form or in approved processed forms. The formulations should be in the form of a liquid like syrup, a suspension, a tablet,

a capsule, a granule, or any other format approved by the Food Authority.

v. The information of plant/herbal ingredients is given in Schedule VI (Part A and B). The files on information of use of these products are to be prepared and maintained by each manufacturer. The files should also contain finished product quality confirmation and the method for testing the presence of the specified ingredient in the food.

vi. The usage level of the specialty food containing some plant-based ingredients shall be within the limits specified in Schedule IV. It may only be exceeded if the ingredient is distributed after delivery or formulated for delivery in more than one portion daily.

vii. When a manufacturer seeks prior approval for using any other plant-based material or its ingredient not specified in Schedule IV, the application must be accompanied by: (a) a file of the product containing the basic information of the material used; (b) a method of quality confirmation to demonstrate the presence of the ingredient in the food; (c) the previously published date on scientific and other details of the material or product and its safety aspects; and (d) the previously conducted studies on humans related to the efficacy of the product.

viii. The file related to all the above information shall be immediately presented to the Food Authority for inspection and review whenever it is asked for. This file should contain the technical and scientific information covering the following facts: (a) the basic information about the ingredients of the product with their scientific names; (b) details of steps of its formulation; (c) data about its shelf life; (d) quality specifications of the product and its determination methods; (e) studies related to the safety and pharmacology of the product; (f) findings of any human study conducted; (g) status of regulation in other countries, if any; and (h) any other relevant information about the product.

ix. The product should only contain those additives that are specified in Schedule VE or VF of the FSSAI Act for the preparation of specialty foods manufactured from plant or botanical ingredients.

The FSSAI Act also provides two explanations on what can be referred as "specialty food containing plant or botanical ingredients." These explanations are:

Explanation 1: The plants and botanicals specified in Schedule IV of the FSSAI in normal and naturally occurring form shall not constitute a food supplement, nutraceutical, or food for special medical or dietary use.

Explanation 2: Mere food forms such as vegetables, cereals, fruits, spices, and legumes minimally processed (cleaned, deweeded, and sorted) either in raw or cooked form or juice shall not constitute food for special medical or health purposes or as nutraceuticals.

In India, the FSSAI regulations of 2011 require the following for the approval of any nutraceutical: The registration of manufacturing company is required (according to Schedule 1b), and an import license has to be issued from the Central Licensing Authority along with complete documents: the layout of the manufacturing unit, the processing unit, the list of machines used in manufacturing, the list of food categories that are to be manufactured, the Food Safety Management System plan or certificate (if any), the source of raw material, the pesticide residue report of water, etc.[13]

8.5.2 REGULATORY GUIDELINES FOR PLANTS-BASED NUTRACEUTICALS IN AUSTRALIA

In Australia, all nutraceuticals are regulated under the TGA, that is, the Therapeutic Goods Act of 1989, as they are referred to as "Therapeutic Goods." In this category, all herbal medicines, vitamins, minerals, nutritional supplements, and plant-based biologics are included regardless of their use as medicine or food (Australian Government, Department of Health Report I–III, 2020). The TGA defines therapeutic goods as beneficial foods that work by modifying the body's functioning. The regulation of therapeutic goods is a must in order to protect public health. In Australia, a two-tiered system is followed for regulating the therapeutic foods that are considered medicines: (1) Low-risk medicines, which only require enlistment on the Australian Register of Therapeutic Goods (ARTG) and (2) High-risk medicines, which must be registered on the ARTG. The high-risk medicines that are registered are evaluated by the TGA for safety, efficacy, and quality, whereas the low-risk medicines that are enlisted are evaluated only for quality and safety; however, their efficacy is not evaluated (Australian Government, Department of Health Report I–III, 2020). There are "listed complementary medicines" containing low-risk ingredients to be used in permissible amounts in the listed medicines. Herbal substances containing medicines and products are "listed" in the current regulatory system only when their safety is well-demonstrated and their manufacturing is done in accordance with the principles of GMP. By 2019, in Australia, there were about 90,988 products listed/registered in the ARTG. Out of them, only ~40 complimentary medications are registered and evaluated at ARTG, while the rest are listed on the register (Australian Government, Department of Health Report I–III, 2020 and Ref. [20]).

Plant-based nutraceuticals and medicines are considered in the low-risk group and relatively benign, so the TGA regulations for them allow their sponsors for self-assessment of their products from time to time. Some of the regulations for listed medicines (including plant-based) are as follows:

1. The medicines that are enlisted on the ARTG must display their unique AUST L number on the medicine label.
2. The listed medicines or products are not evaluated prior to inclusion in the ARTG register, so that the product may have rapid market access.
3. The sponsor of the product, while submitting the application, has to provide a certificate (under Section 26A of the Act) that their product meets all the requirements that are applicable to safe, qualitative, and effective use of it.
4. The listed products may only contain low-risk active ingredients.
5. The import of the listed products must not be prohibited as per the Customs Act.
6. The listed products must not be required to be sterile.

In the current regulatory system, the listed medicines contain a wide range of plant-based substances that are demonstrated as safe. They include the following types:

- **Herbal materials (excluding extracts):** They include whole plant material or its cut parts like leaves, roots, flowers, seeds, bark, etc. The material should be in unprocessed form or like herbal powders (herbal materials that are dried and ground into powders). The material should be properly identified by any three of the given specific tests: (1) Tests for its macroscopic characters; (2) tests for its microscopic characters; (3) tests for ingredients by chromatographic procedures; and (4) tests of chemical reactions. The consigned herbal material must be compared with current scientific literature and reference specimens. The comparative analysis must be conducted by a botany expert to determine any difference between the consigned material and already available data (Australian Government, Department of Health Report I–III, 2020).

- **Herbal extracts:** The herbal extracts are also tested by comparison with current scientific literature and by chromatographic analysis against an authenticated reference specimen. Generally, well-established chemical marker tests are not considered valid tests for the identification of the herb because they might be unspecific to the herb species and give misleading information. For example, the chemical product caffeine is not a suitable marker for *Paullinia cupana* for identification of this species.

8.5.3 REGULATORY GUIDELINES FOR PLANTS-BASED NUTRACEUTICALS IN THE USA

The nutraceuticals in the USA are termed as "dietary supplements (DS)," which are defined as "a product taken orally, containing a dietary ingredient like vitamins, minerals, herbs, or other botanicals intended to supplement the diet." They are made in the form of extracts, liquid concentrates, or metabolic constituents. However, they are not consumed as diet and only labelled as a DS. As per the USA Food and Drug Administration (FDA), currently about 85,000 products are being used as DS. The FDA in the USA has different regulations for DS than for ordinary food products and medicines. The Dietary Supplement Health and Education Act (DSHEA) of 1994 provides a regulatory framework for manufacturing, regulating health claims and labeling, and ensuring the safety of DS, including plant-based nutraceuticals. The most important features of the Act are:

1. According to DSHEA, the manufacturer is solely responsible for ensuring the safety of a nutraceutical before its marketing.
2. The claims related to the safety and efficacy of a nutraceutical should be verified by the Academy of Sciences or other federal authorities after notifying the FDA at least 4 months prior to the product's market launch.[21]
3. In the USA, a drug's safety and efficacy is proven before being approved, whereas DS are primarily regulated through postmarket surveillance and are not approved at all.
4. DS in the USA are primarily regulated through surveillance activity conducted by governmental agencies. The manufacturer

is responsible for ensuring that all the information provided on the product label is not misleading or incorrect.[22]

5. A dietary supplement should not be intended to be used through any other means (like applied topically, injected, inhaled, etc.) than ingestion.
6. As per the FDA requirement, the label of any nutraceuticals or dietary supplement products should have a disclaimer about the fact that the product has not been evaluated by the FDA, and it is not intended for prevention, cure, or treatment of any ailing condition.
7. The food products that are marketed before the implementation of DSHEA are considered safe; however, those which were launched in the market after this date are considered novel ingredients and they have to be assessed by the FDA.
8. DS should not contain ingredients that were previously approved as drugs unless they were marketed as DS prior to being approved as drugs.
9. In 2019, the FDA launched an online tool named "Dietary supplement ingredient advisory list," which is a quick response tool to alert the public when some unlawful ingredient is found in DS based on preliminary determination.

8.5.4 REGULATORY GUIDELINES FOR PLANTS-BASED NUTRACEUTICALS IN CANADA

All nutraceuticals, including plant-based ones, are considered under Natural Health Product (NHP) under the Canadian regulatory system (CRS), which regulates them under the NHP regulations since January 2004. In the NHP, all health products containing vitamins, micronutrients, plant-based medicines, probiotics, amino acids, and essential fatty acids are included (Canada blog ref). The NHP addresses 53 recommendations of the common standing committee on the regulation of NHP. Some of the features of NHP regulations are as follows:

1. All the NHPs only be sold in Canada after obtaining a product license, and the sites where these products are manufactured, packaged, labeled, and imported should also have site licenses.
2. The license can only be obtained after following GMP in product labeling and packaging. The product should be accompanied by safety and efficacy data.

3. These requirements of licensing of the *NHP regulations* apply to all individuals and companies that manufactures, packages, labels, and/or imports NHPs for commercial sale in Canada.
4. To get a license, applicants must give detailed information to Health Canada about the product's ingredients, its source, dose, potency, nonmedicinal ingredients, recommended use(s), and proposed claims. The applicant must also provide evidence about the safe and effective use and quality of the product.
5. After successful assessment by Health Canada on safety and effective uses of the product, a product license along with an eight-digit Natural Product Number (NPN) is provided. This number should be printed on the label of the product. The number implies that the product has been approved and licensed by Health Canada.
6. The licensing application to Health Canada must contain evidence for clinical trial data, studies about its effectiveness, pharmacopeia of the product, and traditional resources. The supporting evidence should back up the proposed health claim of the product and its overall risks.
7. Another post-approval safety surveillance authority, referred to as "The Marketed and Health Products Directorate (MHPD)" is functional in Canada to assess safety trends and health risks (if any) of all regulated marketed health products, including NHPs.
8. A "Licensed natural health products database" is also prepared by Health Canada for consumers to check if a product has been licensed before use.

8.5.5 REGULATORY GUIDELINES FOR PLANTS-BASED NUTRACEUTICALS IN JAPAN

In Japan, food regulation issues were raised voluntarily by stakeholders like manufactures and consumers.[24,25] This legislation came into effect in 1991 and was later renamed as "Health Promotion Law" in 2003.[26] The Japanese Ministry of Health and Welfare is the main regulatory body for all drugs and health supplements and maintains strict regulations for medicines and food supplements.

In Japan, nutraceuticals are known as "Food with Health Claims," which has three distinct categories:

- "Food with Nutrient Function Claims" (FNFC) that are mainly vitamins and minerals,

- "Food for Specified Health Uses" (FOSHU) for other functions, and
- FFC

Food products with botanicals fall under the categories of FOSHU and FFC. Following are important features of the regulation of these products:

1. The products go through strict clinical testing, and they can claim health benefits on their labels because they are proven effective and safe for consumption as claimed.
2. If the product does not have a validated health-beneficial effect but the ingredient meets the requirements on safety, nutrition, ingredient content, etc. set by FOSHU, the product can be approved.
3. If a product shows a reduction in disease risk, it should be limited nutritionally to one ingredient of the product.
4. In 2015, the FFC category was introduced to separate DS used for health promotion and disease prevention from FOSHU. All FFC category products must be submitted to the Consumer Affairs Agency (CAA) for approval.
5. In FFC, it is not the government's responsibility to evaluate the safety and efficacy of functional claims of the food product; in fact, the manufacturing companies are responsible for submitting a premarket approval report.
6. The manufacturing companies are also required to label their products as per the Food Labelling Standards and Guidelines on Notifications of FFC. Due to the FFC regulation system, the growth in the market for functional foods has been accelerated since 2015.
7. In Japan, "herbal medicine" is regulated separately from "Food with Health Claims." In China and Japan, herbal medicines are referred to as "Kampo Medicine." Products that fall into this category are classified as medicines instead of functional foods and are considered to be the same as government-regulated prescription drugs. However, Japan does have over the counter Kampo medicines that must meet certain approval standards to be marketed.

8.5.6 REGULATORY GUIDELINES FOR PLANT-BASED PRODUCTS IN THE UNITED KINGDOM AND OTHER EUROPEAN COUNTRIES

All food supplements are regulated under the Food Law in the United Kingdom and other European countries. The Food Law in the United

Kingdom was formed by the European Union (EU), and it was enacted into the UK law. One more regulation, that is, The European Food Safety Authority (EFSA) under the EU regulates the risk management and evaluates the health claims under the new nutrition and health claims regulation by the food supplement. EFSA applies to all relatively new food ingredients that were not consumed to a significant degree in the EU prior to 1997. The EU food supplements Directive 2002/46 came into force on August 1, 2005 and is implemented in the United Kingdom by the Food Supplements (England) Regulations 2003, and its equivalent regulations are implemented in Scotland, Wales, and Northern Ireland. The Food Standard Agency (FSA) has the responsibility to regulate food-based products, including nutraceuticals, in the United Kingdom. As per FSA, it is the responsibility of the manufacturer of the food supplement to comply with all of the relevant legislations. In the United Kingdom, the legislation which covers food supplements includes: Food Safety Act (FSA) and European Commissions (EU) directives and amendments of 2002/46/EC. Similar legislations are separately into force in England, Scotland, Wales, and Northern Ireland. Some of the regulatory guidelines for nutraceuticals are as follows:

1. The manufacturer has to first register as a Food Business Operator (FBO) with the local authority to sell food supplements.
2. It is the responsibility of the manufacturer to ensure that food supplements are safe for consumption. The food supplier should get itself registered as a business with their local authority and complete all invoices and receipts.
3. The supplier should keep all the records of order invoices and delivery notes on record to be produced when asked by enforcement authorities.
4. The correct labeling of food supplements is very important. The product labeling must be as "food supplement," which must also include all ingredients of the product, their allergen use (if any), conditions of use, and storage. The label must also contain a warning to keep away from children, best before date information, and a nutritional declaration.
5. The herbal-based medicine and ingredients classified as medicine in the United Kingdom come under the jurisdiction of the Medicines and Healthcare Products Regulatory Agency (MHRA), under the Department of Health. MHRA is considering a proposal for a Herbal Medicines Advisory Committee that would report to the proposed Human Medicine Commission. Lately, MHRA has proposed that

the Herbal Medicines Advisory Committee should have members representing Western, Chinese, and Ayurvedic herbalism, along with conventional medicines, and should report directly to the ministry on the regulation of herbal medicines.

6. For medicines containing unproven herbal products, the sole responsibility would be of the manufacturer's for their safety and efficacy in the absence of any scientific evidence.[27]

Changes in the EU from January 1, 2021: On January 1, 2021, the EU Exit Regulations 2019 and the Nutrition EU Exit (nutrition amendment) 2020 came into force. As per these amendments, the responsibilities of regulation of food supplements transferred from the EU organizations to appropriate authorities and bodies in Great Britain and reflected the Protocol on Ireland/Northern Ireland. The list of food supplements in the United Kingdom came into effect on January 1, 2021 that contains the vitamins and minerals for use in food supplements and their modified products.

8.5.7 REGULATION OF PLANT-BASED NUTRACEUTICALS IN RUSSIA

In Russia, the nutraceuticals are known as "Biologically Active Supplements (BAS)" and are defined as, "natural and/or identical to natural biologically active ingredients, intended for consumption with food or inclusion in products to enrich the diet with them." The main aspects of BAS in Russia that are under regulation are state registration, requirements for manufacturing, quality, and safety, and nutritional and health claims testing. In Russia, the regulation that deals with all of them is "Technical Regulations about the Safety of Foods" (Article 24, TP TC 021/2011) and BAS. Some of the salient features of these regulations are as follows:

1. It is mandatory to register by the manufacturer and seller of BAS (introduced in 2004). The registration certificate is issued by the Federal Service for Supervision over Consumer Rights Protection and Human Welfare. The certificate is for both customs clearance and distribution in Russia.
2. Russia, Belarus, and Kazakhstan have established the Customs Union (CU) because the technical regulations developed by the three countries must be consistent with each other.
3. The technical regulations of BAS are the same in all above three countries and are conducted in three stages: (1) Analysis of provided

documents; (2) Microbiological and toxicological testing for BAS-containing herbs, tests for pesticides presence, and authenticity tests to determine the actual presence of ingredients listed; (3) Issuance of certificate of state registration.

4. Some of the documents which are to be submitted for getting a certificate of state registration in Russia are (Russia 2014):

    - Certificate of manufacture confirming that this product is classified as BAS and is not a drug, and also proving its safety for human beings (Free Sale Certificate).
    - Full list of ingredients.
    - If BAS is based on herbs, their name in Latin as well as the method of preparation.
    - Documents on the manufacturing process.
    - Consumer label or its draft in the Russian language as well as a sample of the original label used in the country of manufacture.
    - Usage instructions, duration, doses, and side effects.
    - Explanatory note containing the scientific grounds for BAS compositions.
    - Materials on toxicological, hygienic, and biological efficiency, and other protocols, if requested.
    - Declaration stating that BAS does or does not contain genetically-modified components. Declaration verifying that this BAS does not contain drug or psychoactive-related components or substances.
    - Copy of the trademark certificate, legalized properly, in case a trademark exists; protocols of tests and trials wherever applicable.
    - For BAS used by sportsmen, document from the manufacturer proving that they do not contain components attributed to steroids.
    - Certificates confirming that BAS has the ingredients listed.
    - A document, properly legalized, confirming that the manufacturing process for BAS complies with national and international standards, for example GMP or ISO 9000–9002.

8.5.8 REGULATION OF PLANT-BASED NUTRACEUTICALS IN OTHER REGIONS OF THE WORLD

In some other parts of the world including Colombia, Brazil, and Argentina, an easy and effective registration-based regulation approach has been adopted for determining the quality, safety, and efficacy of nutraceuticals.[29] In Mexico and Chile, a notification is issued to the local authority for production and approval of nutraceuticals. However, some countries have strict requirements, like animal or human clinical study data, as in Brazil, China, and Taiwan.[22] Based on the above information, it is very evident that the regulation of nutraceutical safety, efficacy, and quality is mandatory in all parts of the globe before it enters the market.

8.6 CONCLUSION

The use of nutraceuticals based on herbs and their extracts is well-established in health-promoting and disease-preventing effects, but there are several challenges associated with the development of nutraceuticals that are often ignored because of a lack of control on their manufacturing, labeling, distribution, packaging, and quality. In this chapter, we have discussed mainly three basic global approaches in bringing nutraceutical products to market. (1) The registration-based approach is a flexible approach in which market demand is primary. (2) The notification-based approach, which maintains a balance between premarket resources, consumer access, and consumer safety. (3) The postmarket surveillance approach, which is the most robust as it ensures the safety and quality of the product in the marketplace, regardless of the premarket requirements.

The most important challenge in the regulation of nutraceuticals worldwide is the lack of a common global consensus on defining the category of products. Different countries have different standards and regulations for nutraceuticals; for example, a country like Canada has rigid regulations, whereas the regulations in the USA are the most flexible, with other countries falling in between. It is very necessary for a competent approval authority to substantiate the claim of nutraceutical benefit backed by a complete clinical study showing its effectiveness, safety, and absence of undesirable effects. The regulations of food products have been a concern for many years, and their stringent regulation is very difficult to achieve. However, improved techniques to assess the quality of botanical supplements, especially their purity, bioavailability, and safety aspects, are now available. The improvement of DS in plant-based products like fruits and vegetables, the need for domestication and cultivar development for cultivated production of these plants, and more refined postharvest processing technologies that will allow material to remain active and hygienic also need to be worked out for a large majority of these plants.

KEYWORDS

- **nutraceuticals**
- **herbal food**
- **regulation of nutraceuticals**
- **FSSAI**
- **FDA**

REFERENCES

1. Brower, V. Nutraceuticals: Poised for a Healthy Slice of the Healthcare Market? *Nat. Biotechnol.* **1998,** *16*, 728–731.
2. Pandey, N; Meena, R. P; Rai, S. K; Rai, S. P. Medicinal Plants Derived Nutraceuticals: A Re-emerging Health Aid. *Int. J. Pharma Biosci.* **2011,** *2* (4), 419–441.
3. Bickford, P.C; Tan, J; Shytle, R. D; Sanberg, C. D; El-Badri, N; Sanberg, P. R Nutraceuticals Synergistically Promote Proliferation of Human Stem Cells. *Stem Cells Dev.* **2006,** *15*,118–23.
4. Devi, V. K.; Rehman, F. Nutraceutical Antioxidants-An Overview. *Ind. J. Pharma. Ed.* **2002,** *36* (1), 3–8.
5. Seehafer, M. Bartels Meat 2.0. The Regulatory Environment of Plant-Based and Cultured Meat. *Eur. Food Feed Law Rev.* **2019,** *14*, 323–33.
6. Petetin, L. Frankenburgers, Risks and Approval. *Eur. J. Risk Regul.* **2014,** *2*, 168–186.
7. Dulloo, A. G.; Duret, C.; Rohrer. D. Efficacy of a Green Tea Extract Rich in Catechin Polyphenols and Caffeine in Increasing 24-h Energy Expenditure and Fat Oxidation in Humans. *Am. J. Clin. Nutr.* **1999,** *70*, 1040–1045.
8. Bell, S. J.; Goodrick, G. K. A Functional Food Product for the Management of Weight. *Crit. Rev. Food Sci. Nutr.* **2002,** *42*, 163–178.
9. Brouns, F. Soya Isoflavones: A New and Promising Ingredient for the Health Foods Sector. *Food Res. Int.* **2002,** *35*, 187–193.
10. Sirtori, C. R.; Galli, C. Fatty Acids and the Omega 3. *Biomed. Pharm.* **2002,** *56*, 397–406.
11. Market Survey Report: Nutraceuticals Market Size, Share & Trends Analysis Report by Product (Dietary Supplements, Functional Food, Functional Beverages), by Region, and Segment Forecasts, 2020 – 2028. Published Date: Oct, 2021.
12. Yadav, R.; Mehta, R. The Growing Nutraceuticals Market in India [Online]. https://www.investindia.gov.in/team-india-blogs/growing-nutraceuticals-market-india (accessed Jan10, 2020).

13. Jain, P. N.; Rathod, M. H.; Jain, V.C.; Vijayendra Swami, S. M. Current Regulatory Requirements for Registration of Nutraceuticals in India and USA. *Int. J. Drug Regul. Aff.* **2018,** *6* (2), 22–29.
14. Parikh, D. In *Handbook of Pharmaceutical Granulation Technology*, 3rd ed.; CRC Press: Florida, 2009.
15. Santini, A.; Cammaratta, S. M.; Capone, G.; Lanaro, A.; Tenore, G. C.; Pani, L.; Novellino, L. Nutraceuticals: Opening the Debate for a Regulatory Framework. *Br. J. Clin. Pharmacol.* **2018,** *84* (4), 659–672.
16. Bailey, R. L. Current Regulatory Guidelines and Resources to Support Research of Dietary Supplements in the United States. *Crit. Rev. Food Sci. Nutr.* **2020,** *60* (2) 298–309.
17. Australian Government, Department of Health (I) How Therapeutic Goods are Regulated in Australia: Therapeutic Goods Administration [Online]. https://www.tga.gov.au/how-therapeutic-goods-areregulated-australia (accessed Dec 1, 2020).
18. Australian Government, Department of Health (II) Listed Complementary Medicines. Therapeutic Goods Administration [Online]. https://www.tga.gov.au/listedcomplementary-medicines (accessed Dec 1, 2020).
19. Australian Government, Department of Health. Australian Register of Therapeutic Goods. Therapeutics Goods Administration [Online]. https://www.tga.gov.au/australian-registertherapeutic-goods (accessed Dec 1, 2020).
20. Thakkar, S.; Anklam, E.; Xu, A.; Ulberth, F.; BoLi, J. L. Regulatory Landscape of Dietary Supplements and Herbal Medicines from Aglobal Perspective. *Regul. Toxicol. Pharmacol.* **2020,** *114*, 104647.
21. Dietary Supplement Health and Education Act (DSHEA). United States Food and Drug Administration (FDA). U.S. Department of Health and Human Services. United States. Public Law 103–417 [Online]. http://www.fda.gov; https://ods.od.nih.gov/About/DSHEA_Wording.aspx (accessed Mar 2017).
22. Hasler, C. M. In *Regulation of Functional Foods and Nutraceuticals: A Global Perspective*; Hasler, C. M., Ed.; Blackwell Publishing: Ames, IA, 2008; p 389.
23. https://www.canada.ca/en/health-canada/services/drugs-health-products/natural-non-prescription/regulation.html
24. Shimizu, T. Health Claims on Functional Foods: The Japanese Regulations and an International Comparison. *Nutr. Res. Rev.* **2003,** *16*, 241–252.
25. Saito, M. Role of FOSHU (Food for Specified Health Uses) for Healthier Life. *Yakugaku Zasshi* **2007,** *127*, 407–416.
26. Yamada, K.; Sato-Mito, N.; Nagata, J.; Umegaki, K. Health Claim Evidence Requirements in Japan. *J. Nutr.* **2008,** *138*, 1192–1198.
27. Regulating Herbal Medicine in UK (Editorial Article) *BMJ* **2005,** *33*, 62–63.
28. Russia: Biologically Active Supplements [Online]. http://files.export.gov/x_4455523.pdf (accessed June 2014).
29. Mansour, M. 22 Codex and its Competitors: The Future of the Global Regulatory and Trading Regime for Food and Agricultural Products. In *Regulation of Functional Foods and Nutraceuticals: A Global Perspective*; Hasler, C. M., Eds.; Blackwell Publishing: Ames, IA, 2008; pp 377–389.

CHAPTER 9

IN VITRO AND IN VIVO STUDIES ON NUTRACEUTICALS

JOEL AUGUSTINE, RAVEENA ANN ALEX, and JAYANTHI ABRAHAM

Microbial Biotechnology Laboratory, School of Biosciences and Technology, Vellore Institute of Technology, Vellore, Tamil Nadu, India

ABSTRACT

Nutraceuticals are products with bioactive compounds, which are used as medicine other than nutrition. They are scientifically proven for their nontoxic food supplement and loaded with health benefits. Consumers are greatly solicitous about their health management, administration, and price of drugs. People seek alternative methods for their well-being and to be disease-free, thereby being attracted toward high-technology-based modern medicine. These factors lead to the increasing demand for nutraceuticals. The bioactive phytochemicals have become an important source for nutraceutical ingredients. Scientific research brings out the biological activity of food phytochemicals; however, the health claims attributed to the final marketed nutraceutical products often have little or dubious clinical basis. As less scientific proof is derived from evaluating the efficacy of studies through in vitro assays, human scientific trials are scarce and inconclusive. Numerous papers have reported their ability in treating many diseases such as allergies, diabetes, degenerative diseases, Alzheimer's disease, cardiovascular diseases, eye disorders, inflammation, cancer, and many other diseases. These were confirmed by using in vitro and in vivo studies, further revealing the properties of nutraceuticals in preventing diseases.

Formulations, Regulations, and Challenges of Nutraceuticals.
Tingirikari Jagan Mohan Rao, Rajesh K. Kesharwani, Raj K. Keservani, Anil Kumar Sharma (Eds.)

9.1 INTRODUCTION

Nutraceuticals are products which are used as medicines to provide health benefits other than its nutritious values. In 1989, Stephen Defelice, founder and chairman of the Foundation for innovation in medicine, coined the term nutraceutical from the terms nutrition and pharmaceutical.[48] Nutraceutical products are substances which have physiological benefits or provide protection against persistent diseases. These products are used in improving health, delaying the aging process, regulating blood glucose and cholesterol levels, inhibiting cancer development, reducing inflammation, and providing therapeutic and physiological benefits. They prevent cardiovascular diseases, support the structure and function of the living system, and are found to increase life expectancy. Due to their potential nutritional value, cost-efficiency, convenience, safety, and healing properties, nutraceuticals have a considerable demand.[46] They serve as bioactive compounds that are found in many dietary sources. Food products in nutraceuticals are categorized as dietary fibers, prebiotics, probiotics, unsaturated fatty acids, antioxidants, and herbal plants.[17] When compared to other pharmaceutical substances, these products have not been patented. In recent years, many food bioactive components have been marketed in the form of pharmaceutical products, such as capsules, pills, powders, granulates, etc., which comprise food extracts or phytochemical-enriched extracts to which a beneficial physiological role has been directly or indirectly assigned. The risk of toxicity or adverse effects of drugs led scientists to formulate safer nutraceuticals for health management. This paved the way for a global-wide nutraceutical revolution, creating a new technology concerning medicine and health, which in turn, transformed the food industry into a research-oriented one, similar to the pharmaceutical enterprise. The current population, life trends, and health issues make the nutraceuticals market thrive globally.

9.2 IN VIVO STUDIES OF NUTRACEUTICALS

Nutraceuticals can be orally consumed in the form of tablets, capsules, gels, powders, liquids, etc. In chewable form, they break down into small particles mixed with saliva, thus initiating digestion and supporting better digestion in the stomach. Most nutraceuticals are not completely digested in the stomach, and if unabsorbed, they are excreted out of the body. The lipophilic compounds can undergo intestinal adsorption, fractional compounds pass through epithelial cells, and the rest reach the circulatory system. Later, it is followed by chemical structural alteration, and in the stomach, the pH drops to 3 from 5–6. Most nutraceuticals are stable under low pH condition; some go through irreversible chemical transformation and completely lose their activity. The digestion, followed by absorption and metabolism in the stomach, acts as physicochemical and biochemical barriers for the efficient uptake of nutraceuticals. Intestine is the site where most of the nutrients are digested and absorbed. Here, the digested nutrients are mixed with bile salts, sodium bicarbonates, phospholipids, etc. by dynamic conversion. The lipid digestion and absorption take place in the intestine; the presence of lipophilic compounds increases the lipid absorption. The pH, temperature, and ionic strength affect the bioavailability of nutrients. The final digestion takes place in the colon, where the water and residual nutrients are absorbed by epithelial cells, and then they reach the circulatory system. After the digestion and absorption, only a considerable amount of nutraceutical compounds circulates in the body. The ingested nutraceuticals are poorly available for circulation and make it difficult to meet the daily requirements. The bioavailability depends upon the aqueous solubility, particle size, gastric retention time, diffusion rate across the intestinal epithelial wall, etc. These physical and chemical characteristics are responsible for the rate at which the nutraceuticals are absorbed in the intestine. The protection of labile compounds, extension of gastric retention time, increase in aqueous solubility, controlled release of the compounds, modulation of metabolic activity, etc. can increase the bioavailability of nutraceuticals.

The phospholipid bilayer acts as a carrier material, which enhances the efficiency of the nutraceuticals as they do not engage with the aqueous environment. The nutraceuticals can also be delivered as emulsions, which is a preferred method for oral delivery. Emulsion increases the bioavailability of the nutraceuticals. Micro- and nanoemulsions bring about three times higher bioavailability than conventional methods. They enhance aqueous solubility, passive diffusion rate, and direct uptake by the intestinal lymphatic system. Another class of emulsion is solid lipid nanoparticles. It is slow-digestible and has excellent controlled release of the bioactive compounds. Mainly retinol, peptides, citral, etc. are given as solid lipid nanoparticles.[4,67] Dibenzoylmethane is prepared as a nanoemulsion and has anticancer and antiproliferation activities. Solid lipid nanoparticle-based triptolide has immunosuppressive, antifertility, and antineoplastic activities.[42,68]

The chemical degradation in the stomach causes poor bioavailability of many bioactive compounds. The chemical modification of nutraceuticals developed enhances the bioavailability of the orally ingested nutraceuticals. In this method, the nutraceuticals are coupled with a bioreversible conjugate, which undergoes enzymatic chemical transformation, and a carrier, which improves the biofunctionalities by increasing the physicochemical properties. It increases the absorption of the compounds and improves aqueous stability, membrane permeability, gastric stability, cell permeability, etc. The nutraceutical resveratrol is chemically modified into a prodrug (3,5,40-tri-O-acetylresveratrol) and is used for protection against cardiovascular disease, which possesses antioxidant, anti-inflammatory, and antitumor properties. Tricin hybrid with amino acids has antioxidant, antiviral, anti-inflammatory, antihistaminic, and anticancer properties.[7,41]

The choices of animal models to detect the bioavailability of bioactive compounds and check for the possible side effects are Albino rats, Sprague–Dawley rats, Beagle dogs, etc.[63] Nutraceuticals can be combined with other drugs for treatments of cancer, cardiovascular diseases, etc. Nutraceuticals are used to reduce side effects and also serve as a good nutrient source. In chemotherapy-induced peripheral neuropathy (CIPN), the chemotherapy drugs produce toxic side effects; however, with the combination of nutraceuticals, the neurotoxic side effects were reduced. Schloss et al.[58] conducted a study by delivering nutraceuticals along with chemotherapeutic drugs, which confirmed anti-CIPN effects. Nutraceutical vitamin E with cisplatin and omega-3 fatty acids with paclitaxel when given orally, and glutathione with oxaliplatin administered intravenously, showed promising results while treating CIPN.

Examples of in vivo studies include: In a study, diabetic rats were orally administered with the extracts of *Zingiber officinale*, which resulted in the normalization of the retinal vessel diameter and the reduction of basement membrane thickness.[55]

Silva et al. (2013) demonstrated oral administration of green tea to diabetes-induced rats, which produced a neuroprotective action in the retina. This was characterized by a reduction in neuronal death and a restoration of glutamate uptake.

9.3 IN VITRO STUDIES OF NUTRACEUTICALS

In vitro studies of nutraceuticals have been conducted to identify the activity of biological compounds and their effect on the functionalities of cells. The nutraceuticals that are performed in vitro are conducted at a micromolar (μM) concentration, which is a bit higher than the plasma concentration, that is, <1 μM, obtained from normal dietary intake. In the gastrointestinal absorption study conducted by Ting et al.,[63] the physicochemical condition was designed similarly to that of the human body. To evaluate the bioavailability

TABLE 9.1 Bioactivities of Nutraceuticals and Their In Vivo Methods.

Nutraceuticals	Bioactivities	Techniques	References
Curcumin	Antitumor, antioxidant, anti-inflammatory, anti-HIV, and anticoagulant	Liposomes phospholipid complex Self-emulsifying drug delivery system	[61,62]
Silybin	Hepatoprotective and antioxidant	Phospholipid complex	[63,28]
α-Tocopherol	Antioxidant and neuroprotective	Nanoemulsion	[63]
Dibenzoylmethane	Anticancer and antiproliferation	Nanoemulsion	[42]
Curcumin	Antitumor, antioxidant, and anti-inflammatory	Organogel-based nanoemulsion	[68]
Oridonin	Antioxidant, anti-inflammatory, antibacterial, and antitumor	Self-emulsifying drug delivery system	[63,67]
EGCG	Antioxidant, antiviral, anti-inflammatory, anticancer, cardioprotective, and neuroprotective	Peracetylation (chemical modification)	[63]
Resveratrol	Antioxidant, anti-inflammatory, cardioprotective, and antitumor	Prodrug (2,3,4'-tri-O-acetylresveratol)	[41]

of the nutraceutical, it was designed similarly to the GI tract in terms of pH, temperature, enzyme activity, etc.; chemical tests were conducted to detect the stability of the nutraceutical at 37°C, pH varying between 2.5 and 7, and ionic composition to find out the efficiency. The GI tract is more complex than the designed chemical environment, due to the enzyme activity of the stomach. In a single-factored chemical method, the invigilator screened the dosing efficiency, which allowed more samples to be performed at a given time. It was simple to perform and cost-effective. In a multifactored model, two or more bioaccessibility factors are used. In vitro digestion model consists of physiochemical conditions of the stomach and small intestine. This model can predict the oral efficiency of the nutraceutical, but unable to reproduce dynamic physical condition of human body factors, such as peristalsis motion of GI tract, integration of all the influential physiological factors, transitional change in physiological environment, and elimination from the absorption site. An instrument TIM-1 gastrointestinal simulation system designed by TNO Quality Life, the Netherlands (a research organization), features the physiological and physicochemical factors such as body temperature, gastric and intestinal residence time, pH, luminal composition, and peristalsis movement. It is automated by a computer system, and the residues are collected, which allows the elimination of loss due to the degradation of nutraceuticals. The bioavailability can be calculated by subtracting the initial concentration from eliminated concentration. This system is the most precise one to estimate the bioaccessibility of nutraceuticals.[9,62,69]

After the digestion, the absorption takes place through the permeable cellular membrane. In the gastrointestinal model, the bioaccessibility is determined by the help of a dialysis membrane. From this, the transport rate of dissolved compounds can be estimated. It can predict the functionalities of compound permeability and absorption based on the physical, chemical, and biological events in the intestine. This model can efficiently determine the lymphatic transport of lipid compounds. It is a monolayer technique and has disadvantages such as the fact that the study was conducted in different laboratories, the transport coefficient is different in different laboratories, and its performance depends upon the quality of the layer.[25] An in situ animal model in single-pass intestinal perfusion (SPIP) study provides the difference in the absorption behavior at different intestinal regions. In this study, controlling the parameters, such as the compound concentration and the flow rate through the perfused section, for intra- and inter-experimental comparability, the volume decreases due to the water absorption when perfusates normalize the difference between inlet and outlet concentrations, and then the compound permeation efficiency can be calculated.[21,62]

In cell lines: An in vitro study conducted by Anusree et al.[5] reveals the nutraceutical ability of punicic acid (a polyunsaturated fatty acid) relevant to diabetes through enhanced GLUT4 expression and adiponectin secretion. Punicic acid, a major bioactive compound present in pomegranate seed oil, contains PPARγ agonist activity by different groups, which plays a role in the control of adipocyte differentiation and insulin sensitivity. It increased the adiponectin (adipokine secreted from adipocytes) secretion and upregulated GLUT4 expression and translocation in adipocytes. 3T3-L1 cell lines from mice were used in research on adipose tissues. Different concentrations of punicic acid were treated with fully differentiated 3T3-L1 cells for 24 h and the release of adiponectin from the cells into the medium was studied using a mouse adiponectin ELISA kit.[5] Lanthascreen time-resolved fluorescent energy transfer was used to analyze the in vitro binding of punicic acid to PPARγ; competitive binding assay along with other assays such as 2-deoxy glucose uptake assay, immunofluorescence imaging of GLUT4, inhibition of reactive oxygen species (ROS) were studied.

9.4 SOURCES OF NUTRACEUTICALS

Different types of natural plants and animals, along with marine sources and minerals, have been examined for biologically active metabolites as sources that can be used for developing new nutraceuticals. For the production of nutraceuticals as an emerging and economically viable concept, microorganisms are more widely used.[18] Bacteria, yeast, and microalgae also can act as producers for the production of nutraceuticals and food ingredients.[34] Some nutraceuticals thus obtained are listed below:

i. Shark liver oil is rich in bioactive substances that are used in therapeutic and dietary products. It contains alkylglycerols and squalene, which show immunogenic activities such as antimicrobial and antitumor activities. And the compound n-3 polyunsaturated fatty acid (PUFA) is used to treat cardiovascular disease and has antiallergic properties. Shark liver oil is rich in fatty acids and is used as a dietary supplement.[1]
ii. Marine bacteria thrive in harsh oceanic environments and thus result in the production of valuable and unsaturated secondary metabolites. Wang et al.[66] demonstrated that an aerobic, nonpigmented bacterium strain, KMM 3042 produces bioactive substances such as pyrroles, pseudopeptide pyrrolidinedione, phloroglucinol, phenazine, benzaldehyde, quinoline, quinolone, phenanthrene, phthalate, andrimid, moiramides, zafrin, and bushrin. Some of these compounds have antimicrobial properties and some are cathepsin-B inhibitors.[31] The engineered *E.coli* produces nutraceutical β-carotene by improving ATP and NADPH supplies.
iii. Marine fungi, due to the presence of some unique carbon frameworks in them, produce a wide variety of bioactive metabolites that possess antiplasmodial, anticancer, antibacterial, anti-inflammatory, and antiviral activity. These novel compounds are used as new lead structures for medicine and for plant protection. *Penicillium* sp., isolated from deep ocean sediment, exhibits antitumor activities in their alkaloid-rich extracts due to the presence of meleagrin D and E and roquefortine H and I. But, they exhibit weak cytotoxicity.[31]
iv. The marine cyanobacteria, *Spirulina*, produce bioactive compounds that have high nutritional value. These bioactive compounds contain essential nutrients, such as provitamins, minerals, polyunsaturated fatty acids, and proteins. *Spirulina platensis* possesses the antioxidant activity of phycobiliproteins. These phenolics, proteins, and fatty acids can be used to produce nutraceuticals and dietary compounds.[15]

v. Macroalgae or seaweeds are well-known type of algae for the production of nutraceuticals with many bioactive compounds with disease-preventive effect. From macroalgae, about 2400 natural products have been isolated belonging to the classes of Rhodophyceae, Phaeophyceae, and Chlorophyceae. *Gracilaria opuntia* from Rhodophyceae class posesses many biological properties. Seaweed produces vitamins and minerals, such as vitamin A, D, E, K, B_1, B_2, B_9, B_{12} and minerals mainly calcium, potassium, zinc, manganese, etc. The red seaweed *Palmaria palmata* contains iron; *Ulva lactuca* is rich in calcium. Kim et al.[38] showed in their clinical trial study with humans that daily intake of seaweeds such as *Undaria pinnatifida* and *Sacchariza polyschides* helps in balancing blood glucose levels, decreases serum triglyceride concentrations, and increases high-density lipoprotein cholesterol in type-2 diabetes subjects.

vi. Plant source: The phytochemicals are plants' bioactive compounds, which provide health effects for humans and animals. The vegetables and fruits have free radical scavenging effects against cancer of the stomach, lungs, oral cavity, pharynx and esophagus, colon, and pancreas. These phytochemicals include polyphenolics, terpenoids, alkaloids, nitrogen-containing compounds, fibers, etc. The polyphenolics were considered as antinutrient because some compounds like tannins decrease the digestive enzyme activity, amino acid availability, mineral uptake, and have other toxic activities. It was later discovered that these compounds have antioxidant activity and provide health benefits.[11,56] Terpenes have a wide range of antioxidant properties; two of them are tocotrienols and tocopherols, and they are found in grains. α-d-Tocopherol has a similar effect to that of vitamin E. These compounds suppress the growth of tumor cell by initiating apoptosis. The carotenoid terpenes are pigmented compounds present in fruits and vegetables; α-carotenoid, β-carotenoid, and ε-carotenoid have vitamin A activity. The carotenoid γ-carotene, lycopene, and lutein have anticancer properties against prostate, uterine, breast, colorectal, and lung cancers.[8] Limonoids are the monoterpenes found in orange peel oil, which have chemotherapeutic activity and inhibit pancreatic carcinogenesis. Phytosterol is another class of terpenes, which is similar to cholesterol. The unsaturated phytosterol is more effective in decreasing the cholesterol concentration and it reduces the serum LDL.[35]

The phenolic compound flavonoid from grapes induces free radical-mediated cell signaling pathways, allergies, ulcer, inflammation, etc. It is given orally to prevent platelet aggregation, reduce thrombosis, and treat cardiac diseases. The isoflavonoids and daidzein from soy inhibit the hormone-dependent and independent cancer cells. They are also used to treat osteoarthritis and have cardioprotective effects. Glucosinolates are the plant metabolites seen in cruciferous plants. They have many health effects, such as activating the liver detoxification enzymes and protecting against carcinogens, mutagens, ROS, etc. These glucosinolates are mainly seen in cruciferous plants like broccoli, cauliflower, cabbage, etc.[19,24]

TABLE 9.2 Common Nutraceuticals and Their Therapeutic Effects.[14]

Source	**Constituent**	**Therapeutic effects**
Aloe vera	Aloins and aloesin	Anti-inflammatory, wound healing properties, and dilates capillaries
Garlic	Alliin and allicin	Antibacterial, antifungal, anti-inflammatory, and fibrinolytic
Liquorice	Glycyrrhizin and liquitin	Tropical antibacterial and antiviral
Onion	Allicin and alliin	Antibiotic and antiatherosclerotic
Turmeric	Curcumin	Anti-inflammatory, antiarthritic, anticancer, and antiseptic
Ginger	Zingiberene and gingerols	Stimulant, chronic bronchitis, and throat ache
Grains	Tocotrienols and tocopherols	Suppress tumor growth
Fruits and vegetables	α-carotenoid, β-carotenoid, ε-carotenoid, γ-carotene, lycopene, and lutein	Anticancer properties against prostate, uterine, breast, colorectal, and lung cancers
Orange peel oil	Limonoids	Inhibit pancreatic carcinogenesis
Grapes	Flavonoids	Induce free-radical-mediated cell signaling pathways, allergies ulcer, and inflammation
Soy	Isoflavonoids and daidzein	Used to treat osteoarthritis and have cardioprotective effects
Cruciferous plants	Glucosinolates	Activate the liver detoxification enzymes, protection against carcinogens, mutagens, and ROS

9.5 NUTRACEUTICALS AS MEDICINES

9.5.1 ALZHEIMER'S DISEASE

Alzheimer's is a most common form of dementia, which is a neurodegenerative disease. It is a progressive decline in memory and other mental functions. Several studies suggest that oxidative stress can be related to Alzheimer's and many other neurodegenerative disorders. Antioxidants, such as curcumin, lutein, turmerin, etc., can exert positive effects by combating oxidative stress, and thereby postpone the development of Alzheimer's.[46]

9.5.2 DIABETIC RETINOPATHY

Diabetic retinopathy is a complication of diabetes. It is caused by damage to the blood vessels of the retina resulting in vision loss and blindness. This damage is the result of a high blood sugar level. Various natural dietary compounds, such as polyphenols, carotenoids, and saponins, have been analyzed for potential treatments to prevent the retinal oxidative stress typical of diabetic retinopathy. These compounds are present in various natural products like fruits, vegetables, herbs, beverages, and spices, and are extremely effective in fortifying the endogenous antioxidant defenses through a scavenger activity or by the stimulation of antioxidant enzyme expression.[55] An antioxidant, lipoic acid supplement, is taken for the treatment of diabetic neuropathy and is also effective in people with complications of diabetes (long-term dietary supplement).

9.5.3 LIPID LOWERING EFFECT

Combination of garlic with olive oil in diabetic patients efficiently regulates serum cholesterol and triglycerides levels along with dyslipidemia.[44]

9.5.4 CARDIOVASCULAR DISEASES

Vitamins, antioxidants, dietary fibers, and omega-3 polyunsaturated fatty acids, along with physical exercise, are suggested for the cure and treatment of cardiovascular diseases.[46]

9.5.5 *LOWER BLOOD CHOLESTEROL AND HYPERTENSION*

Buckwheat seeds are whole grains similar to cereals that have phytosterols, flavones, flavonoids, proteins, thiamin-binding proteins, etc. Taking these seeds can lower blood cholesterol levels and hypertension in people.[40]

9.6 BIOAVAILABILITY OF NUTRACEUTICALS

One of the most important characteristics of bioactive compounds is bioavailability, which plays a major role in the manufacture and design of nutraceuticals. It is used to assess the involvement of nutraceuticals and its bioefficiency. There are some factors affecting the bioavailability like bioaccessibility, absorption, and transformation.[47] To determine the bioavailability of a compound, studies are conducted using in vivo and in vitro methods with bioactive compounds encapsulated within various carriers. The oral bioavailability depends on the stability of gastrointestinal fluid and absorption through intestinal epithelial layer. The biopharmaceutical compounds that are orally taken have various matrices, and they are grouped into four based on their solubility and permeability.

1. High permeability and high solubility.

   These biopharmaceuticals consist of high permeability and solubility; they get digested and absorbed easily, and no factor is limiting the bioavailability.

2. High permeability and low solubility.
   This comprises lipophilic compounds which have high permeability in the epithelial cells and less soluble in gastrointestinal fluid. Solubility of the compound is limiting the bioavailability.

3. Low permeability and high solubility.
   The high solubility in GI fluids and low permeability to epithelial cells are mostly hydrophilic compounds; the limiting factor is permeability through the GI membrane. To increase the permeability of fatty acids, inclusion agents are added.

4. Low permeability and low solubility.
   The low permeability and solubility of compounds results in poor bioavailability; these compounds' bioavailability is improved by self-emulsifying drug systems or microemulsion.[50]

A nutraceutical that has greater than 75% bioavailability is known as a highly bioaccessible nutraceutical, and one with less than 75% bioavailability is relatively low bioaccessible. In vitro studies involve many stages, including digestion, which is done in mouth, colon, stomach, and intestine models, gastrointestinal peristalsis, motility, and its specifications. These stages assess the bioaccessibility and bioavailability of the nutraceuticals.[20]

9.7 STUDIES OF NUTRACEUTICALS IN HUMANS, ANIMALS, AND CELL LINES

9.7.1 IN HUMANS

i. A study conducted by Proksch et al.[51] demonstrated the role of the nutraceutical on skin wrinkles in humans in double-blind, placebo-controlled study with a bioactive collagen peptide (BCP) called VERISOL®. Total of 114 subjects, with 57 subjects in each treatment group received 2.5 g of VERISOL or placebo for a period of 8 weeks. Before and after (4 and 8 weeks) treatment, skin wrinkles of the subjects were measured. When compared to placebo, a significant reduction of eye wrinkle volume was observed after a period of 4–8 weeks on intake of BCP. Higher content of procollagen type I and elastin was also observed in addition.[49,51]
ii. A study involved 84 healthy Asian subjects (double-blind, randomized, and placebo-controlled study) taking galacto-oligosaccharides (GOS) capsules (1 g in a capsule) twice a day for 12 weeks. It was observed that in those subjects who had intake of the GOS, their corneometer values from baseline to 12 weeks were higher, and their percentage of wrinkle areas and transepidermal water loss (TEWL) was significantly reduced.[32]
iii. In a single-blind study on 150 hypercholesterolemia patients, enteric-coated garlic powder tablets (400 mg garlic and 1 mg allicin) two times per day showed a reduction in cholesterol and LDL levels.[39]

9.7.2 IN ANIMALS

i. Studies on resveratrol, a nonflavonoid found mostly in grapes or blueberries, showed that it can improve learning and memory process. Oral administration of resveratrol in 50 mg/kg/day to aged male Wistar rats for a period of 12 weeks halted the special and emotional memory loss along with neuroinflammatory signal inhibition.[29]

ii. Feeding cod liver oil rich in PUFAs to a mouse model with rheumatoid arthritis substantially resulted in lower serum levels of interleukins (IL-6, IL-10, and IL-12) and also in tumor necrosis factor-α (TNF-α), prostaglandin E2, thromboxane B2, and leukotriene B4.[16,65]

iii. Curcuminoids present in turmeric act as a potential cognitive nutraceutical product. Studies on 14-month-old mice treated with 50 mg/kg for 21 days and SAMP-8 model mice through intraperitoneal and oral administration of curcumin resulted in enhanced cognitive performance. Oral treatment with curcumin (480 mg/kg) over a long-term period of 12 weeks increased spatial and recognition memory capacities.[57,60]

9.7.3 IN CELL LINES

In vitro studies were performed by Fernández-Bedmar et al.[27] using cytotoxicity and DNA fragmentation assays, in HL60 promyelocytic cell lines, to evaluate the nutraceutical value of different *Capsicum* sp. The HL60 cells were tested with different concentrations of capsicum compounds (capsaicin, capsanthin, and lutein) and cell growth and viability of the cells were observed. The results showed that capsaicin induced decreases in CDK1, cyclin E, and cyclin A, leading to G_0/G_1 arrest, and also increased the ROS and Ca^{2+} production. It also decreased cytochrome c release, leading to caspase-3 activity, which induced apoptosis in HL-60 cells.[27,64]

9.8 AN EXAMPLE OF IN VITRO AND IN VIVO STUDIES OF CARNITINE

Carnitine is a naturally occurring endogenous metabolite, which can be found in nearly all cells of the body in most mammals. It is a water-soluble molecule, accounted to have numerous biological activities, and can be considered a restrictively fundamental supplement for its significance in human physiology.[2,22] It exists in two enantiomers, such as D-carnitine and L-carnitine. Each of them is biologically active; but only L-carnitine naturally occurs in animals, and D-carnitine is toxic because it inhibits the activity of the L-form. Carnitine showed in vitro anticancer activity by means of preventing colon cancer cells (Caco-2 cells) proliferation through a reduction in prostaglandin E2 synthesis and induction of colon cancer cell apoptosis. Acetyl-l-carnitine acts as an antiprostate cancer agent by inhibiting the production of chemokines (CXCL12 and CCL2), chemokines ligand receptor (CXCR4), and proinflammatory cytokines (IFN-γ and TNF-α) in many prostate cancer cell lines.[6] In vitro studies of carnitine's anticancer role have been conducted by Huang et al.[33] in different samples of HepG2 tumor-bearing mice, human SMMC-7721, primary cultured thymocytes, and hepatoma HepG2 cell lines. Following the L-carnitine treatment, the main conclusions observed have been prevention of cancer cell growth, inhibition of histone deacetylases HDAC I/II activities caused by L-carnitine binding to active sites of HDAC, elevation of histone acetylation and acetylated lysine accumulation, and induction of the p21^{cip1} gene, mRNA, and protein expression in cancer cells.[22,33]

A study by Farahzadi et al.[26] showed that the use of L-carnitine (0.2 mM) for 2 days prolonged the lifespan of human mesenchymal stem cells, obtained from aged participants, by lengthening telomerase and increasing hTERT gene expression. Another study in blood leukocytes showed that administering L-carnitine at 30 and 150 M prevented DNA damage induced by L-2-hydroxyglutaric and D-2-hydroxyglutaric in L-2-hydroxyglutaric-aciduria-affected people.[54]

9.8.1 IN VITRO STUDIES

9.8.1.1 IN ANIMAL STUDIES

Studies showed that the use of L-carnitine has many health effects in animals. It reduced the DNA damage, histopathological abnormalities, and also increased the activity of antioxidant enzymes during cadmium-induced damage in male mice[3] and controlled contrast-induced nephropathy through a preventive function.[10] In a study conducted by Kelek et al.,[36] there was an elevation of total antioxidant properties and scavenging of oxygen free radicals in rats treated with L-carnitine (50 mg/kg/day) for 7 months by regulating Bax and Bcl-2, dropping caspase-3 activity.

Chickpea (*Cicer arietinum L.*) is a widely cultivated legume and characterized for high content of carbohydrates, fats, proteins, vitamins, and fats.[52] It can lower the chance of incidents of chronic diseases,[45] change lipid metabolism, decrease the formation of lipid droplets during 3T3-L1 adipocyte differentiation,[59] and lower insulin resistance in type-2 diabetes.

9.8.2 *IN VIVO STUDIES*

9.8.2.1 IN HUMAN

Daily intake of L-carnitine (two or three 500 mg tablets) decreased the imatinib-induced muscle cramps in gastrointestinal stromal tumor sick individuals.[13] When female patients suffering from osteoarthritis are prescribed a dosage of 750 mg/day for 8 weeks, it decreased the intensity of pain and reduced the serum inflammatory mediators, such as matrix metalloproteinases-1 and interleukin-1b.[43] In a study by Kido et al.[37] in elderly hemodialyzed patients, L-carnitine supplementation gives last-stage kidney problem accelerated amino acid metabolism, fatty acid metabolism, blood act L-carnitine levels, and energy synthesis in skeletal and heart muscles.

KEYWORDS

- **nutraceuticals**
- **in vivo**
- **in vitro**
- **bioavailability**
- **encapsulation**

REFERENCES

1. Akutsu, K.; Tanaka, Y.; Hayakawa, K. Occurrence of Polybrominated Diphenyl Ethers and Polychlorinated Biphenyls in Shark Liver Oil Supplements. *Food Addit. Contam.* **2006,** *23* (12), 1323–1329.
2. Alesci, S. Ed.; In *Carnitine: The Science Behind a Conditionally Essential Nutrient*; New York Academy of Sciences, 2004; vol 1033.
3. Alharthi, W. A.; Hamza, R. Z.; Elmahdi, M. M.; Abuelzahab, H. S.; Saleh, H. Selenium and L-carnitine Ameliorate Reproductive Toxicity Induced by Cadmium in Male Mice. *Biol. Trace Elem. Res.* **2020,** *197* (2), 619–627.
4. Almeida, A. J.; Souto, E. Solid Lipid Nanoparticles as a Drug Delivery System for Peptides and Proteins. *Adv. Drug Deliv. Rev.* **2007,** *59*, 478–490.
5. Anusree, S. S.; Priyanka, A.; Nisha, V. M.; Das, A. A.; Raghu, K. G. An *in vitro* Study Reveals the Nutraceutical Potential of Punicic Acid Relevant to Diabetes via Enhanced GLUT4 Expression and Adiponectin Secretion. *Food Funct.* **2014,** *5* (10), 2590–2601.

6. Baci, D.; Bruno, A.; Cascini, C.; Gallazzi, M.; Mortara, L.; Sessa, F.; ... & Noonan, D. M. Acetyl-L-Carnitine Downregulates Invasion (CXCR4/CXCL12, MMP-9) and Angiogenesis (VEGF, CXCL8) Pathways in Prostate Cancer Cells: Rationale for Prevention and Interception Strategies. *J. Exp. Clin. Cancer Res.* **2019,** *38* (1), 1–17.
7. Belluti, F.; Fontana, G.; Bo, L. D.; Carenini, N.; Giommarelli, C.; Zunino, F. Design, Synthesis and Anticancer Activities of Stilbene–Coumarin Hybrid Compounds: Identification of Novel Proapoptotic Agents. *Bioorg. Med. Chem.* **2010,** *18*, 3543–3550.
8. Bendich, A. J. Carotenoids and the Immune Response. *J. Nutr.* **1989,** *119*, 112–115.
9. Blanquet, S.; Zeijdner, E.; Beyssac, E.; Meunier, J. P.; Denis, S.; Havenaar, R.; Alric, M. A Dynamic Artificial Gastrointestinal System for Studying the Behavior of Orally Administered Drug Dosage Forms Under Various Physiological Conditions. *Pharm. Res.* **2004,** *21* (4), 585–591.
10. Boyacioglu, M.; Turgut, H.; Akgullu, C.; Eryilmaz, U.; Kum, C.; Onbasili, O. A. The efficient of L-carnitine on oxidative stress responses of experimental contrast-induced nephropathy in rat. *J. Vet. Med. Sci.* **2013,** 76(1), 1–8.
11. Bravo, L. Polyphenols: Chemistry, Dietary Sources, Metabolism, and Nutritional Signicance. *Nutr. Rev.* **1998,** *56*, 317–333.
12. Centrone, M.; Gena, P.; Ranieri, M.; Di Mise, A.; D'Agostino, M.; Mastrodonato, M.; Tamma, G. *In vitro* and *In vivo* Nutraceutical Characterization of Two Chickpea Accessions: Differential Effects on Hepatic Lipid Over-Accumulation. *Antioxidants* **2020,** *9* (3), 268.
13. Chae, H.; Ryu, M. H.; Ma, J.; Beck, M.; Kang, Y. K. Impact of l-carnitine on Imatinib-Related Muscle Cramps in Patients with Gastrointestinal Stromal Tumor. *Investig. New Drugs* **2020,** *38* (2), 493–499.
14. Chauhan, B.; Kumar, G.; Kalam, N.; Ansari, S. H. Current Concepts and Prospects of Herbal Nutraceutical: A Review. *J. Adv. Pharm. Technol. Res.* **2013,** *4* (1), 4.
15. Colla, L. M.; Reinehr, C. O.; Reichert, C.; Costa, J. A. V. Production of Biomass and Nutraceutical Compounds by Spirulina Platensis Under Different Temperature and Nitrogen Regimes. *Bioresour. Technol.* **2007,** *98* (7), 1489–1493.
16. Curtis, C. L.; Harwood, J. L.; Dent, C. M.; Caterson, B. Biological Basis for the Benefit of Nutraceutical Supplementation in Arthritis. *Drug discov. Today* **2004,** *9* (4), 165–172.
17. Das, L.; Bhaumik, E.; Raychaudhuri, U.; Chakraborty, R. Role of Nutraceuticals in Human Health. *J. Food Sci. Technol.* **2012,** *49* (2), 173–183.
18. Dewapriya, P.; Kim, S. K. Marine Microorganisms: An Emerging Avenue in Modern Nutraceuticals and Functional Foods. *Food Res. Int.* **2014,** *56*, 115–125.
19. Dillard, C. J.; German, J. B. Phytochemicals: Nutraceuticals and Human Health. *J. Sci. Food Agric.* 2000, *80* (12), 1744–1756.
20. Dima, C.; Assadpour, E.; Dima, S.; Jafari, S. M. Bioavailability and Bioaccessibility of Food Bioactive Compounds; Overview and Assessment by *in vitro* Methods. *Compr. Rev. Food Sci. Food Saf.* **2020,** *19* (6), 2862–2884.
21. Ding, X. X.; Kaminsky, L. S. Human Extrahepatic Cytochromes P450: Function in Xenobiotic Metabolism and Tissue-Selective Chemical Toxicity in the Respiratory and Gastrointestinal Tracts. *Annu. Rev. Pharmacol.* **2003,** *43*, 149–173.
22. Durazzo, A.; Lucarini, M.; Nazhand, A.; Souto, S. B.; Silva, A. M.; Severino, P.; Santini, A. The Nutraceutical Value of Carnitine and its use in Dietary Supplements. *Molecules* **2020,** *25* (9), 2127.
23. Espín, J. C.; García-Conesa, M. T.; Tomás-Barberán, F. A. Nutraceuticals: Facts and Fiction. *Phytochemistry* **2007,** *68* (22–24), 2986–3008.

24. Fahey, J. W.; Zhang, Y.; Talalay, P. Broccoli Sprouts: An Exceptionally Rich Source of Inducers of Enzymes that Protect Against Chemical Carcinogens. *Proc. Natl. Acad. Sci. USA* **1997,** *16*, 10367–10372.
25. Failla, M. L.; Huo, T.; Thakkar, S. K. *In vitro* Screening of Relative Bioaccessibility of Carotenoids from Foods. *Asia Pac. J. Clin. Nutr.* **2008,** *17* (S1), 200–203.
26. Farahzadi, R.; Mesbah-Namin, S. A.; Zarghami, N.; Fathi, E. L-carnitine Effectively Induces hTERT Gene Expression of Human Adipose Tissue-Derived Mesenchymal Stem Cells Obtained from the Aged Subjects. *Int. J. Stem Cells* **2016,** *9* (1), 107.
27. Fernández-Bedmar, Z.; Alonso-Moraga, A. *In vivo* and *in vitro* Evaluation for Nutraceutical Purposes of Capsaicin, Capsanthin, Lutein and Four Pepper Varieties. *Food Chem. Toxicol.* **2016,** *98*, 89–99.
28. Filburn, C.; Kettenacker, R.; Griffin, D. Bioavailability of a Silybin, A Phosphatidylcholine Complex in Dogs. *J. Vet. Pharmacol. Ther.* **2007,** 30, 132–138.
29. Gocmez, S. S.; Gacar, N.; Utkan, T.; Gacar, G.; Scarpace, P. J.; Tumer, N. Protective Effects of Resveratrol on Aging-Induced Cognitive Impairment in Rats. *Neurobiol. Learn. Mem.* **2016,** *131*, 131–136.
30. Goñi, I.; Valentín-Gamazo, C. Chickpea Flour Ingredient Slows Glycemic Response to Pasta in Healthy Volunteers. *Food Chem.* **2003,** *81* (4), 511–515.
31. Gupta, C.; Prakash, D. Nutraceuticals from Microbes of Marine Sources. In *Nutraceuticals-Past, Present and Future*, 2019; p 99.
32. Hong, Y. H.; Chang, U. J.; Kim, Y. S.; Jung, E. Y.; Suh, H. J. Dietary Galacto-oligosaccharides Improve Skin Health: A Randomized Double Blind Clinical Trial. *Asia Pac. J. Clin. Nutr.* **2017,** *26* (4), 613–618.
33. Huang, H.; Liu, N.; Guo, H.; Liao, S.; Li, X.; Yang, C.; Liu, J. L-carnitine is an Endogenous HDAC Inhibitor Selectively Inhibiting Cancer Cell Growth *in vivo* and *in vitro*. *PloS One* **2012,** *7* (11), e49062.
34. Hugenholtz, J.; Smid, E. J. Nutraceutical Production with Food-Grade Microorganisms. *Curr. Opin. Biotechnol.* **2002,** *13* (5), 497–507.
35. Jones, P. J. H.; Ntanios, F. Y.; Raeini-Sarjaz, M.; Vanstone, C. A. Cholesterol-lowering ef®cacy of a Sitostanol-containing Phytosterol Mixture with a Prudent Diet in Hyperlipidemic Men. Am. J. Clin. Nutr. **1999,** *69*, 1144–1150.
36. Kelek, S. E.; Afşar, E.; Akçay, G.; Danışman, B.; Aslan, M. Effect of Chronic L-carnitine Supplementation on Carnitine Levels, Oxidative Stress and Apoptotic Markers in Peripheral Organs of Adult Wistar Rats. *Food Chem. Toxicol.* **2019,** *134*, 110851.
37. Kido, J.; Inoue, H.; Shimotsu, H.; Yoshida, Y.; Suzuki, Y.; Nakamura, K.; Matsumoto, S. Effect of l-carnitine on Amino Acid Metabolism in Elderly Patients Undergoing Regular Hemodialysis. *Blood Purif.* **2020,** *49* (5), 614–621.
38. Kim, M. S.; Kim, J. Y.; Choi, W. H.; Lee, S. S. Effects of Seaweed Supplementation on Blood Glucose Concentration, Lipid Profile, and Antioxidant Enzyme Activities in Patients with Type 2 Diabetes Mellitus. *Nutr. Res. Pract.* **2008,** *2* (2), 62–67.
39. Kojuri, J.; Vosoughi, A. R.; Akrami, M. Effects of Anethumgraveolens and Garlic on Lipid Profile in Hyperlipidemic Patients. *Lipids Health Dis.* **2007,** *6* (1), 1–5.
40. Li, S. Q.; Zhang, Q. H. Advances in the Development of Functional Foods from Buckwheat. *Crit. Rev. Food Sci. Nutr.* **2001,** *41* (6), 451–464.
41. Liang, L.; Liu, X.; Wang, Q.; Cheng, S.; Zhang, S.; Zhang, M. Pharmacokinetics, Tissue Distribution and Excretion Study of Resveratrol and its Prodrug 3,5,40 -tri-Oacetylresveratrol in Rats. *Phytomedicine* **2013,** *20*, 558–563.

42. Lin, W.; Hong, J. L.; Shen, G.; Wu, R. T.; Wang, Y.; Huang, M. T.; Newmark, H. L.; Huang, Q.; Khor, T. O.; Heimbach, T.; Kong, A. N. Pharmacokinetics of Dietary Cancer Chemopreventive Compound Dibenzoylmethane in Rats and the Impact of Nanoemulsion and Genetic Knockout of Nrf2 on its Disposition. *Biopharm. Drug Dispos.* **2011,** *32*, 65–75.
43. Malek Mahdavi, A.; Mahdavi, R.; Kolahi, S. Effects of l-carnitine Supplementation on Serum Inflammatory Factors and Matrix Metalloproteinase Enzymes in Females with Knee Osteoarthritis: A Randomized, Double-Blind, Placebo-Controlled Pilot Study. *J. Am. Coll. Nutr.* 2016, *35* (7), 597–603.
44. Memon, A. R.; Ghanghro, A. B.; Shaikh, I. A.; Qazi, N.; Ghanghro, I. H.; Shaikh, U. Effects of Olive Oil and Garlic on Serum Cholesterol and Triglycerides Levels in the Patients of Type–II Diabetes Mellitus. *J. Liaquat Univ. Med. Health Sci.* **2018,** *17* (02), 101–105.
45. Mirmiran, P.; Yuzbashian, E.; Asghari, G.; Sarverzadeh, S.; Azizi, F. Dietary Fibre Intake in Relation to the Risk of Incident Chronic Kidney Disease. *Brit. J. Nutr.* **2018,** *119* (5), 479-485.
46. Nasri, H.; Baradaran, A.; Shirzad, H.; Rafieian-Kopaei, M. New Concepts in Nutraceuticals as Alternative for Pharmaceuticals. *Int. J. Prev. Med.* **2014,** *5* (12), 1487.
47. Ozturk, B.; Argin, S.; Ozilgen, M.; McClements, D. J. Nanoemulsion Delivery Systems for Oil-Soluble Vitamins: Influence of Carrier Oil Type on Lipid Digestion and Vitamin D3 Bioaccessibility. *Food Chem.* **2015,** *187*, 499–506. https://doi.org/10.1016/j.foodchem.2015.04.065
48. Pandey, M.; Verma, R. K.; Saraf, S. A. Nutraceuticals: New Era of Medicine and Health. *Asian J. Pharm. Clin. Res.* **2010,** *3* (1), 11–15.
49. Pérez-Sánchez, A.; Barrajón-Catalán, E.; Herranz-López, M.; Micol, V. Nutraceuticals for Skin Care: A Comprehensive Review of Human Clinical Studies. *Nutrients* **2018,** *10* (4), 403.
50. Pouton, C. W.; Porter, C. J. H. Formulation of Lipid-Based Delivery Systems for Oral Administration: Materials, Methods and Strategies. *Adv. Drug Deliv. Rev.* **2008,** *60*, 625–637.
51. Proksch, E.; Schunck, M.; Zague, V.; Segger, D.; Degwert, J.; Oesser, S. Oral Intake of Specific Bioactive Collagen Peptides Reduces Skin Wrinkles and Increases Dermal Matrix Synthesis. *Skin Pharmacol. Physiol.* **2014,** *27* (3), 113–119.
52. Rachwa-Rosiak, D.; Nebesny, E.; Budryn, G. Chickpeas—Composition, Nutritional Value, Health Benefits, Application to Bread and Snacks: A Review. *Crit. Rev. Food Sci. Nutr.* 2015, *55* (8), 1137–1145.
53. Rao, P. J.; Naidu, M. M. Nanoencapsulation of Bioactive Compounds for Nutraceutical Food. In *Nanoscience in Food and Agriculture*; Springer: Cham, 2016; vol *2*, pp 129–156.
54. Rodrigues, D. G. B.; de Moura Coelho, D.; Sitta, Â.; Jacques, C. E. D.; Hauschild, T.; Manfredini, V.; ... & Vargas, C. R. Experimental Evidence of Oxidative Stress in Patients with l-2-hydroxyglutaric Aciduria and that l-Carnitine Attenuates *in vitro* DNA Damage Caused by d-2-Hydroxyglutaric and l-2-hydroxyglutaric Acids. *Toxicol. In Vitro* **2017,** *42*, 47–53.
55. Rossino, M. G.; Casini, G. Nutraceuticals for the Treatment of Diabetic Retinopathy. *Nutrients* **2019,** *11* (4), 771.

56. Salunkhe, D. K.; Jadhav, S. J.; Kadam, S. S.; Chavan, J. K. Chemical, Biochemical, and Biological Signi®cance of Polyphenols in Cereals and Legumes. *CRC Crit. Rev. Food Sci. Nutr.* **1982,** *17*, 277–305.
57. Sarker, M. R.; Franks, S. F. Efficacy of Curcumin for Age-associated Cognitive Decline: A Narrative Review of Preclinical and Clinical Studies. *Geroscience* **2018,** *40* (2), 73–95.
58. Schloss, J. M.; Colosimo, M.; Airey, C.; Masci, P. P.; Linnane, A. W.; Vitetta, L. Nutraceuticals and Chemotherapy Induced Peripheral Neuropathy (CIPN): A Systematic Review. *Clin. Nutr.* **2013,** *32* (6), 888–893.
59. Shinohara, S.; Gu, Y.; Yang, Y.; Furuta, Y.; Tanaka, M.; Yue, X.; ... & Kimura, H. Ethanol Extracts of Chickpeas Alter the Total Lipid Content and Expression Levels of Genes Related to Fatty Acid Metabolism in Mouse 3T3-L1 Adipocytes. *Int. J. Mol. Med.* **2016,** *38* (2), 574–584.
60. Singh, P.; Sivanandam, T. M.; Konar, A.; Thakur, M. K. Role of Nutraceuticals in Cognition During Aging and Related Disorders. *Neurochem. Int.* **2021,** *143*, 104928.
61. Takahashi, M.; Uechi, S.; Takara, K.; Asikin, Y.; Wada, K. Evaluation of an Oral Carrier System in Rats: Bioavailability and Antioxidant Properties of Liposome-Encapsulated Curcumin. *J. Agric. Food Chem.* **2009,** *57*, 9141–9146.
62. Ting, Y.; Zhao, Q.; Xia, C.; Huang, Q. Using *in vitro* and *in vivo* Models to Evaluate the Oral Bioavailability of Nutraceuticals. *J. Agric. Food Chem.* **2015,** *63* (5), 1332–1338.
63. Ting, Y.; Jiang, Y.; Lan, Y.; Xia, C.; Lin, Z.; Roger, M. A.; Huang, Q. Use of Viscoelastic Emulsions to Increase the Oral Bioavailability of Tangeretin. *Mol. Nutr. Food Res.* **2015.**
64. Tsou, M. F.; Lu, H. F.; Chen, S. C.; Wu, L. T.; Chen, Y. S.; Kuo, H. M.; ... & Chung, J. G. Involvement of Bax, Bcl-2, Ca2+ and Caspase-3 in Capsaicin-Induced Apoptosis of Human Leukemia HL-60 Cells. *Anticancer Res.* **2006,** *26* (3A), 1965–1971
65. Venkatraman, J. T.; Chu, W. C. Effects of Dietary ω-3 and ω-6 Lipids and Vitamin E on Serum Cytokines, Lipid Mediators and Anti-DNA Antibodies in a Mouse Model for Rheumatoid Arthritis. *J. Am. Coll. Nutr.* **1999,** *18* (6), 602—613.
66. Wang, J.; Guleria, S.; Koffas, M. A.; Yan, Y. Microbial Production of Value-Added Nutraceuticals. *Curr. Opin. Biotechnol.* **2016,** *37*, 97–104.
67. Yang, X.; Tian, H.; Ho, C. T.; Huang, Q. Stability of Citral in Emulsions Coated with Cationic Biopolymer Layers. *J. Agric. Food Chem.* **2011,** *60*, 402–409.
68. Yu, H.; Huang, Q. Improving the Oral Bioavailability of Curcumin Using Novel Organogel-Based Nanoemulsions. *J. Agric. Food Chem.* **2012,** *60*, 5373–5379.
69. Zhao, Q.; Lan, Y.; Xia, C.; Rogers, M. A.; Huang, Q. Comparison of *in vitro* Models in 415 Determining Bioaccessibility of Nutraceuticals: The Case of Coenzyme Q10. *J. Agr. Food Chem.* **2015.**
70. Zulet, M. A.; Macarulla, M. T.; Portillo, M. P.; Noel-Suberville, C.; Higueret, P.; Martínez, J. A. Lipid and glucose utilization in hypercholesterolemic rats fed a diet containing heated chickpea (Cicerаretinum L.): a potential functional food. *Int. J. Vitam. Nutr. Res.* **1999,** *69* (6), 403–411.

CHAPTER 10

REGULATORY FRAMEWORK FOR THE DEVELOPMENT OF NUTRACEUTICALS

MANSAVI BHARDWAJ

PhD Scholar, School of Biotechnology, Shri Mata Vaishno Devi University Katra, Jammu and Kashmir, India

ABSTRACT

The therapeutic role of nutraceuticals and dietary supplements in improving the quality of health is becoming more widely recognized. They have been shown to have many health advantages and their use will surely keep us away from the ailments and help humans to maintain a good life style. Because these items are intended for human consumption, the regulatory bodies around the world are focusing on the product quality and safety. Maintaining safety and quality standards as per numerous regulatory rules set by the respective government has become increasingly crucial as food goods travel from one country to another; this may be the significant catalyst for industrial growth. This chapter provides a snapshot of regulatory framework for the development of nutraceuticals in India as wells as other countries. This will include all the legal and necessary requirements for the product to be approved as a nutraceutical.

10.1 INTRODUCTION

Hippocrates, the legendary father of modern medicine explicated the deep relationship between the usage of proper health foods and their therapeutic advantages, validating the fact that this concept is too old and he also quoted "*Let food be thy medicine, and medicine be thy food*". It wasn't until late

Formulations, Regulations, and Challenges of Nutraceuticals.
Tingirikari Jagan Mohan Rao, Rajesh K. Kesharwani, Raj K. Keservani, Anil Kumar Sharma (Eds.)

1980s and early 1990s that a formal strategy and knowledge of nutraceuticals began to emerge and it become a new word of our dictionary in 1989 coined by Dr. Stephen DeFelice, founder of the Foundation of Innovation in Medicine. Despite the fact that the term "nutraceuticals" is now widely understood and accepted, it still lacks a well-defined definition. The food or drinks that deliver medical- or health-related benefits including the disease treatment and prevention come under "nutraceuticals" according to Dr. DeFelice.[1] The main idea behind inventing the nutraceuticals was to make people aware that apart from the source of energy and other important nutrients, food also provides numerous health benefits.[2] Still sometimes "functional foods," "dietary integrators," and "nutraceuticals" are used incorrectly or as overlapping terms. Due to drastic change in today's lifestyle and eating habits, the concept of nutraceuticals has now been recognized as major means of disease prevention.[3] They seem to be more attractive to many people than any other prescribed drugs due to the fact that it is a more natural approach. Dietary supplements, they believe, will make them feel healthier, stronger, and energetic and help them to avoid many diseases. Also, when established medicaments for their particular conditions fail, some patients seek these products. Isolated nutrients, various herbal products, processed products, and many other designer foods are grouped under nutraceuticals possessing important physiological and biological activities.[4] Nutraceuticals are another name for bioactive compounds, present in human diet that give health advantages in addition to the product's basic nutritional value.[5] These bioactive compounds are generally present in minute quantities in our food as natural elements providing numerous health advantages in addition to the basic nutritional value and also have therapeutic potential.[6] The link between the food, physiological function, and the diseases has been better understood over the past few decades. On the other hand, the present nutrition science is significantly revealing more details on roles and actions of particular food elements in promoting health. Offering medical and health benefits, the production of these nutraceuticals and related products must strongly be controlled by a powerful regulatory structure. In developing a good nutraceutical policy, a clear understanding of nutraceuticals in a regulatory system would surely decrease the uncertainty.

10.2 EXPANSION OF GLOBAL NUTRACEUTICAL MARKET

The interest of not just department of research and development and various multinational corporations but also consumers have increased in nutraceuticals. This is due to the large sums of money spent on healthcare. The Global Health Expenditure database (GHED) generally supervised by the World Health Organization, compiles statistics on health spending for 192 nations during the last two decades. According to its report, India has 7% of its GDP whereas the United States of America (USA) has 17% of GDP into healthcare. These data provide us a clear picture that the USA has now world's largest and fastest growing market for functional foods and nutraceuticals.[7] Despite its rapid growth, the nutraceuticals industry is still in its early stages, as evidenced by the number of marketed goods and nutraceutical's contribution to the country's GDP. In 2017, the worldwide market of nutraceuticals was worth $379.061 billion, and by 2026, it is predicted to be worth $734.601 billion.[8] Diverse variety of spices, medicinal herbs, plants, and trees are native to India and have a significant domestic market size.[8] These items are also being exported globally in huge amount from India with Japan and USA as major clients.[9] In the recent years, the market of herbal supplements is also expected to show a tremendous growth from 0.6 billion to 1.7 billion US $.[10] Between the year 2010 and 2020, the number of published research articles related to both dietary supplements or nutraceuticals has been increased to 70,000 in PubMed, indicating massive research in this field.

In November 2019, severe acute respiratory syndrome corona virus two (SARS-CoV2) occurred, resulted in an even bigger increase in dietary and nutraceutical sales in early 2020. Consumers seek further protection from viral infection and sickness during a severe worldwide pandemic on the idea that supplement goods would provide health advantages or "immune boosting" effects if ingested.[11–13] In comparison to the previous year, the sales of nutraceuticals or dietary supplements increase by 5% in 2019 but further 44% increase in sales were seen during the same period of pandemic.[13] Different countries have shown sudden rise in the sales of supplements or nutraceuticals during the first and the second wave of COVID-19 pandemic just to boost the immune functioning of the body.

Interestingly, several nutraceutical candidates and dietary supplements such as probiotics, zinc, Vitamin D, Vitamin C, omega-3 polyunsaturated fatty acids, and selenium are being further explored for their role as anti-viral or immunomodulators in the treatment of COVID-19.[14,15] Among all, Vitamin D had grabbed the attention of scientists in recent years especially in this pandemic period as its deficiency has been significantly associated with the risk of COVID-19 infection.[16] Vitamin D has been shown to increase the antiviral efficacy of epithelial cells of bronchi, reduce the multiplication and further release of rhinovirus (non-enveloped virus), linked to respiratory infections.[17] This inhibitory effect of Vitamin D is relatively more against enveloped viruses such as corona virus, as shown by many studies.[18] That being a case, the link between COVID-19 and Vitamin D is currently the topic of extensive research and multiple therapeutic trials for the treatment of such respiratory infections. The nutraceuticals and supplements market's remarkable growth trends are unlikely continue with pandemic period but in near future, the market is predicted to return prepandemic levels.[18] It's hardly surprise that the regulatory authorities are monitoring the sector more closely now that global consumption of nutraceuticals and dietary supplements is on the rise and because of findings made during the pandemic.

10.2.1 NUTRACEUTICAL MARKET IN INDIA AT GLANCE

As per the Ministry of Food Processing Industries, India's current share in global nutraceutical market is 2% and it is expected to account for 3% in upcoming years.

FIGURE 10.1 Schematic pie diagram is showing the India's share in global nutraceutical market.

The estimated market value of Indian nutraceutical market is 4 USD billion by 2020 and 7% expected rate of growth in 2019–2024 period.[19] Many domestic firms and multinational companies currently hold the Indian nutraceutical market such as Dabur, ITC, Patanjali, Himalaya, and Baidya Nath.[20]

10.3 REGULATORY FRAMEWORK FOR THE DEVELOPMENT OF NUTRACEUTICALS

10.3.1 REGULATION OF NUTRACEUTICALS IN THE UNITED STATES OF AMERICA

The population of America in majority falls sick due to the consumption of dietary supplements, as shown by many studies. These supplements were majorly consumed in the form of different juices and snacks to provide much healthier life. This headline resulted in the formulation of regulations regarding the development and the use of nutraceuticals and other dietary supplements in the USA.

USFDA stands for the United States Food and Drug Administration is basically a federal agency of Department of Health and Human Services. Under numerous sets of regulations, USFDA regulates the dietary supplements and ingredients in the USA. In 1994, DSHEA which stands for the Dietary Supplement Health Education Act came into existence and created certain standards for any product to be categorized as dietary supplements. Some of the major norms set under this act are as listed below:

i. No need to register with the FDA before producing or selling the products.
ii. Safety of the dietary supplement is of top priority. It is illegal for manufacturers and distributors of dietary supplements and other related components to advertise adulterated products. Therefore, companies are responsible for assessing the safety of their products prior to marketing.
iii. In any case, it is the responsibility of FDA to take necessary actions against any unsafe dietary supplement products in the market.
iv. Manufacturers must guarantee that the information on their product labels is accurate and not misleading.

10.3.1.1 PROCESS OF REGISTRATION OF NUTRACEUTICALS IN THE UNITED STATES OF AMERICA

Foremost, only those dietary supplements can be registered that are according to the definition of DSHEA.

The regulation under 21 CFR 190 Subpart B where CFR stands for the Code of Federal Regulations and deals with dietary supplements states that at least 75 days before the introduction of a dietary supplement, the notification regarding the new dietary ingredient must be submitted by the manufacturer to the Office of Nutritional Products, Labeling and Dietary Supplements (HSF-820), Center for Food Safety and Applied Nutrition, Food and Drug Administration, 5100 Paint Branch Pkwy, College Park, MD 20,740. Also, the information that may include citations to the published articles, that supports the manufacturer's conclusion that the dietary supplement containing that particular ingredient is relatively safe for consumption by the consumers. Below are the documents required to be submitted during the registration process:

i. Name and address of the manufacturer or applicant
ii. Both common and scientific name of the new dietary ingredient
iii. Full description of the dietary supplement or supplements that incorporate the new ingredient. This may include:
   a. The amount of new dietary ingredient in the dietary supplement
   b. Recommended conditions of use on the dietary supplement's label, if any. In other cases, normal conditions of usage of the supplement would be preferred.
   c. Evidences related to safety measures, if any when used under recommended conditions, will expected to be safe
   d. Signature of the manufacturer and the distributor of the dietary supplements containing new dietary ingredient.[21,22]

10.3.1.2 PROCESS OF CLEARANCE FOR NUTRACEUTICALS IN THE UNITED STATES OF AMERICA

i. After successful submission of all the related documents, FDA would further notify the submitter about the date of receipt of registration or notification. The filing date of notification would be considered the day when the agency would receive the documents.
ii. The manufacturer shall not be allowed to deliver the dietary supplement containing a new dietary ingredient for introduction into the interstate commerce for 75 days from the date of filing.
iii. In case, if additional information in support of notification already filed for new dietary ingredient is provided by the manufacturer or distributor, then the agency will examine all the submissions related to that notification if they are meaningful. In that case, the agency

will also look into whether there is any need to reset the time frame of 75 days.

iv. FDA will assign a new filing date, if the new submission is determined to be a significant modification. The manufacturer of the dietary supplement will be notified with new filing date by the agency itself. In this case, the new filing date will be the date of receipt when the agency receives the additional information for further substantial revision.

v. All the information contained in the notification related to new dietary supplement will be released by the FDA after 90 days for public display except for any confidential information.[23]

10.3.2 INDIAN REGULATORY FRAMEWORK FOR THE DEVELOPMENT OF NUTRACEUTICALS

India is the second largest country both in terms of population and also for the fruits and vegetables production. Being the second largest producer of fresh fruits and vegetables, India still lags in the processing of these fruits and vegetables in comparison to the USA which holds the largest proportion in the world's nutraceutical market. Studies have shown that the term "nutraceuticals" was introduced after the year 2015 in the Indian regulatory framework of food products. In India, "foods for special dietary uses" are known as nutraceuticals'.[24] Earlier, there was no single autonomous body at the national level for the regulation of food products. Different food processing laws were made both at states and central levels for the regulation of food products. Among them, major central laws include the following:

i. Food Products Order, 1995
ii. Milk and Milk products Order, 1992
iii. Edible Oils packaging (Regulation) Order, 1988
iv. Meat Food Products Order, 1973
v. The Solvent Extracted Oil, De-oiled Meal and Edible Flour (Control) Order, 1967
vi. Fruit Products Order, 1955
vii. Prevention of Food Adulteration Act, 1954
viii. Vegetable Oil Products (Control) Order, 1947.[24]

Finally, FSSAI stands for the Food Safety and Standards Authority of India came into existence in 2008 under the Food Safety and Standards Act,

2006 consolidating all the existing laws and order related to food products. The main idea behind the establishment of FSSAI is to create certain scientific standards for the regulation of food products to guarantee safe and healthy food for human consumption.

10.3.2.1 MAJOR REQUIREMENTS FOR REGULATING NUTRACEUTICALS IN INDIA

For maintaining the proper channel of different rules and regulations for regulating nutraceutical markets in India, there are certain claims required before registration. These may include:

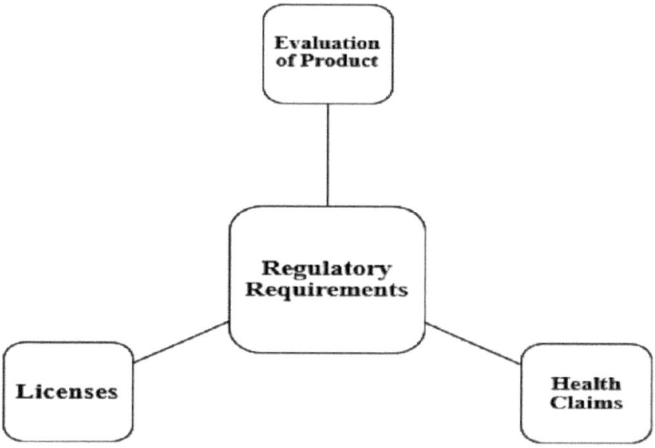

FIGURE 10.2 Schematic diagram is showing three different regulatory requirements in India.

10.3.2.1.1 Licenses

Depending on the status of product, various types of licenses are required to get the product registered in India. These might include:

i. License for Manufacturing
ii. Import License
iii. License for Marketing
iv. Other major clearances both on state as well as national level, by the regulatory side before the launching of product in Indian market.

10.3.2.1.2 Evaluation of Products

Major steps involved in the evaluation of products are as follows:

i. Documentation
ii. Collection of sample or product in the presence of witnesses
iii. Dispatch of sample to the authority
iv. Analysis of the food product within timeframe
v. Other Adjudication formalities such as appeal and hearing.

10.3.2.1.3 Health Claims

Different countries have different regulatory frameworks for nutraceuticals so it is very important to look into each detail of the regulation to avoid any kind of chaos or difficulty. India has figured out how to keep its own label claim content with numerous claims. Any depiction that suggests a relationship between a food or a component of that particular food and health is termed as "Health Claim".[25] To wish to get their product registered, a foreign person encounters various difficulties in doing so, these underlined "health claims" may surely provide solutions:

i. India has its own sets of packaging and labeling regulations
ii. Packaging must also display the components of the consignment as well as the need for the sample product to be packed according to the Indian norms
iii. The information that must be provided in the claim
iv. Structural Claim
v. Functional Claim.

10.3.2.2 PROCESS OF REGISTRATION IN INDIA

To get entry into the Indian market, registration from the FSSAI is mandatory. Different protocols are to be followed by companies to get their product registered. For public safety, the rules and regulations are quite stringent, sometimes creating difficulties for the companies to get approval for their dietary supplement products. Below are the details for the process of registration:

i. Multiple forms like FORM A, FORM B and FORM C must be filled by the manufacturer
ii. Import License, if required to be issued by the Central Licensing Authority
iii. Self-Attested Declaration Form.[24]

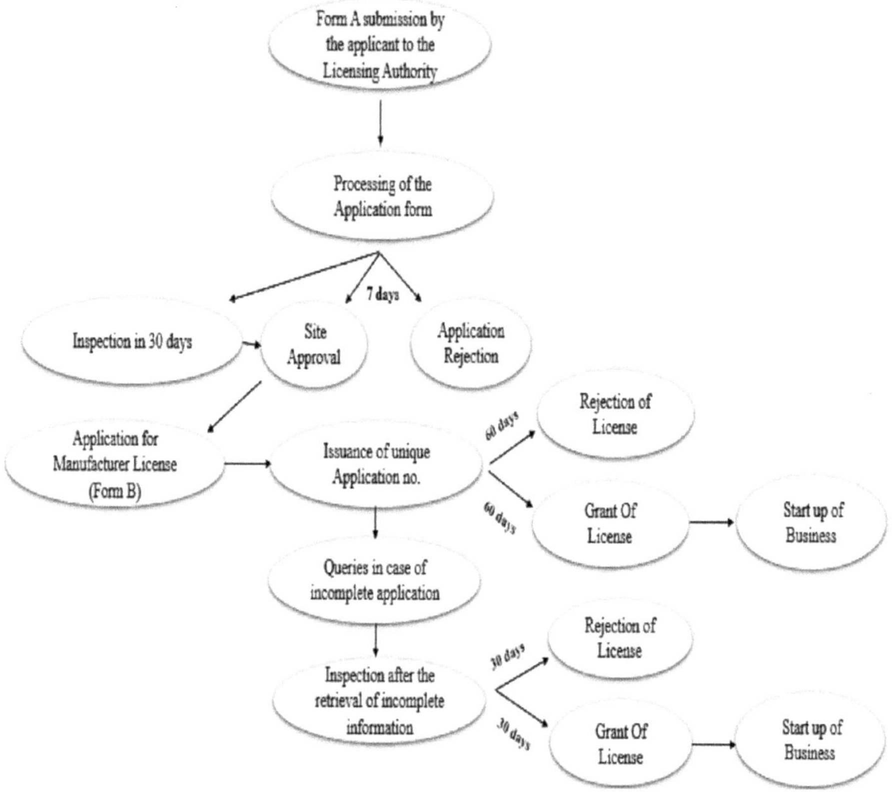

FIGURE 10.3 Flowchart showing the registration process in India.

10.4 CONCLUSION

Globally, the nutraceutical market is evolving rapidly. The demand for nutraceuticals is increasing day-by-day, especially since the pandemic year when people have started releasing the value of good health. With the tremendous increase in the demand for nutraceuticals, the rules and regulations of nutraceuticals and other dietary supplements are now strengthened. Different countries have their own regulatory framework for nutraceuticals.

It is all because of the establishment of FSSAI in 2008 in India, a strong and powerful regulatory framework has been developed that has transitioned a multi-level control to a single line of control with a focus on self-compliance rather than being entirely a regulatory authority. Likewise, in other countries, such as the USA sharing the largest proportion in the nutraceutical market also have their own regulatory framework to deal with the nutraceuticals. Updation of the Dietary Supplement and Health Education Act (DSHEA) in 1994 has provided a roadmap for the registration of nutraceuticals or dietary supplements in the USA for marketing purposes. Thus, it is critical for the market participants to be aware of all the recent nutraceutical rules and regulations in order to get their products approved by the high authorities. The most successful nutraceutical players in the next years are likely to be those whose functional products are merely one aspect of a larger line of goods that meet both standard and health value points. An understanding of the relationship between food and disease by consumers will determine future nutraceutical demands.

KEYWORDS

- **nutraceuticals**
- **dietary supplements**
- **regulatory framework**

REFERENCES

1. Brower, V. Nutraceuticals: Poised for a Healthy Slice of the Healthcare Market? *Nat. Biotechnol.* **1998,** *16* (8), 728–731.
2. Chauhan, B.; Gopal, K.; Nazia, K.; Shahid H. A. Current Concepts and Prospects of Herbal Nutraceutical: A Review. *J. Adv. Pharma. Technol. Res.* **2013,** *4* (1), 4.
3. Sumi, Y. O. S. H. I. H. I. K. O. Research and Technology Trends of Nutraceuticals. *Sci. Technol. Trends Q. Rev.* **2008,** *28*, 10–21.
4. Andlauer, W.; Peter, F. Nutraceuticals: A Piece of History, Present Status and Outlook. *Food. Res. Int.* **2002,** *35* (2–3), 171–176.
5. Kalra, E. K. Nutraceutical-definition and Introduction. *Aaps. Pharmsci.* **2003,** *5* (3), 27–28.

6. Biesalski, H.-K.; Lars O. D.; Ibrahim E.; Rolf G.; Michael M.; Dieter S.; et al. Bioactive Compounds: Definition and Assessment of Activity. *Nutrition* 2009, *25* (11–12), 1202–1205.
7. World Nutraceuticals, Industry Study with Forecasts to 2010 & 2015. *The Freedonia Group*; Cleveland, OH USA; 2006.
8. https://www.globenewswire.com/newsrelease/2019/01/29/1706747/0/en/Global-Nutraceuticals-MarketReport-2018-Market-Accounted-for-379-06-Billion-in-2017-andExpected-to-Reach-734-60-Billion-by-2026.html
9. Patwardhan, B.; Dnyaneshwar, W.; Palpu, P.; Narendra B. Ayurveda and Traditional Chinese Medicine: A Comparative Overview. *Evid. Based Complement. Alternative Med.* **2005,** *2* (4), 465–473.
10. Sharma, A.; Pramod, K.; Pankaj, S., Birendra, S. A Comparative Study of Regulatory Registration Procedure of Nutraceuticals in India, Canada and Australia. *Int. J. Pharma. Qual. Assur.* **2013,** *4* (4), 61 66.
11. Lordan, R.; Halie, M. R.; Casey, S. G. Dietary Supplements and Nutraceuticals Under Investigation for COVID-19 Prevention and Treatment. *Msystems* **2021,** *6* (3), e00122–21.
12. Furlong, C. *Food and Beverage Trends in Europe During COVID-19*; 2020.
13. Lordan, R. Dietary Supplements and Nutraceuticals Market Growth During the Coronavirus Pandemic–Implications for Consumers and Regulatory Oversight. *Pharmanutrition* **2021,** *18*, 100282.
14. Tsoupras, A.; Ronan, L.; Ioannis, Z. Thrombosis and COVID-19: The Potential Role of Nutrition. *Front. Nutr.* **2020,** *7*, 177.
15. Zabetakis, I.; Ronan, L.; Catherine, N.; Alexandros, T. COVID-19: The Inflammation Link and the Role of Nutrition in Potential Mitigation. *Nutrients* **2020,** *12* (5), 1466.
16. Mez, J.; Daniel, H. D.; Patrick, T. K.; Bobak, A.; Victor, E. A., Bertrand R. H.; et al. Clinicopathological Evaluation of Chronic Traumatic Encephalopathy in Players of American Football. *Jama* **2017,** *318* (4), 360–370.
17. Beard, J. A.; Allison, B.; Rob S. Vitamin D and the Anti-viral State. *J. Clin. Virol.* **2011,** *50* (3), 194–200.
18. Lordan, R. Notable Developments for Vitamin D amid the COVID-19 Pandemic, but Caution Warranted Overall: A Narrative Review. *Nutrients* **2021,** *13* (3), 740.
19. https://foodprocessingindia.gov.in/sectors/Nutraceuticals
20. http://mrssindia.com/uploads/reports/pdf/assocham-knowledge-report-on-nutracuticals-released-at-3rd-national-symposium-6.pdf.
21. Kessler, R. C.; Roger, B. D.; David F. F.; Maria, I. V. R.; Ellen, E. W.; Sonja A. W.; et al. Long-term Trends in the Use of Complementary and Alternative Medical Therapies in the United States. *Ann. Inter. Med.* **2001,** *135* (4), 262–268.
22. Jain, P. N.; Meera H. R.; Vineet C. J.; Vijayendraswamy S. M.. Current Regulatory Requirements for Registration of Nutraceuticals in India and USA. *Int. J. Drug Regul. Aff.* **2018,** *6* (2), 22–29.
23. https://www.govinfo.gov/content/pkg/CFR-2011-title21-vol3/pdf/CFR-2011-title21-vol3-part190.pdf
24. Verma, B.; Harvinder, P. Regulations of Nutraceuticals in India & US. *Pharma. Innov.* **2018,** *7* (7), 811–816.
25. Health claims in India, http://www.fssai.gov.

PART III

Opportunities and Challenges of Nutraceuticals

CHAPTER 11

OPPORTUNITIES AND CHALLENGES IN THE DEVELOPMENT OF FUNCTIONAL FOODS AND NUTRACEUTICALS

BHUSHAN R. RANE, AKASH J. AMKAR, VINIT S. PATIL, PRASAD K. VIDHATE, and ABHIJIT R. PATIL

Shri D.D. Vispute College of Pharmacy & Research Center, Panvel, Maharashtra, India

ABSTRACT

Functional foods and nutraceuticals are entire, fortified, enriched, or improved foods that give health advantages in addition to basic nutrients such as vitamins and minerals whenever ingested in effective amounts as a portion of a diversified diet on a routine basis. Functional foods and nutraceuticals are currently one of the most intensely researched and widely promoted fields of food and nutrition science. Nutritional supplements are important components of functional foods and nutraceuticals for sustaining health and combating diseases such as cancer, obesity, cardiovascular diseases (CVDs), and diabetes. It also has an economic importance for farmers and consumers have been discussed in this review. According to our current situation globally, regional regulation authorities have their own regulations for functional foods and nutraceuticals. Due to this, developmental issues are frequently underestimated due to a lack of authoritative control is also highlighted in this review, and maintaining the quality of these products is also a major challenge.

Formulations, Regulations, and Challenges of Nutraceuticals.
Tingirikari Jagan Mohan Rao, Rajesh K. Kesharwani, Raj K. Keservani, Anil Kumar Sharma (Eds.)

11.1 INTRODUCTION

The father of modern medicine, Hippocrates, expressed, "Let food be the medicine, and the medicine shall be the food," which has heightened consumer and food and nutrition scientist interest in the health benefits of specific foods.[73] In Japan in the 1980s, the idea of functional food came into existence for the nutritional quality and increased the life span of elderly people to save or control healthcare expenditures.[38] Stefen De Felice coined the phrase nutraceuticals in 1989, combining the words nutrition and pharmaceutical. He demonstrated that nutraceuticals are foods/food ingredients that provide medicinal or health benefits, such as restriction and/or treatment to cure disease.[26] According to Shahidi, food or a component of food is used in the production of nutraceuticals and functional foods. These may resemble traditional foods in appearance.[92] Nutraceuticals, on the other hand, are typically thought to be pure or separated forms of foods that are offered as medicine in the market.[46,49–51]

11.2 FUNCTIONAL FOOD AND NUTRACEUTICALS

11.2.1 FUNCTIONAL FOODS OVERVIEW

Beyond providing vital nutrients, whole, enriched, fortified, processed, or enhanced food has been employed as a functional food. One food is functional since it delivers nutrients as well as has a wide range of physiological impacts (Guangchang et al., 2012). Several natural dietary components, such as fruits, vegetables, meat, fish, poultry, cereals, and grains, are thought to provide health advantages in addition to fundamental nutrients. Many scientists believe that individuals believe that foods sold in stores are fortified with particular nutrients before processing and hence qualify as functional foods. Some people believe that all traditional foods are functional foods that we consume on a daily basis (Istva et al., 2008). Because of the advancement of food for reducing the risk of chronic and cardiovascular disease (CVD) and illness, all processed foods in supermarkets are not considered functional foods unless there is scientific proof that they contain the most essential nutrients and that their presence can reduce health deficiencies. Due to mental health issues, stress, sadness, memory loss, and poor focus, modern living has a significant detrimental impact on human life (Kwak and Jukes, 2001). To prevent noncommunicable diseases, scientists are working to find suitable synergies by developing new food products with the help of food scientists and with the help of consumers to reduce diseases such as obesity, chronic disease, and various types of diabetes, musculoskeletal disease, gastrointestinal (GI) diseases, anxiety, depression, and stress caused by aging effect (Mattioli et al, 2016). People in most European countries are accustomed to eating fast food as a result of their hectic lifestyles; the global food business is estimated to be worth US$33 billion. Minerals and trace elements such as copper, manganese, and selenium are responsible for serving as a cofactor against antioxidant enzymes, which helps to protect our bodies from oxidative stress (Gormely, 2013). Some vitamins, such as vitamin C and vitamin E, behave as free radicals and are utilized as antioxidant quenchers. As a result, this research is being looked at to see what foods, which have functional food features, we consume in our daily lives. The current study is based on the following goals.

11.2.1.1 DEFINITION OF FUNCTIONAL FOOD

A functional food typically incorporates added, scientifically created substances that have a specific health benefit. Despite the fact that the word "functional food" has been defined multiple times,[87] there is no universally agreed definition for this category of foods.[4] In most countries, there is no legal definition of the phrase, and even nutritionists and food specialists find it difficult to draw a line between conventional and functional meals (Mark-Herbert, 2004).[72] Several national agencies, academic institutions, and the food industry have suggested definitions for functional foods so far. The two approaches are "Foods that may provide health benefits beyond basic nutrition" and "Foods that resemble normal food in appearance and are meant to be consumed as part of a regular diet but have been modified to subserve physiological roles beyond the provision of simple nutrient requirements."[9] A Concerted Approach by the European Commission toward Functional Food Science in Europe (FuFoSE) with the coordination of the International Life Science Institute (ILSI) Europe, according to the following definition of functional food, "A food product can only be termed functional if, in addition to the fundamental nutritional effects, it has beneficial effects on one or more body functions that are important to a human being, either improving general and physical conditions or decreasing the risk of disease evolution. The amount of functional food consumed and the form in which it is consumed should be as usual for dietary purposes. As a result, it couldn't be in tablet or capsule shape like regular food."[21] In Japan, "foods for specified health uses" (FOSHU) goods have been available in capsules and tablets since 2001, despite the fact that the vast majority of items are still available in more traditional formats.[74] Functional foods, on the other hand, are not considered discrete food categories in European legislation, but rather a concept (Station et al., 2005).[18] As a result, the rules that must be followed are complex and vary depending on the type of food. All foods are covered under the General Food Law Regulation. Furthermore, based on the product's characteristics and its intended purpose, legislation governing dietetic foods, genetically modified organisms (GMOs), dietary supplements, or novel foods may also apply to functional foods. Rather than regulating the product category as a whole, legislative efforts in the European Union (EU) are currently focused on limiting the use of health claims on packages and in marketing. A list of authorized claims must be published for all member states, and nutrient profiles must be prepared for goods carrying health claims (EC, 2006),[72] In accordance with EU legislation on food nutrition and health claims.[23] "Function claims" and "reduction of disease risk claims" are two types of health claims. Health claim can be "function claims" and "reduction of disease claims." Detailed aspect of relevant legislation is thoroughly discussed for example by Coppens et al. (2006), Asp (2007) and Mathioudakis (2007).[6,18,61] Kwak and Jukes (2001a, 2001b) examined the regulatory concept of functional food in several countries including Japan, Korea, UK, USA, and Australia.[54,55]

11.2.1.2 APPLICATION AND EXAMPLES OF FUNCTIONAL FOOD

Sr. No.	Functional Food	Bioactive Component	Health Benefits	References
1.	Fortified margarine	Plant sterol and stanol ester	Reduce total and low-density lipoprotein (LDL) cholesterol	Weingartner et al., 2009
2.	Psyllium	Soluble fiber		
3.	Soy	Protein		
4.	Whole oat product	β-Glucan	Reduce total and LDL cholesterol	22
5.	Cranberry juice	Proanthocyanidins	Reduce UTI	
6.	Fatty fish	n-3 fatty acids	Risk of heart disease is reduced	30
7.	Garlic	Organosulfur compounds	Reduce the risk of heart and cancer disease	88
8.	Green tea	Catechin	Reduce the risk of various types of cancer	40
9.	Spinach, kale, collard green	Lutein, zeaxanthin	Reduce the risk of age-related blindness (macular degeneration)	1
10.	Tomatoes and processed	Lycopene	May reduce the risk of prostate cancer and inhibit LDL oxidation.	97
11.	Cruciferous vegetables	Glucosinolates, indole	Reduce the risk of various types of cancer	93
12.	Fermented dairy	Probiotics	Enhance immunity and GI health	68

11.2.2 NUTRACEUTICALS

A new diet-health approach has emerged recently, emphasizing the positive elements of nutrition. The new lifestyle that people have adopted has caused changes in their basic eating habits. Junk food consumption has skyrocketed, contributing to a slew of ailments caused by poor nutrition. Obesity is now acknowledged as a global problem. With economic development, the standard of living has improved in terms of earnings, expenditure, and lifestyle. However, it has also presented a significant problem in the shape of "lifestyle diseases." Food habits were the first to suffer as a result of this lifestyle shift. Food and health have long been linked. "Let food be thy medicine and medicine be thy food," Hippocrates reportedly stated. Traditional medicine in pre-Columbian America, Europe, Asia, and Africa is full of examples of foods that have been used to prevent and heal sickness. Food, on the other hand, has come to be considered primarily as a resource of nutrients (i.e., energy, protein, and fat) influenced by Western medicine's rationalistic ideology, to the exclusion of all other uses. However, as the prevalence of chronic diseases rises due to changing demographics, a significant and growing amount of research indicates that focused nutrition utilizing naturally occurring chemicals may be able to stabilize or perhaps cure many of the most difficult health problems.[67] Heart disease, followed by cancer, osteoporosis, arthritis, and a host of other disorders, is the main cause of death in the majority of developing nations globally. Consumers who are dissatisfied with current medicine's pricey, high-tech disease-treatment approach are looking for supplementary or alternative beneficial items and the bureaucracy of managed care promotes the attractiveness of nutritional supplements (Raut et al., 2020). Dietary supplements, functional foods, and functional beverages are the three primary types of nutraceuticals.[49,50] Dietary supplements can also be divided into nutritional and mineral supplements, homegrown supplements, plant concentrates, and protein supplements (Rizk et al., 2020; Hanan et al., 2020).

11.2.2.1 DEFINITION OF NUTRACEUTICALS

Stephen DeFelice, MD, founder and chairman of the Foundation for Innovation in Medicine (FIM), Cranford, NJ, coined the term "nutraceutical" in 1989 by combining the words "nutrition" and "pharmaceutical." According to DeFelice, nutraceutical is described as "a food or component of a food that delivers medicinal or health advantages, including the prevention and treatment of a disease." However, the term nutraceutical, as it is widely used in marketing, has no regulatory definition. He suggests that functional foods and nutraceuticals should be redefined. Food is referred to as "functional food" when it is cooked or prepared to utilize "scientific intelligence," with or without an understanding of how or why it is employed. Consequently, functional food provides the vitamins, lipids, proteins, carbs, and other nutrients it requires for optimal existence. A nutraceutical is a functional food that helps with preventing disease and/or treatment of the disorder (Ekta et al., 2003). nutraceuticals are food-derived products that are claimed to provide additional health advantages over and beyond the fundamental nutritional content of foods.[17] Since the initial definition of nutraceuticals was given, a variety of other definitions have been presented; by either limiting the dose types covered by the phrase (e.g., pills and capsules), enforcing some sort of processing constraints limiting the health benefits criteria, or establishing their relationship to other product categories. Several definitions concur or cross-reference the original DeFelice definition; some of them are as follows: Any substance that is food or a component of fruit and offers medicinal or health advantages, including the treatment and prevention of diseases, is referred to as a nutraceutical. These goods might include everything from isolated nutrients, nutritional supplements, and particular diets to foods that have been genetically modified, herbal products, and processed foods including cereals, soups, and beverages. "Any nontoxic food ingredient with scientifically demonstrated health advantages, including disease treatment and prevention," is referred to as a nutraceutical. Nutraceuticals are products made from food sources that advertise additional health advantages over and above the fundamental nutritional value of the food they are produced from.[75]

11.2.2.2 APPLICATIONS AND EXAMPLES OF NUTRACEUTICALS

- Nutraceuticals in different diseases

Sr. No.	Diseases	Examples	Reference
1	CVD	n-3 PUFAs, Tannins, Anthocyanins, octacosanol	12
2	Eye disorder	Lycopene, β-carotene, green tea, vitamin C, vitamin E, astaxanthin.	79
3	Alzheimer's disease	Curcumin, lutein, lycopene, turmerin	(Grammatikos et al., 2021)
4	Diabetes	Isoflavones, omega-3 fatty acid, psyllium	56
5	Cancer	Lycopene, biochanin, daidzein, β-carotene	13
6	Allergy	Quercetin	(Merseglia et al., 2021)
7	Immunity problems	Astragalus, garlic, *Echinacea angustfolia*	(Mannuci et al., 2021)
8	Obesity	Capsaicin, psyllium fiber, green tea, ma-huang guarana	86

- **Nutraceuticals as dietary supplements**

Sr no	Examples	Application	Reference
1	Folic acid	Preventing neural tube defects, helps in red blood cells formation	(Aliali et al., 2021)
2	Vitamin A	Antioxidant, growth, prevent some skin conditions	44
3	Calcium	Bone, muscles contraction, teeth nerve health, prevention of osteoporosis	77
4	Iron	Carry oxygen, produce energy	(Aliali et al., 2021)
5	Vitamin D	Musculoskeletal health, calcium absorption assistance, and bone and tooth health	(Nastri et al., 2021)

11.3 CURRENT SCENARIO AND MARKETED CONDITION

The statement "Let food be your medicine, your medicine will be your food," made by Hippocrates, the father of modern medicine, is gaining interest today as nutrition scientists and consumers recognize the health benefits of certain foods (El Sohaimy, 2017). The concept of functional food was introduced in Japan. In the 1980s, health authorities in Japan recognized the need for improved health care and increased the life expectancy of a growing number of elderly people in the community to control the expense of basic health care. The concept of a diet specifically designed to improve health or reduce the risk of disease is introduced and has been termed a "specific healthcare diet" (FOSHU). To increase physical and mental well-being and which may also reduce the risk of disease Such diets include fortified foods, processed foods, rich foods, dietary supplements, and healthy foods (Keservani et al., 2019). The term "nutraceutical" was coined by the words, "nutrition" and "medicine" in 1989 by Stephen DeFelice, chairman and founder of the Founding Foundation for Medical Initiation (FIM), Cranford. He described nutraceuticals as "foods (or part of a diet) that provide medical or health benefits, which include the prevention and/or treatment of a disease." It is usually a form of pills, supplements, and pills. Nutraceuticals include a wide range of products including beverages, dietary supplements, broken-down nutrients, genetically modified foods, herbal products, and processed foods.

11.3.1 NUTRACEUTICALS MARKET OVERVIEW

The global population of more than 60 years is expected to reach 1 million by 2020, 70% of whom will be living in developed countries promoting antiaging products so that the product can be seen at a higher level (Global Industry Analysts, 2018). The global market for food and nutrition in 2013 was nearly US$175 billion in different countries. The nutritious beverage market alone is expected to have the highest growth, with a Combined Annual Growth Rate (CAGR) of 8.8%. The sector is expected to have a net worth of about US$87 billion in 2016. The Nutrition Market is the second largest market, expected to reach US$67 billion in 2016, at a CAGR of 6.4%.

11.3.2 MOTIVATORS OF FUNCTIONAL FOODS AND NUTRACEUTICAL MARKET

Asia–Pacific Nutraceutical product market is an emerging market for food ingredients and active food components. Japan is the largest consumer of nutritious food in Asia, while China is the second largest consumer of nutritious products as it has the largest number in the world and people pay close attention to their eating habits.[49] The Japanese Nutrition Industry includes a wide range of products in a variety of fields including active and nutritional age and nutritional supplements and diets. Rapid population growth and healthy eating habits among people have been a 13-year market force (Shimizu, 2018). According to a report by (Ken Research, 2019), a very high percentage of sales are recorded for solid and sporting products. The food supplement market has also shown healthy growth over the years. Strength foods and beverages are in great demand with vitamins that lead to dietary supplements. Major players such as Yakult, Pocari Sweat, and Aquarius are giants in the well-established Japanese food and beverage market, while Otsuka Pharmaceuticals, Miki, Amway, and Taisho Pharmaceuticals are giants in the food market.[50] Due to the highly developed nature of the Japanese nutrition market, nutrition industries are expected to produce specific products that will meet the needs of different ages in the country. Revenue from the nutrition market is expected to grow at a CAGR of 1.5% in 2017. The active food and beverage market in India is expected to have a growth of 70.74% compared to the food supplement in 2017 (Keservani et al., 2014). The Indian nutraceuticals market has shown remarkable growth during the previous 7 years. The increased health awareness about the different types of nutritious foods available on the market and people's willingness to spend on healthy beverage food supplements during the years 2009, 2011, 2013, 2015, and 2016. The global nutraceutical market—food, beverages, and supplements, 2009–2016 ($ Millions).

11.3.3 INTERNATIONAL REGULATIONS FOR LABELING AND HEALTH-RELATED CLAIMS

In the USA, functional and nutritious foods are regulated by the US Food and Drugs Authority (FDA) under the auspices of the Federal Food, Drug, and Cosmetic Act. Since they are not legally defined, they are regulated under the same rules as other foods and food products (Yamada et al., 2017). The FDA is also in charge of reviewing the quality, safety, and labeling of dietary supplements to ensure that they meet all Dietary Supplement Health and Education requirements and FDA regulations prior to marketing (El Sohaimy et al., 2018). By January 2018, food labs in the USA should have information that provides size and new labeling requirements for specific package sizes, further understanding of food science, and packages should have renewed designs as recommended by the FDA.[49] Manufacturers wishing to market new food supplements should notify the FDA and add information indicating that the article is suitable as a new food supplement and is safe under the terms of use as set out in its label before marketing a new product. Similarly, functional food is still a "real" category according to EU or Irish food law but is regulated by existing food law (Global Industry Analysts et al., 2020). In Europe, the European Commission writes recommendations based on EFSA safety assessments and decides whether or not to accept a new food product. EU law stipulates that only those ingredients (vitamins and minerals) in the list are marketable as of August 1, 2005, while new additives must be fully tested for safety (Sonmez et al., 2018). The European Commission makes the labeling of nutrition mandatory and instructs food producers to provide information on energy levels and six nutrients as well as food allergies. According to the FDC Act, Section 201 (g) (1), articles that treat, reduce, cure, or prevent any disease are drugs. However, the Health and Healthcare Labelling (NLEA) Act of 1990 authorizes the FDA to allow certain claims to reduce the risk of disease, known as "health claims," from appearing on food labels. Health claims are exempt from a medical condition, provided that all applicable requirements for each type of claim are met (Sonmez et al., 2014). If this condition isn't fulfilled, the food will be marked as unethical and illegal (Yamada et al., 2020).

11.3.4 THE KEY FACTORS RESPONSIBLE FOR MARKET GROWTH ARE

- Development of work-related antioxidants

- Increasing income
- Increased demand for health care prevention
- Increased awareness of health and wellness
- Constantly innovating
- Increased lifestyle of lifestyle-related diseases
- A growing number of gymnasiums in developing regions
- The rise of the e-commerce industry

11.3.5 COVID-19'S EFFECT ON THE INTERNATIONAL FOOD MARKET

Nutraceutical supply chains are struggling worldwide to keep up with the rapid spread of the COVID-19 virus. Most companies have their own production units in Asia. The disruption of procurement due to the COVID-19 epidemic has led to a decline in the availability of nutritional supplements worldwide (Yamada et al., 2020). The disruptive effects of the new coronavirus epidemic have placed a great strain on the supply of nutritious products worldwide, thereby increasing the risk of scarcity.[26] Although production in all the various industries in Asia has begun slowly again, U.S. manufacturers pharmacists, and nutritionists, who receive the most directly and indirectly from Asia, are at greater risk for shortages (Global Industry Analysts et al., 2018). However, the COVID-19 epidemic has also created growth opportunities for nutritional suppliers as the focus is on the use of supplements such as many vitamins, omega-3 fatty acids, and protein supplements to maintain immunity.

11.3.6 RULES GOVERNING HEALTHY EATING

There are a variety of laws governing Nutraceuticals with different geographical names. The name even varies from region to region from Nutraceuticals to dietary supplements and there are some countries that include nutraceuticals under the food umbrella (Degnan et al., 2018). A dietary supplement is usually an oral supplement made up of a food ingredient intended to be an addition to a diet (Coppens et al., 2018). A number of definitions and terms are used around the world to describe nutraceuticals, such as dietary supplement in the USA; in Canada, it is called Natural Health Product; Australia uses the name "complementary medicines"; EU explains as "food supplements"; and in India, it is known to be "foods used special foods."

11.3.7 REGIONAL DEFINITIONS LEGAL DEFINITIONS ARE BASED ON TWO FACTORS

One is the source or source; another benefit I can give the buyer. The nomenclature of nutraceuticals also differs from noninduced countries: the US describes them as "Food Ingredients," in Canada, they are referred to as "Natural Health Products" and Japan classifies them as "Special Health Food Use (FOSHU)". These definitions could be broad or specific; there are different definitions and regulations of dietary supplements and active foods in the United States, Canada, and Europe, and in Japan, both dietary supplements and active foods are regulated under a set of similar regulations (Coppens et al., 2017). The US and Canada list the product specifications you should have. It is called nutraceutical, and Europe and Japan simply provide general guidelines on the structures by which the product should be named. In Canada, the definition of food and nutrition supplements includes conventional and folk medicines. While the US includes herbs and botanicals in its definition of FOSHU, Japan does not mention traditional medicine under FOSHU (Pettman et al., 2016). The Indian definition (in terms of the Safety and Safety and Food Security Act, 2016) describes the components that a product should contain- and the general nutritional features found in traditional medicines are not included in the description.

11.3.7.1 US REGULATIONS

The US Food and Drug Administration regulates food, supplements, drugs, and cosmetics. Unlike prescription and dietary drugs, nutraceuticals are not subject to FDA approvals; are regarded as "dietary supplements" in terms of the Dietary Supplement, Health and Education Act (DSHEA) 1994. The FDA no longer has the power to regulate such "dietary supplements" as a result of DSHEA and considered them "a separate dietary control category" (Coppens et al., 2016). They do not belong in the food or medication category. A dietary supplement is, according to the DSHEA, "a product (apart from tobacco) intended to supplement foods containing or contain one or more of the ingredients listed: vitamins, minerals, amino acids, herbs, or other botanicals; or a concentrate, metabolite, constituent, extract, or combination of the ingredients listed." (Pettman et al., 2016). In addition, it should also be accompanied by the following methods: intended for the absorption of a capsule, pill, powder, tablet, or liquid that should not be consumed as a regular food or as a single food/food item labeled diet (Degnan et al., 2018).

11.3.7.2 EU REGULATION

In EU food law, the regulatory framework for "active foods" or "nutraceuticals" does not exist. The rules are many and depend on the diet (Yamada et al., 2018). The principles of General Food Law Regulation, which include the obligation for food safety, tracking, recall, and information, apply to all foods (Pettman et al., 2017). A nutrient product must adhere to the regulatory standards for medical products with relation to safety, efficacy, quality inspection, and marketing authorization processes when making a claim about its therapeutic effects and medical applications." For example, claims should not imply that with a nutraceutical diet/drink, the disease will be prevented or treated, only that it can help improve health, perhaps helping to prevent the onset of the disease. Alternative "active food" may be used (Degnan et al., 2020). Novel Food Regulation works on "active play" and "new". In accordance with Directive 2002/46/EC of the European Parliament and Council of 2002 regarding the standardization of related foods in member states, nutraceuticals are specifically regulated as "food ingredients" (Coppens et al., 2019). This establishes harmonious rules for labeling ingredients and introduces certain rules for minerals and vitamins in dietary supplements. The regulations are regulated by the European Food Safety Authority (EFSA)

11.3.7.3 INDIAN REGULATION

As in other countries, Indian law does not provide any legal conditions for the treatment of various diseases using nutraceuticals (Pettman et al., 2017). The Indian government's dietary regulations on the market include the Food and Standards Act (FSSA), passed in 2006, has not yet been put into effect. It consists of eight laws that the state's food security commissioner enforces (Coppens et al., 2020). According to the FSSA, "special diets" are specially designed or designed to meet specific nutritional needs, which are due to a specific physical or living condition or certain diseases and disabilities (Degnan et al., 2019). These foods are presented in such a way that your composition should be very different from the Indian standard (IS) food of the same natural diet, and may include one or more of the ingredients listed below.

- Plants or plant extracts or parts thereof are powdered, concentrated or extracted from water, hydroalcoholic extract or ethyl alcohol, single or compound (Yamada et al., 2018).

- Amounts of minerals, vitamins, metals, proteins, their compounds, or amino acids that do not exceed the Recommended Daily Indigent Grant or enzymes (within permissible limits).
- Items from animals.
- Consumer foods that add to food by increasing the total amount of food.

11.3.7.4 CANADIAN REGULATION

In Canadian law, as "Natural Health Products," nutraceuticals are included in this category. "According to the Natural Health Products Act" introduced in 2004 by Health Canada (Degnan et al., 2017). The law is regulated by the Food and Drugs Authority of Canada (Coppens et al., 2016). The description includes the following products:

- Vitamins and minerals
- Herbs
- Homeopathic remedies
- Traditional Chinese medicine
- Probiotics
- Other substances such as fatty acids and amino acids

11.3.7.5 JAPAN REGULATIONS

Japan is a pioneer in the area of effective food/nutrition control. The concept of Specific Health Food Use (FOSHU) was established in 1991 (Yamada, 2018). The Department of Health and Welfare must approve food that has been certified as FOSHU after receiving complete scientific documentation to back up the claim as a regular element of a diet (Pettman et al., 2016). As a FOSHU rule, a functional diet can have three functions:

- Nutrition
- Sensory satisfaction
- Physically better

11.3.7.6 AUSTRALIAN REGULATIONS

In Australia, medical/food products are called "complementary medicines" and are regulated drugs under the Medical Supplies Act, 1989, which came into effect in 1991 (Coppens et al., 2016).

11.3.7.7 RUSSIA REGULATION

According to a directive from the Russian Ministry of Health and Social Development 1898, dated 15 April 1997, entitled "On the Procedure for the Testing and Certification of Active Dietary Supplements," nutraceuticals are administered under the name Biologically Active Dietary Supplements (BADS) (Coppens et al., 2016). Definitions include nutraceuticals (vitamins, dietary fiber, minerals, and amino acids) and para-pharmaceuticals (bioflavonoids, alkaloids, essential oils, and polysaccharides) (Yamada et al., 2018). Legally, BADS is "a clinically proven diet." They are commended for preventing and preventing the side effects of drug treatment and achieving complete remission (Pettman, 2016). The focus on Russia is largely based on antiretroviral drugs rather than part of a person's daily diet (Degnan et al., 2018).

11.4 OPPORTUNITIES

Human health and well-being are largely determined by the eating of nutritious foods. The emergence of foods with health advantages is an excellent chance to enhance public health; as a result, the scientific community, users, and food processors have paid close attention to these compounds in recent years.

A high intake of fruits, vegetables, and whole grains is associated with a decreased risk of chronic illnesses such as cancer and cardiovascular disorders, which are the leading causes of mortality in Europe, the United States[57] and many other nations.

11.4.1 CANCER

Epidemiological studies have established that dietary variables can influence carcinogenesis. In addition, experimental studies have shown that many bioactive dietary ingredients or natural items can help to avoid cancer.[8,57] Many food ingredients have been discovered to have anti-mutagenic and anti-carcinogenic capabilities, despite the fact that their nutritional benefits are still unknown. Such encouraging research strengthens the case for the use of bioactive dietary components as chemopreventative drugs.[76]

Plants such as garlic, ginger, curcumin, ginseng, ginger, and green tea extract have a variety of anti-oncogenesis activities. Inhibition of DNA alkylation, tumor initiation, proliferation, and metastasis, as well as the stimulation of autophagy and intrinsic apoptosis, are examples of such pathways to get the anticancerous effect of the above examples (Alali et al., 2020).[36]

TABLE 11.1 Some of the Examples of Nutraceuticals which are used for Specific Cancer Cure.

Type of cancer	Nutraceutical	References
Breast cancer	Allicin in garlic	65,98
	Curcumin	52
	Vitamin D	104
Colon cancer	Polyphenols	100
	Terpenoids	41
	Alkaloids	66,111
	Micronutrients	16
Oral cancer	Strawberry	16
	Rosemary	Prinz-langenohl et al., 2001
	Geraniol	112
Prostate cancer	Vitamin D	43
	Catechins in green tea	109

11.4.2 OBESITY

An excessive accumulation of bodily fat distinguishes obesity as a medical condition.[80] According to numerous researches higher amounts of body fat are linked to an increased risk of a number of health problems. Weight loss is becoming more widely recognized as having significant health benefits for overweight persons, and it has been linked to an improvement in the life expectancy of those suffering from obesity-related problems.[34] Because of the presence of arachidonic acid (ALA)/conjugated linolenic acid (CLA) and arachidonic acid (ALA)/docosahexaenoic acid (DHA)/eicosapentaenoic acid (EPA), flaxseed and fish oil are considered useful foods for obesity. The antiobesity qualities of ALA, CLA, DHA, and EPA are far more essential than their other benefits.[103] Asparagus, garlic, leeks, onions, bananas, Jerusalem artichoke, chicory, wheat bran, and barley are all good sources of prebiotics that have promising antiobesity properties.[28] Foods that contained probiotic bacteria as functional ingredients with the goal of improving digestive health. Some particular variants of *Lactobacillus* and *Bifidobacterium* have shown continuous antiobesogenic characteristics throughout most human and animal studies, such as *Lacticaseibacillus casei*, *Lacticaseibacillu rhamnosus*, *Lactobacillus gasseri*, *Lactiplantibacillus plantarum*, *Bifidobacterium infantis*, *Bifidobacterium longum*, and *Bifidobacterium breve*.[24,33]

11.4.3 CARDIOVASCULAR DISEASE

The risk of CVD is enhanced not only by poor dietary habits, but also by lifestyle choices such as smoking and alcohol drinking.[85] It has been discovered that those who consume nutritious foods, live active lifestyles, those who not smoke and not consume excessive alcohol have a lower incidence of CVDs. It is also recognized that diets that increase serum total cholesterol, LDL cholesterol, and triacylglycerol concentrations causes CVDs and their related problems while decreasing HDL cholesterol concentrations reduce the risk of coronary artery disease. When consumed in sufficient quantities, functional foods can help reduce the risk of CVDs through a variety of ways. Lowering blood cholesterol levels, lowering plaque accumulation, minimizing lipoprotein oxidation, enhancing arterial compliance, scavenging free radicals and limiting platelet aggregation are all examples.[34] Ginger extract increased the antioxidant enzyme and decreased oxidative stress in the blood.[71] Beetroot includes betalain and phenolic chemicals, which boost LDL resistance to oxidation, protects the liver from damage, and lower blood pressure.[14,20,99]

11.4.4 DIABETES

Functional foods present within the Mediterranean diet containing polyphenols, terpenoids, flavonoids,[45,47,48] alkaloids, sterols, pigments, and unsaturated fatty acids play a significant role in maintaining wellness, and make a significant contribution to prevent much more disease.[102] More specifically For type 2 diabetes, such foods have been reported to have improved antioxidant, anti-inflammatory, and anticholesterol qualities, as well as improved insulin sensitivity and lowered resistance, all of which are important factors in the prevention, management, and treatment of T2DM.[3]

11.4.5 ECONOMIC SIGNIFICANCE OF NUTRACEUTICALS AND FUNCTIONAL FOODS

Over the past decade, the market for functional foods has expanded and gaining market share that is tentatively predicted to exceed that of organic foods.

Furthermore, demand for functional foods and nutraceuticals is increasing in developing countries, giving a potential opportunity to expand local markets. Nutraceuticals and Functional foods economic returns have the potential to benefit each point of the supply chain, from raw material farmers and processors to retailers. Many poor countries with rich biodiversity and traditional knowledge of the health impacts of specific indigenous plant species may find functional foods to be an opportunity for economic prosperity.

Farming for the functional foods sector can benefit primary producers and rural communities in other ways than diversifying and high-value production. Growing functional food markets through domestication of wild plant species; improved links to the private sector, such as through contract farming employment or business prospects from processing functional foods; and work on plantations can all help poor people, If their sustainable usage is carefully regulated, functional qualities can raise the value of otherwise rare plant species, which can contribute in biodiversity conservation.[107]

11.5 CHALLENGES IN FORMULATING NUTRACEUTICALS

11.5.1 POOR AQUEOUS SOLUBILITY

Aqueous solutions are insoluble for most lipophilic bioactive substances. As a result, the body has a tough time absorbing them. As a result, nutraceuticals' positive properties are mostly ignored. Phytosterols, phytostanols, carotenoids, oil-soluble vitamins, curcumin, and other similar substances are examples of such substances.[110] Most of these cannot be utilized in meals or beverages since they are not water soluble. The most critical challenge in today's high-priced nutraceuticals is complex procedures are necessary to make lipid-based nutraceuticals soluble in watery food products. This effectively raises the cost of producing edible-grade nutraceuticals, making their purchase more expensive. As a result, cost-effective solutions that improve nutraceutical solubility must be developed. As a result, extensive study has already been conducted in that regard, and the water solubility of nutraceuticals has been improved using specialized technology (Salvia-Trujillo et al., 2015).

11.5.2 HIGH MELTING POINT OF NUTRITIONAL SUPPLEMENTS

At room temperature, compounds such as carotenoids, phytosterols, and fatty alcohols are crystalline. This makes incorporating such ingredients into foods difficult, particularly at room temperature. If they are forced into foods, their stability and shelf life are significantly reduced, resulting in an unpleasant appearance, irritating odor, and mouthfeel, and so impacting the market value and customer desire for such goods. Furthermore, such foods have lesser bioavailability, reducing their value. In these situations, the nutraceuticals are often converted to a liquid phase by dissolving them in acceptable edible-grade solvents. After that, they are added to foods. They can also be added as suspended nanocrystals to the relevant food product in a way that does not compromise its shelf life or degrade its flavor or quality (McClements et al., 2012).

11.5.3 CHEMICAL INSTABILITY

The most significant difficulty for the food and nutraceutical industries is the instability of hydrophilic and lipophilic bioactive chemicals, which results in the rapid degradation of these foods under standard storage settings.[82] This makes long-term marketing, storage, and transportation of such products extremely challenging. This is notably true for molecules containing omega-3 fatty acids, such as fish oils, flaxseed oil, and cod liver oils; carotenoids, such as lycopene or curcumin; and so on (McClements and Decker, 2000; Boon et al., 2010; Waraho et al., 2011; Sen Gupta and Ghosh, 2013; Heger et al., 2014). The level of chemical degradation is entirely dependent on the bioactive product's composition, ambient circumstances such as temperature, pH, pressure, and so on, as well as the presence of metals or other oxidation-promoting factors. To protect such molecules from degradation, the creation of nanoscale products is critical.

11.5.4 RAW MATERIALS

If the raw materials are manufactured outside of the United States, several important factors should be kept in mind. The following are the specifics: Importing items from the outside necessitates pre-approval from the US Customs. Once production is set and the manufacturing process is in a steady state of operation, preapproved customs clearance is advantageous since it aids in seamless product flow. It is crucial to understand foreign laws, rules, and regulations so that product imports do not come to a halt and no unexpected complications arise in the middle of the cycle. It's critical to be aware of the product's list of forbidden ingredients that are not permitted in the USA. It is indeed critical to compare the costs of manufacturing locally against manufacturing in another country and then importing the final goods.

Since the products are imported, customs duty may be an additional cost element. Check for import limits—not only on the type of food, but also on the form; for example, depending on whether the product is liquid, powder, or solid, restrictions differ. The vendor's available capacity for quantity and quality production, as well as a specified response time the total time required from placing an order to receiving the semifinished goods from the overseas supplier. The time it takes for items to arrive from overseas and from local sellers, as well as the difference. Imported vs domestic item's total life. Imported goods must outlast domestic goods.[7]

11.5.5 WAREHOUSING

The warehousing department will ensure that the following requirements are completed before storing the material: Verify that the material was received in accordance with the purchase order. Make sure you have all you need for efficient material storage. Materials should be stored in such a way that they are easy to handle and the distance between the handler and the object must be kept to a minimum. Ensure that thorough physical verification is performed on a regular basis. All materials must be labeled, and they must be visible at all times. In the warehouse, at all times, all coverage insurance must cover the entire risk and value of any claim. The insurance should begin as soon as the goods are acquired to ensure that no things are left uninsured. Hazardous, toxic substances, as well as extremely combustible and temperature-sensitive products, must be stored with extreme attention. Storage department personnel must pay close attention to all criteria. Essential for product and staff safety it is necessary to keep objects in a way that prevents the absorption of moisture. The material can only be stored once it has been cleared by the quality assurance department. This section is also in charge of ensuring inventory levels are maintained in the management of numerous incoming materials it's possible that the level will be set based on the information you provide based on a variety of variables, such as overall lead time, EOQ, and transit time for each customer material, and the cost of keeping an inventory on hand. Recognize the items that are slow to move and those that are stationary.[7]

11.5.6 QUALITY

For checking incoming material, the quality department must have a manual. The following is an example of a comprehensive checklist that might be used. For all incoming material, a manual or handbook must be developed, including the procedure for each parameter, process, and method of checking. As soon as the incoming item is reported to the factory/warehouse location, security officers must immediately begin the verification process, which should begin with a check to see if the material is in fact accompanied by the purchase order. Quality, safety, and health report cards, as well as the amount specified in the purchase order, must all be properly examined. After that, the quality control department is involved. Physically inspect the material, then draw a sample at random or according to the quality control manual's requirements, and perform an exhaustive and thorough check based on numerous characteristics. If the inbound material passes all of the manual's standards, the quality control team will give the warehouse team permission to store the merchandise. If the material does not meet the specifications, it should be returned to the vendor with a rejection note explaining reasons.[7]

11.6 CONCLUSION

Nutraceuticals have demonstrated their health advantages and illness prevention capability, and they should be consumed in accordance with their acceptable suggested consumption. Nutraceuticals have the potential to deliver significant health advantages, particularly in the prevention and/or treatment of acute and chronic human disorders, promoting optimal health, lifespan, and quality of life. However, the quality, safety, long-term side effects, and toxicity of the product as well as studies on dietary supplements and human clinical trials will all be factors in its development. Better-characterized and research-proven products can assist boost customer confidence in nutraceutical products around the world. As we should be aware, the functional food industry does not consistently review or apply existing standards. However, diet is merely one component of a total lifestyle approach to good health this should include regular exercise, cigarette avoidance, stress reduction, healthy body weight maintenance, and other positive health habits. Functional foods would not be a part of a successful strategy for enhancing health and reducing disease risk until all of these issues have been resolved.

11.7 FUTURE PERSPECTIVE

The vast majority of the world's population is concerned with maintaining good health, and nutraceuticals play a vital role in this regard. Furthermore, the negative side effects of pharmaceuticals on the market have enhanced their propensity toward such items. Currently, a wide range of nutraceuticals formulations with varied bioactive ranging from vitamins to plant bioactive are available in the global market. The list of nutraceutical goods is growing over time as a result of supportive research linking their use to the prevention of complex clinical illnesses such as cancer, diabetes, cardiovascular disease, and so on. The fewer health risks compared to synthetic medications account for the high consumer acceptance. Customers are expressing a strong interest in various sorts of nutraceuticals for stress relief, boosting energy, mental alertness, preventing metabolic disorders, battling oxidative stress, and so on. To compete and sustain in the global market, nutraceutical companies are focusing on developing new products and attracting customers through advertising in print and online media (Zaden et al., 2014). Countries such as the United States, Germany, China, and the United Kingdom are investing heavily in nutraceutical research in the hopes of reaping large profits in the future, which will be dominated by critical issues such as antibiotic resistance, for which these products may only serve as simple solutions. Currently, in industrialized countries such as the United States, more than half of the population relies on nutraceuticals for day-to-day health maintenance. Nutraceuticals are being developed using technological advances such as nutrigenomics and real-time cellular imaging tools. Such technology interventions in the creation of nutraceutical formulations and the allied food business would have a significant positive impact on healthcare management. Encapsulation technology advancements have enabled many chemicals to be encapsulated in a single capsule. Encapsulation in liquid pellets, oils, granules, and powders can now be included in hard gelatin capsules thanks to advances in nutraceutical formulation technology. This method has the potential to be expanded to isolate and purify nutraceuticals. Various novel delivery techniques, such as nanoparticles, phytosomes, liposomes, and lipid nanoparticles, are gradually emerging to define the future of next-generation nutraceuticals with high efficacy. Enzymes are a unique class of nutraceuticals with enormous potential for future usage. Enzyme-oriented nutraceuticals enabled by modern technologies such as nanotechnology will undoubtedly hold a significant market share in the future. Unique and sophisticated characterization methodologies based on genomes, proteomics, and metabolomics will be investigated for the improvement of nutraceutical goods. Because of developments in nutraceutical formulation technology, encapsulation in liquid pellets, oils, granules, and powders can now be contained in hard gelatin capsules. This technology could be expanded to identify and purify nutraceuticals. Novel delivery strategies, such as nanoparticles, phytosomes, liposomes, and lipid nanoparticles, are gradually emerging to determine the future of high-efficacy next-generation nutraceuticals. Enzymes are a unique class of nutraceuticals with huge future potential. Enzyme-oriented nutraceuticals, made possible by modern technologies such as nanotechnology, will likely have a sizable market share in the future. For the enhancement of nutraceutical products, novel and sophisticated characterization approaches based on genomics, proteomics, and metabolomics will be examined (Jha et al., 2021).

KEYWORDS

- **cancer**
- **regulations**
- **dietary supplements**
- **cardiovascular disease**
- **functional food**
- **nutraceuticals**

REFERENCES

1. Abdel-Aal, E.-S.; Akhtar, H.; Zaheer, K.; et al. Dietary Sources of Lutein and Zeaxanthin Carotenoids and Their Role in Eye Health. *Nutrients* **2013,** *5* (4), 1169–1185. https://doi.org/10.3390/nu5041169
2. AlAli, M.; Alqubaisy, M.; Aljaafari, M. N.; et al. Nutraceuticals: Transformation of Conventional Foods into Health Promoters/Disease Preventers and Safety Considerations. *Molecules* **2021,** *26* (9), 2540 https://doi.org/10.3390/molecules26092540
3. Alkhatib, A.; Tsang, C.; Tiss, A.; et al. Functional Foods and Lifestyle Approaches for Diabetes Prevention and Management. *Nutrients* **2017,** *9* (12), 1310.
4. Alzamora, S. M.; Salvatori, D.; Tapia, M. S.; et al. Novel Functional Foods from Vegetable Matrices Impregnated with Biologically Active Compounds. *J. Food Eng.* **2005,** *67* (1–2), 205–214.
5. Aronson, J. K. Defining Nutraceuticals: Neither Nutritious Nor Pharmaceutical. *Br. J. Clin. Pharmacol.* **2017,** *83*, 8–19.
6. Asp, N. G. International Symposium on Functional Foods in Europe "International Developments in Science and Health Claims". *Nutr. Metab. Cardiovasc. Dis.* **2007,** *17* (2), 170.

7. Bagchi, D.; Sreejayan, N. *Developing New Functional Food and Nutraceutical Products*; Elsevier: Amsterdam, 2017.
8. Balsano, C.; Alisi, A. Antioxidant Effects of Natural Bioactive Compounds. *Curr. Pharm. Des.* **2009,** *15* (26), 3063–3073.
9. Bech-Larsen, T.; Grunert, K. G. The Perceived Healthiness of Functional Foods. *Appetite* **2003,** *40* (1), 9–14.
10. Borchers, A. T.; Keen, C. L.; Gerswin, M. E. The Basis of Structure/Function Claims of Nutraceuticals. *Clin. Rev. Allergy Immunol.* **2016,** *51*, 370–382.
11. Bousquet, J.; Bewick, M.; Cano, A.; et al. Building Bridges for Innovation in Ageing: Synergies between Action Groups of the EIP on AHA. *J. Nutr. Health Aging* **2017,** *21* (1), 92–104.
12. Sosnowska, B.; Penson, P.; Banach, M. The Role of Nutraceuticals in the Prevention of Cardiovascular Disease. *Cardiovasc. Diagn. Ther.* **2017,** 7 (1), S21–S31S2S31.
13. Calvani, M.; Pasha, A.; Favre, C. Nutraceutical Boom in Cancer: Inside the Labyrinth of Reactive Oxygen Species. *Int. J. Mol. Sci.* **2020,** *21* (6), 1936. https://doi.org/10.3390/ijms21061936.
14. Cao, C.; Pathak, S.; Patil, K. In *Antioxidant Nutraceuticals: Preventive and Healthcare Applications*; CRC Press/Taylor & Francis Group: Boca Raton, 2018.
15. Cencic, A.; Chingwaru, W. The Role of Functional Foods, Nutraceuticals, and Food Supplements in Intestinal Health. *Nutrients* **2010,** *2* (6), 611–625.
16. Chang; H.-F.; Yang; L.-L. Gamma-Mangostin, a Micronutrient of Mangosteen Fruit, Induces Apoptosis in Human Colon Cancer Cells. *Molecules* **2012,** *17* (7), 8010–8021.
17. Chaudhari, S. P. Nutraceuticals: A Review. *World J. Pharm. Pharm. Sci.* **2017,** 681–739. https://doi.org/10.20959/wjpps20178-9825.
18. Coppens, P.; da Silva, M. F.; Pettman, S. European Regulations on Nutraceuticals, Dietary Supplements and Functional Foods: A Framework Based on Safety. *Toxicology* **2006,** *221* (1), 59–74.
19. Dadhania, V. P.; Trivedi, P. P.; Vikram, A.; et al. Nutraceuticals Against Neurodegeneration: A Mechanistic Insight. *Curr. Neuropharmacol.* **2016,** *14*, 627–640.
20. Danwilai, K.; Konmun, J.; Sripanidkulchai, B.; et al. Antioxidant Activity of Ginger Extract as a Daily Supplement in Cancer Patients Receiving Adjuvant Chemotherapy: A Pilot Study. *Cancer Manag. Res.* **2017,** *9*, 11–18.
21. Diplock, A. T.; Aggett, P. J.; Ashwell, M.; et al. Scientific Concepts of Functional Foods in Europe: Concensus Document. *Br. J. Nutr.* **1999,** *81* (suppl. 1), S1–S27.
22. Dragsbæk, K.; Neergaard, J. S.; Laursen, J. M.; et al. Metabolic Syndrome and Subsequent Risk of Type 2 Diabetes and Cardiovascular Disease in Elderly Women. *Medicine* **2016,** *95* (36).
23. EC. Regulation (EC) No. 1924/2006 of the European Parliament and of the Council of 20 December 2006 on Nutrition and Health Claims Made on Foods. *Official Journal of the European Union*, 2006, *L 12*, 3–18.
24. Ejtahed, H.-S.; Angoorani, P.; Soroush, A.-R.; et al. Probiotics Supplementation for the Obesity Management; A Systematic Review of Animal Studies and Clinical Trials. *J. Funct. Foods* **2019,** *52*, 228–242.
25. El Khoury, D.; Cuda, C.; Luhovyy, B. L.; et al. Beta Glucan: Health Benefits in Obesity and Metabolic Syndrome. *J. Nutr. Metab.* **2012,** 1–28. https://doi.org/10.1155/2012/851362.
26. El Sohaimy, S. A. Functional Foods and Nutraceuticals-modern Approach to Food Science. *World Appl. Sci. J.* **2012,** *20* (5), 691–708.

27. Fernando, G., Hernández, A. F., Martín Domingo, M. C. Toxic Contamination of Nutraceuticals and Food Ingredients In: *Nutraceuticals: Efficacy, Safety and Toxicity*; Gupta, R. C. Ed.; Academic Press: Boston, MA; Chapter 58, 2016; pp 1–13.
28. Florowska, A.; Krygier, K.; Florowski, T.; et al. Prebiotics as Functional Food Ingredients Preventing Diet-related Diseases. *Food Funct.* **2016,** *7*, 2147–2155.
29. Freedonia Group. World Nutraceutical Ingredients, Industry Study with Forecast for 2013 and 2018; 2009: 1–487.
30. Gebauer, S. K.; Psota, T. L.; Harris, W. S.; et al. N-3 Fatty Acid Dietary Recommendations and Food Sources to Achieve Essentiality and Cardiovascular Benefits. *Am. J. Clin. Nutr.* **2006,** *83* (6 Suppl), 1526S1535S. https://doi.org/10.1093/ajcn/83.6.1526S.
31. Gormley, R. Fish as a Functional Food: Some Issues and Outcomes. *Sea Health UCD* 2013.
32. Grammatikos, A. P. The Genetic and Environmental Basis of Atopic Diseases. *Ann. Med.* **2008,** *40* (7), 482–495. https://doi.org/10.1080/07853890802082096.
33. Green, M.; Arora, K.; Prakash, S. Microbial Medicine: Prebiotic and Probiotic Functional Foods to Target Obesity and Metabolic Syndrome. *Int. J. Mol. Sci.* **2020,** *21* (8), 2890.
34. Gul, K.; Singh, A. K.; Jabeen, R. Nutraceuticals and Functional Foods: The Foods for the Future World. *Crit. Rev. Food Sci. Nutr.* **2015,** *56* (16), 2617–2627.
35. Gupta, R. C. *Nutraceuticals: Efficacy, Safety and Toxicity*; Academic Press; Boston, MA, 2016.
36. Gupta, S. V.; Pathak, Y. In *Advances in Nutraceutical Applications in Cancer: Recent Research Trends and Clinical Applications*; CRC Press: Boca Raton, 2020.
37. Hallikainen, M. A.; Sarkkines, E. S.; Gylling, H.; et al. Comparison of the Effect of Plant Sterol Ester and Plant Stanol Ester-enriched Margarines in Lowering Serum Cholesterol Concentration in Hypercholesterolaemia Subjects on a Low-fat Diet. *Eur. J. Clin. Nutr.* **2000,** *54*, 715–725.
38. Hasler, C. M. Functional Foods: Benefits, Concerns and Challenges—A Position Paper from the American Council on Science and Health. *J. Nutri.* **2002,** *132* (12), 3772–3781.
39. Heger, M.; van Golen, R. F.; Broekgaarden, M.; et al. The Molecular Basis for the Pharmacokinetics and Pharmacodynamics of Curcumin and Its Metabolites in Relation to Cancer. *Pharmacol. Rev.* **2013,** *66* (1), 222–307. https://doi.org/10.1124/pr.110.004044.
40. Isemura, M. Catechin in Human Health and Disease. *Molecules* **2019,** *24* (3). https://doi.org/10.3390/molecules24030528.
41. Jayaprakasha, G. K.; Murthy, K. N. C.; Uckoo, R. M.; et al. Chemical Composition of Volatile Oil from Citrus Limettioides and Their Inhibition of Colon Cancer Cell Proliferation. *Ind. Crops Prod.* **2013,** *45*, 200–207.
42. Kalra, E. K. Nutraceutical--Definition and Introduction. *AAPS Pharm. Sci.* **2003,** *5* (3), E25. https://doi.org/10.1208/ps050325.
43. Karlsson, S.; Olausson, J.; Lundh, D.; et al. Vitamin D and Prostate Cancer: The Role of Membrane Initiated Signaling Pathways in Prostate Cancer Progression. *J. Steroid Biochem. Mol. Biol.* **2010,** *121* (1–2), 413–416.
44. Kelsey, N. A.; Wilkins, H. M.; Linseman, D. A. Nutraceutical Antioxidants as Novel Neuroprotective Agents. *Molecules* **2010,** *15* (11), 7792–7814. https://doi.org/10.3390/molecules15117792.

45. Keservani, R. K.; Sharma, A. K. Flavonoids: Emerging Trends and Potential Health Benefits. *J. Chin. Pharm. Sci.* **2014,** *23* (12), 815.
46. Keservani, R. K.; Kesharwani, R. K.; Vyas, N.; et al. Nutraceutical and Functional Food as Future Food: A Review. *Der. Pharmacia Lettre* **2010a,** *2* (1), 106–116.
47. Keservani, R. K.; Sharma, A. K.; Kesharwani, R. K. Nutraceutical and Functional Foods for Cardiovascular Health. In *Food Process Engineering*; Apple Academic Press, CRC Press: 2016a; pp 257–278, 291–312. ISBN: 9781771884020.
48. Keservani, R. K.; Sharma, A. K.; Kesharwani, R. K. Medicinal Effect of Nutraceutical Fruits for the Cognition and Brain Health. *Scientifica*, **2016b,** 10. Article ID 3109254. doi:10.1155/2016/3109254.
49. Keservani, R. K.; Sharma, A. K.; Kesharwani, R. K. An Overview and Therapeutic Applications of Nutraceutical and Functional Foods. *J. Drug Deliv. Sci. Technol.* **2017,** 160–201. ISBN:9781522507543.
50. Keservani, R. K.; Sharma, A. K.; Kesharwani, R. K., Eds.; *Nutraceuticals and Dietary Supplements: Applications in Health Improvement and Disease Management*; CRC Press: 2020. ISBN: 9781771888738.
51. Keservani, R. K.; Kesharwani, R. K.; Sharma, A. K.; et al. Nutritional Supplements: An Overview. *Int. J. Curr. Pharm.* **2010b,** *1* (1), 59–75.
52. Kronski, E.; Fiori, M. E.; Barbieri, O.; et al. MIR181B Is Induced by the Chemopreventive Polyphenol Curcumin and Inhibits Breast Cancer Metastasis via down-Regulation of the Inflammatory Cytokines CXCL1 and -2. *Mol. Oncol.* **2014,** *8* (3), 581–595.
53. Kumar Jha, S.; Roy, P.; Chakrabarty, S. Nutraceuticals with Pharmaceutics: Its Importance and Their Applications. *Int. J. Drug Dev. Res.* **2021,** 13.
54. Kwak, N.-S.; Jukes, D. J. Functional Foods. Part 1: The Development of a Regulatory Concept. *Food Control* **2001a,** *12* (2), 99–107.
55. Kwak, N. S.; Jukes, D. J. Functional food. Part 2: The Impact on Current Regulatory Terminology. *Food Control* **2001b,** 12, 109–117.
56. Limer, J. L.; Speirs, V. Phyto-Oestrogens and Breast Cancer Chemoprevention. *Breast Cancer Res.* **2004,** *6* (3). https://doi.org/10.1186/bcr781.
57. Liu, R. H. Potential Synergy of Phytochemicals in Cancer Prevention: Mechanism of Action. *J. Nutr.* **2004,** *134* (12).
58. Maha, Z. R.; Hanan, F. A. Nutraceuticals and Their Nanotechnology-Based Therapeutic Applications in Different Diseases. *Int. J. Biol. Med.* **2020,** 15–32. https://doi.org/10.36811/ijbm.2020.110020.
59. Mannucci, C.; Casciaro, M.; Sorbara, E. E.; et al. Nutraceuticals against Oxidative Stress in Autoimmune Disorders. *Antioxidants* **2021,** *10* (2), 261. https://doi.org/10.3390/antiox10020261.
60. Marseglia, G.; Licari, A.; Giorgio C. Complementary Treatment of Allergic Rhino conjunctivitis: The Role of the NutraceuticalLertal(R). *Acta Biomed. Ateneo Parm* **2021,** *91* (1), 97–106.
61. Mathioudakis, B. In *Regulation on Nutrition and Health Claims: Current Status and Beyond.* International Developments in Science & Health Claims. ILSI Europe International Symposium on Functional Foods, 2007.
62. Mazzanti, G.; Moro, P. A.; Raschi, E.; et al. Adverse Reactions to Dietary Supplements Containing Red Yeast Rice: Assessment of Cases from the Italian Surveillance System. *Br. J. Clin. Pharmacol.* **2017,** *83*, 894–908.

63. McCarville, J. L.; Caminero, A.; Verdu, E. F. Novel Perspectives on Therapeutic Modulation of the Gut Microbiota. *Ther. Adv. Gastroenterol.* **2016,** *9*, 580–593.
64. McClements, D. J. Crystals and Crystallization in Oil-In-Water Emulsions: Implications for Emulsion-Based Delivery Systems. *Adv. Colloid Interface Sci.* **2012,** *174*, 1–30. https://doi.org/10.1016/j.cis.2012.03.00
65. Modem, S.; DiCarlo, S. E.; Reddy, T. R. Fresh Garlic Extract Induces Growth Arrest and Morphological Differentiation of MCF7 Breast Cancer Cells. *Genes & Cancer* **2012,** *3* (2), 177–186.
66. Mondal, A.; Gandhi, A.; Fimognari, C.; et al. Alkaloids for Cancer Prevention and Therapy: Current Progress and Future Perspectives. *Eur. J. Pharmacol.* **2019,** *858*, 172472
67. Borkar, N.; Saurabh, S. S.; Rathore, K. S.; et al. An Insight on Nutraceuticals. *Pharma Tutor* **2015,** *3* (8), 13–23.
68. Nagpal, R.; Kumar, A.; Kumar, M.; et al. Probiotics, Their Health Benefits and Applications for Developing Healthier Foods: A Review. *FEMS Microbiol. Lett.* **2012,** *334* (1), 1–15. https://doi.org/10.1111/j.1574-6968.2012.02593.
69. Nally, J. D. *Good Manufacturing Practices for Pharmaceuticals*, Sixth edn; CRC Press: New Vernon, NJ, 2016.
70. Nastri, L.; Moretti, A.; Migliaccio, S.; et al. Do Dietary Supplements and Nutraceuticals Have Effects on Dental Implant Osseointegration? A Scoping Review. *Nutrients* **2020,** *12* (1). https://doi.org/10.3390/nu12010268.
71. Neha, K.; Haider, M. R.; Pathak, A.; et al. Medicinal Prospects of Antioxidants: A Review. *Eur. J. Med. Chem.* **2019,** *178*, 687–704.
72. Niva, M. 'All foods affect health': Understandings of Functional Foods and Healthy Eating Among Health-oriented Finns. *Appetite* **2007**, *48*, 384–393.
73. Nwosu, O. K.; Ubaoji, K. I. Nutraceuticals: History, Classification and Market Demand. In *Functional Foods and Nutraceuticals*; 2020, 13–22.
74. Ohama, H.; Ikeda, H.; Moriyama, H. Health Foods and Foods with Health Claims in Japan. *Toxicology* **2006,** 22, 95–111.
75. Palthur, M. P.; Palthur, S. S.; Chitta, S. K. Nutraceuticals: A Conceptual Definition. *Int. J. Pharm. Pharm. Sci.* **2010,** *2* (3), 19–27.
76. Pan, M.-H.; Ghai, G.; Ho, C.-T. Food Bioactives, Apoptosis, and Cancer. *Mol. Nutr. Food Res.* **2007,** *52* (1), 43–52.
77. Pandey, M. K.; Gupta, S. C.; Karelia, D.; et al. Dietary Nutraceuticals as Backbone for Bone Health. *Biotechnol. Adv.* **2018,** *36* (6), 1633–1648. https://doi.org/10.1016/j.biotechadv.2018.03.014.
78. Pang, G.; Xie, J.; Chen, Q.; et al. How Functional Foods Play Critical Roles in Human Health. *Food Sci. Hum. Wellness* **2012,** *1* (1), 26–60.
79. Pellegrini, M.; Senni, C.; Bernabei, F.; et al. The Role of Nutrition and Nutritional Supplements in Ocular Surface Diseases. *Nutrients* **2020,** *12* (4), 952. https://doi.org/10.3390/nu12040952.
80. Piché, M.-E.; Tchernof, A.; Després, J.-P. Obesity Phenotypes, Diabetes, and Cardiovascular Diseases. *Circ. Res.* **2020,** *126* (11), 1477–1500.
81. Pinto da Costa, J. A Current Look at Nutraceuticals—Key Concepts and Future Prospects. *Trends Food Sci. Technol.* **2017,** *62*, 68–78.

82. Rao, J.; Decker, E. A.; Xiao, H.; et al. Nutraceutical Nanoemulsions: Influence of Carrier Oil Composition (Digestible versus Indigestible Oil) on β-Carotene Bioavailability. *J. Sci. Food Agric.* **2013,** *93* (13), 3175–3183. https://doi.org/10.1002/jsfa.6215.
83. Raut, H.; Lodhi, G. N.; Kalode, D. P. A General Review on Nutraceuticals: Its Golden Health Impact Over Human Community. *Int. J. Adv. Res.* **2021,** *9* (1), 1121–1125. https://doi.org/10.21474/ijar01/12399
84. Rautiainen, S.; Manson, J. E.; Lichtenstein, A. H.; et al. Dietary Supplements and Disease Prevention—A Global Overview. *Nat. Rev. Endocrinol.* **2016,** *1*, 407–420.
85. Riccioni, G.; Mancini, B.; Di Ilio, E.; et al. Protective Effect of Lycopene in Cardiovascular Disease. *Eur. Rev. Med. Pharmacol. Sci.* **2008,** *12*, 183–190.
86. Rissanen, T. H.; Voutilainen, S.; Virtanen, J. K.; et al. Low Intake of Fruits, Berries and Vegetables Is Associated with Excess Mortality in Men: The Kuopio Ischaemic Heart Disease Risk Factor (KIHD) Study. *J. Nutri.* **2003,** *133* (1), 199–204. https://doi.org/10.1093/jn/133.1.199.
87. Roberfroid, M. B. Global View on Functional Foods: European Perspectives. *Br. J. Nutr.* **2002,** *88*, S133–S138.
88. Ruhee, R. T.; Roberts, L. A.; Ma, S.; et al. Organosulfur Compounds: A Review of Their Anti-Inflammatory Effects in Human Health. *Front. Nutr.* **2020,** 7. https://doi.org/10.3389/fnut.2020.00064.
89. Salvia-Trujillo, L.; Qian, C.; Martín-Belloso, O.; et al. Influence of Particle Size on Lipid Digestion and β-Carotene Bioaccessibility in Emulsions and Nanoemulsions. *Food Chem.* **2013,** *141* (2), 1472–1480. https://doi.org/10.1016/j.foodchem.2013.03.050.
90. Santini, A.; Novellino, E. Nutraceuticals in Hypercholesterolaemia: An Overview. *Br. J. Pharmacol.* **2017,** *174*, 1450–1463.
91. Santini, A.; Tenore, G. C.; Novellino, E. Nutraceuticals: A Paradigm of Proactive Medicine. *Eur. J. Pharm. Sci.* **2017,** *96*, 56–61.
92. Shahidi, F. Nutraceuticals and Functional Foods: Whole versus Processed Foods. *Trends Food Sci. Tech.* **2009,** *20* (9), 376–387.
93. Shida, M.; Hara, M.; Fukino, N.; et al. Glucosinolate Metabolism, Functionality and Breeding for the Improvement of Brassicaceae Vegetables. *Breed. Sci.* **2014,** *64* (1), 48–59. https://doi.org/10.1270/jsbbs.64.48.
94. Siró, I.; Kápolna, E.; Kápolna, B.; et al. Functional Food. Product Development, Marketing and Consumer Acceptance—a Review. *Appetite* **2008,** *51* (3), 456–467.
95. Southan, C.; Sharman, J. L.; Benson, H. E.; et al. The Iuphar/BPS Guide to Pharmacology in 2016: Towards Curated Quantitative Interactions between 1300 Protein Targets and 6000 Ligands. *Nucleic Acids Res.* **2015,** *44* (D1).
96. Stanton, C.; Ross, R. P.; Fitzgerald, G. F.; et al. Fermented Functional Foods Based on Probiotics and Their Biogenic Metabolites. *Curr. Opin. Biotech.* **2005,** *16*, 198–203.
97. Story, E. N.; Kopec, R. E.; Schwartz, S. J.; et al. An Update on the Health Effects of Tomato Lycopene. *Annu. Rev. Food Sci. Technol.* **2010,** *1* (1), 189–210. https://doi.org/10.1146/annurev.food.102308.124120.
98. Talib, W. H. Consumption of Garlic and Lemon Aqueous Extracts Combination Reduces Tumor Burden by Angiogenesis Inhibition, Apoptosis Induction, and Immune System Modulation. *Nutrition* **2017,** *43–44*, 89–97.
99. Tesoriere, L.; Allegra, M.; Butera, D.; et al. Absorption, Excretion, and Distribution of Dietary Antioxidant Betalains in Ldls: Potential Health Effects of Betalains in Humans. *Am. J. Clin. Nutri.* **2004,** *80* (4), 941–945.

100. Umesalma, S.; Sudhandiran, G. Differential Inhibitory Effects of the Polyphenol Ellagic Acid on Inflammatory Mediators NF-KB, Inos, COX-2, TNF-α, and IL-6 in 1,2-Dimethylhydrazine-Induced Rat Colon Carcinogenesis. *Basic Clin. Pharmacol. Toxicol.* **2010,** *107* (2), 650–655.
101. van der Zanden, L. D. T.; van Kleef, E.; de Wijk, R. A.; et al. Knowledge, Perceptions and Preferences of Elderly Regarding Protein-Enriched Functional Food. *Appetite* **2014,** *80,* 16–22. https://doi.org/10.1016/j.appet.2014.04.025.
102. Vasto, S.; Barera, A.; Rizzo, C.; et al. Mediterranean Diet and Longevity: An Example of Nutraceuticals? *Curr. Vasc. Pharmacol.* **2014,** *12*, 735–738.
103. Venkatakrishnan, K.; Chiu, H.-F.; Wang, C.-K. Extensive Review of Popular Functional Foods and Nutraceuticals against Obesity and Its Related Complications with a Special Focus on Randomized Clinical Trials. *Food Funct.* **2019,** *10* (5), 2313–2329.
104. Wang, Q.; Lee, D.; Sysounthone, V.; et al. 1,25-Dihydroxyvitamin D3 and Retonic Acid Analogues Induce Differentiation in Breast Cancer Cells with Function- and Cell-Specific Additive Effects. *Breast Cancer Res. Treat.* **2001,** *67* (2), 157–168.
105. Weingartner, O.; Bohm, M.; Laufs, U. Controversial Role of Plant Sterol Esters in the Management of Hypercholesterolaemia. *Eur. Heart J.* **2008,** *30* (4), 404–409. https://doi.org/10.1093/eurheartj/ehn580.
106. Wildman, R. E.; Wildman, R.; Wallace, T. C. *Handbook of Nutraceuticals and Functional Foods*, CRC Press: 2016.
107. Williams, M.; Pehu, E.; Ragasa, C. In *Functional Foods: Opportunities and Challenges for Developing Countries*; 2006.
108. Yamada, K.; Sato-Mito, N.; Nagata, J.; et al. Health Claim Evidence Requirements in Japan. *J. Nutr.* **2008.**
109. Yang, K.; Gao, Z.-Y.; Li, T.-Q.; et al. Anti-Tumor Activity and the Mechanism of a Green Tea (Camellia Sinensis) Polysaccharide on Prostate Cancer. *Int. J. Biol. Macromol.* **2019,** *122*, 95–103.
110. Yi, J.; Li, Y.; Zhong, F.; et al. The Physicochemical Stability and in Vitro Bioaccessibility of B-carotene in Oil-In-Water Sodium Caseinate Emulsions. *Food Hydrocoll.* **2014,** *35*, 19–27. https://doi.org/10.1016/j.foodhyd.2013.07.025.
111. Zhou, J.; Feng, J.-H.; Fang, L. A Novel Monoterpenoid Indole Alkaloid with Anticancer Activity from Melodinus khasianus. *Bioorg. Med. Chem. Lett.* **2017,** *27* (4), 893–896.
112. Zhu, X.; Xiong, L.; Zhang, X.; et al. Lyophilized Strawberries Prevent 7,12-Dimethylbenz[α]Anthracene (DMBA)-Induced Oral Squamous Cell Carcinogenesis in Hamsters. *J. Funct. Foods* **2015,** *15*, 476–486.

CHAPTER 12

NUTRACEUTICALS MANUFACTURING CHALLENGES IN FAST GROWING FOOD SEGMENT IN INDIA

MANISH KUMAR,[1] SHIV SHANKAR PANDEY,[2] PUSHPENDRA KUMAR,[3] VIMAL KUMAR YADAV,[4] and AJAY KUMAR SHUKLA[4]

[1]*Department of Pharmacy, Madhav University, Pindwara, Rajasthan, India*

[2]*Tahira Institute of Medical Sciences, Gorakhpur, Uttar Pradesh, India*

[3]*Faculty of Pharmacy, Uttar Pradesh University of Medical Sciences, Saifai, Etawah, Uttar Pradesh, India*

[4]*Institute of Pharmacy, Dr. Ram Manohar Lohia Avadh University, Ayodhya, India*

ABSTRACT

Nutraceuticals have exploded in popularity in India in recent years, particularly for protective and positive health care and sickness management. The words "nutriceuticals" are made up of two words: "nutrient and pharmaceutica," nutrient refers to a healthy, useful dietary component, and "pharmaceutical," refers to a therapeutic solution. Nutraceuticals are increasingly finding their way into home cupboards for many reasons, such as preventive healthcare, health management, skincare haircare, beauty, and wellness. The scientific report 2018 predicts that the entire Indian nutraceuticals market will grow at a Compound Annual Growth Rate (CAGR) of 21%, from $4 billion in 2017 to $18 billion in 2025. The current chapter focused on the

Formulations, Regulations, and Challenges of Nutraceuticals.
Tingirikari Jagan Mohan Rao, Rajesh K. Kesharwani, Raj K. Keservani, Anil Kumar Sharma (Eds.)

importance of healthy diet, health issues related to unhealthy diet, the development of new nutraceuticals, functional foods, and food supplements, and understanding of the analytical, formulation, and regulatory requirements of nutraceuticals. This chapter also covered purpose, scope, and significance of nutraceuticals use and challenges to the government for implementing guidelines and regulations over purity of active ingredients, and rationalized use of nutraceutical.

12.1 INTRODUCTION

Hippocrates (founder of western medicine) identified food as medicine in the 5th century BC, saying, "Let food be thy medicine.[4]" Dr. Stephen L. DeFelice, the founder and chairman of the Foundation of Innovation Medicine (FIM) in Crawford, New Jersey, was credited with coining the term "nutraceutical." Since Dr. DeFelice coined the term, Health Canada has changed the definition to include a product removed or purified from foods, often supplied in therapeutic forms not generally related to food, and proven to supply a physiological benefit or protection against chronic disease. Betacarotene and lycopene are two examples.[2,14,47,61,62] The following are the factors that led to the switch to nutraceuticals:

1. Customers are becoming increasingly concerned about growing healthcare expenditures.
2. People who are dissatisfied with drugs' capacity to promote health are turning to nutraceuticals to improve their health and prevent chronic disease.
3. Health-care practitioners realize that our excessively processed food lacks nutrients necessary for optimal health and is made from crops farmed with chemical fertilizers, pesticides, herbicides, and genetically modified seeds.
4. Those who believe that prevention is preferable to cure.
5. People suffering with chronic diseases for which allopathic therapy has not been able to provide a solution.
6. Patients who are financially weak and believe in herbal treatment.

The phrase "Nutraceutical" was coined in 1989 and is derived from the words "nutrition" and "pharmaceutical." Despite being heavily utilized in marketing, it lacks a regulatory definition.

12.1.1 CATEGORIZATION OF NUTRACEUTICALS

Nutraceuticals[57] is a broad word for goods generated from food sources that give considerable health advantages in addition to nutrition. Table 12.1 compares nutraceuticals with pharmaceuticals.

TABLE 12.1 Comparison between Nutraceuticals with Pharmaceuticals.

Nutraceuticals	Pharmaceuticals
Nutraceuticals are primarily concerned with health prevention and maintenance.	Pharmaceuticals are largely concerned with illness, disease, and treatment.
Although it is frequently used in marketing, there is no regulatory (FDA) guideline.	Pharmaceuticals must be monitored by a physician and approved by the FDA (FDA).
Nutraceuticals target the disease or problem's root cause.	Pharmaceuticals treat symptoms and manifestations in a specific portion of the body or the entire body.
Nutraceuticals take a long time to demonstrate their effects.	Pharmaceuticals are usually instantaneous or deliver results in a short period of time.
Nutraceuticals are sold in drug stores, health food stores, supermarkets, online businesses, and the internet without a prescription.	Pharmaceuticals must be dispensed with a prescription from a certified doctor to be marketed (except OTC).
Neutraceuticals have little, if any, negative side effects.	The majority of pharmaceuticals have adverse effects.

Different forms of nutraceuticals exist. Nutraceuticals are nonspecific biological medicines that assist individuals to improve their well-being, preventing cancer and managing symptoms. They can be divided into the three major types listed below.

1. The term "nutrients" refers to compounds with proven nutritional functions, such as fatty acids, minerals, and amino acids.
2. Concentrates and extracts of herbs or botanical items—Botanicals.
3. Reagents from various sources, such as pyruvate, chondroitin sulfate, and steroid hormone precursors. These are used for particular purposes, like meal replacements, weight-loss supplements, and sports nutrition. Figure 12.1 displays the nutraceuticals based on the market availability of foods.

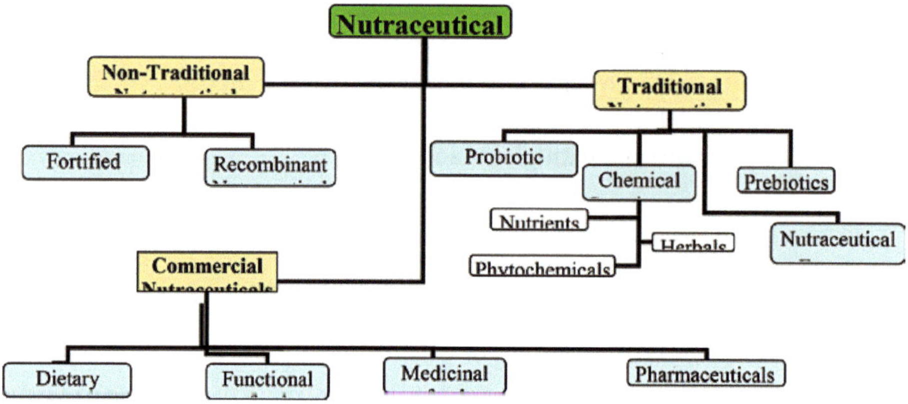

FIGURE 12.1 Nutraceuticals categorized based on foods availability in market.

12.1.1.1 DIETARY SUPPLEMENTS

Dietary supplements are usually sold in pill form such as vitamin B capsules. Commonly dietary supplement capsules contain nutrients derived from concentrated foods. A comprehensive definition of a dietary supplement was created by the Dietary Supplement Health and Education Act (DSHEA), which was passed in 1994. A product that is consumed orally and contains a "dietary ingredient" is known as a dietary supplement. These items contain a variety of "dietary elements," including vitamins, minerals, herbs or other botanicals, amino acids, enzymes, organ tissues, glandulars, and metabolites. Dietary supplements come in a number of forms, such as tablets, capsules, soft gels, gelcaps, liquids, and powders, and are extracts or concentrations."

12.1.1.2 FUNCTIONAL FOODS

Instead of using liquid or capsule-based nutritional supplements, functional foods are designed to allow customers to enjoy enriched foods in their natural state. The process of enriching or fortifying functional meals is referred to as nitrification. This procedure restores a product's nutritional content to levels similar to those before processing. Vitamin D and other additional nutrients are sometimes added to milk.

12.1.1.3 FOOD AS MEDICINE

The founder of Western medicine, Hippocrates, promoted the therapeutic benefits of eating. The use of foods to treat and prevent disease has been documented by a number of civilizations, including the Sumerians, Egyptians, Chinese, and Indians. The medicinal effects of food have been addressed in Ayurveda, a 5000-year-old ancient Indian health science. According to historical records, people have been studying food's therapeutic properties for thousands of years. Considered the father of Western medicine, Hippocrates advised people to "let food be thy medicine."

12.1.1.4 FARMACEUTICALS

Farmaceuticals is a mashup of the phrases farm and pharmaceuticals," according to a report written for the US Congress titled "Agriculture: A Glossary of Terms, Programs, and Laws." It refers to chemicals produced from genetically engineered plants or animals that have medicinal benefit (usually through biotechnology). Advocates claim that employing plants and even animals as pharmaceutical factories might be much more affordable than using currently used techniques.

12.1.2 NUTRACEUTICALS ARE DIVIDED INTO CATEGORIES BASED ON

12.1.2.1 CHEMICAL CONSTITUENTS

A. Nutrients
B. Natural sources
C. Plant constituents

Phytochemicals are essentially plant nutrients with specific biological properties that improve human health.

1. A substrate for biological reactions.
2. Cofactors in enzyme reactions.
3. Enzymatic reaction inhibitors.
4. Absorbents that bind to and remove unwanted gastrointestinal components.
5. Boost the stability or absorption of vital nutrients.
6. Bacterial growth factor that is beneficial
7. Fermentation substrate for beneficial bacteria
8. Selective inhibitors of pathogenic microorganisms in the intestine.
9. Reactive or hazardous chemical scavengers.
10. Ligands that agonist or antagonist receptors on the cell surface or inside the cell.[11]

12.1.2.2 PROBIOTIC MICROORGANISMS

Pathogens including yeasts, other bacteria, and viruses that could normally make people sick are eliminated by the symbiotic interaction with the human gastrointestinal system. By altering the microflora, preventing pathogen adhesion to the intestinal epithelium, competing for nutrients necessary for pathogen survival, producing antitoxin, and reversing some of the effects of infection on the intestinal epithelium, such as secretory changes and neutrophil migration, they fight infection. They combat infection by altering the intestinal microflora, blocking pathogen adhesion to the intestinal epithelium, producing antitoxin, competing for nutrients required for pathogen survival, and reversing the effects of infection on the intestinal epithelium, such as secretory changes and neutrophil migration. Lactose intolerance can be treated with probiotics, which produce a specialized enzyme (ß-galactosidase) that hydrolyzes the harmful lactose into its component sugars. The criteria such as safety, functionality, and technology should be considered while selecting probiotic benchmarks.

12.1.2.3 NUTRACEUTICAL ENZYMES

Enzymes are necessary for life; without them, our body would stop working. Those who have hypoglycemia, blood sugar imbalances, digestive disorders, and obesity may find that adding enzyme supplements to their diet helps to alleviate their symptoms. These enzymes can be found in bacteria, plants, and animals, among other places.

12.1.2.4 PREBIOTICS

Prebiotics are substances that humans do not digest and are a relatively new addition to our lexicon. Rather, they operate as a source of food for healthy probiotic bacteria. This encourages the growth of probiotic bacteria in a favorable environment, reducing the chances of harmful germs infiltrating our digestive tract. Prebiotic inulin is present in a variety of processed foods. It is a type of fiber derived from the roots of plants such as chicory, Jerusalem artichoke, and dandelions.[46]

12.1.2.5 NONTRADITIONAL NUTRACEUTICALS

Is biotechnology employed in the manufacturing of synthetic foods? Bioactive components found in food samples have been used to generate products that promote human health. They are arranged in a specific manner.

1. Nutraceuticals with added vitamins and minerals.
2. Nutraceuticals made from recombinant DNA.

12.1.2.6 FORTIFIED NUTRACEUTICALS

They are fortified with vitamins and minerals, sometimes up to 100% of the Dietary Reference Intake for each item. It consists of essential ingredients like folic acid-fortified ingredients and fortified foods produced via agricultural breeding. Cholecalciferol-fortified milk, for example, is used to treat vitamin D insufficiency.[36]

12.1.2.7 RECOMBINANT NUTRACEUTICALS

Other foods that provide energy include bread, wine, fermented starch, yoghurt, cheese, and vinegar. These foods are all made with the biotechnology method. Biotechnology allows for the production of probiotics as well as the extraction of bioactive components through enzyme/fermentation techniques and genetic engineering.

12.1.2.8 COMMERCIAL NUTRACEUTICALS

Finding an innovative chemical is more difficult, expensive, and harmful than it has ever been. Because of the large and rapidly growing industry, various pharmaceutical companies are now seeking to develop nutraceuticals. Only a few of the therapeutic categories addressed by nutraceuticals comprise anti-arthritis, cold and cough, sleeping difficulties, digestion, certain cancers, osteoporosis, blood pressure, cholesterol management, painkillers, depression, and diabetes prevention.[40–43] One of the most promising findings in human nutrition and health research in the last three decades has been the revelation of health advantages of eating omega-3-rich seafood. Commercial nutraceuticals include followings.

12.1.2.9 MEDICINAL FOOD

Medicinal food is designed to be ingested or delivered internally while being closely supervised by a medical professional. It is intended for the particular nutritional management of a condition or sickness. It requires specific nutritional support. Medicinal food is based on well-established scientific principles and contains a specific nutrient that the body cannot produce normally due to a disease or condition. It is widely prescribed for phenylketonuria, coeliac disease, and lactose intolerance, among other health problems that cause problems with normal meal intake, digestion, absorption, or metabolism.[3]

12.1.2.10 FUNCTIONAL FOODS

"Any food or food ingredient that might offer medical benefits in addition to the usual nutrients is referred to as a functional food," according to a widely accepted definition. Rather than using liquid or capsule-based nutritional supplements, functional foods are created to allow consumers to consume enriched foods in their natural state. The process of enriching or fortifying functional meals is known as nitrification. This procedure restores a product's nutritional content to levels similar to what it was before it was processed. Vitamin D and other additional nutrients are sometimes added to milk. According to Health Canada, "Typical food that has been enhanced with components or substances for purposes beyond than merely nutritional ones, such as a specific physiological or medicinal benefits." All functional meals in Japan must adhere to three principles:

1. They are available in their natural state rather than as a capsule, pill, or powder;
2. Included in the diet daily;
3. Aimed at controlling or preventing disease through the regulation of a biological process.[13]

Dietary fibers are divided into two categories.

1. Water-insoluble fibers.
2. Water-soluble fibers.

A daily intake of 30–40 g is recommended.

12.1.2.11 LIPIDS

Lipids are sources of high energy for the body. Lipids or fats divided into following types are:

(1) Monosaturated (MFA) and (2) polyunsaturated fatty acids (PUFA) like Eicosapentaenoic acid (EPA) Docosahexaenoic acid (DHA).

Palmitic, lauric, and myristic acids are important cholesterol-raising fatty acids in current diets. The high cholesterol and saturated fat of the Eskimos' diet makes them resistant to heart disease. Fish and soybean oils are rich in linolenic acid. Linoleic acid content is too high in corn and soybean oils.

12.1.2.12 NUTRACEUTICALS AND DISEASES

As new treatments emerge for conditions like Parkinson's, Alzheimer's, obesity, diabetes, cancer, osteoporosis, and coronary heart disease, nutraceuticals are swiftly gaining popularity. According to data, a variety of biological functions, including the activation of antioxidant defenses, signal transduction pathways, the expression of genes related to cell survival, cell proliferation and differentiation, and the preservation of mitochondrial integrity, have been linked to the mechanistic actions of natural compounds. These revolutionary ideas could help prevent the signs of many age-related or chronic diseases. It is crucial to remember that the nutrients included in many foods, fruits, and vegetables generate the well-documented health advantages. For instance, lutein and zeaxanthin shelter the eyes against cataracts and macular degeneration; β-carotene and lycopene shield the skin from UV radiation damage. Lycopene may aid in the prevention of prostate cancer, and it may support cardiovascular health. A few of the pharmacological benefits that nutraceutical products provide include antiaging, protection against some chronic diseases, body homeostasis, cardiovascular diseases, neurodegenerative diseases, metabolic disorders like diabetes, degenerative cancer, protein deficiency, ophthalmic complications, allergic problems, and Parkinsonism. Figure 12.2 provides an overview of nutritional supplements' use in disease management.[37]

12.2 SCOPE AND OPPOURTUNITY IN INDIAN NUTRACEUTICAL AND GLOBAL MARKETS

At a CAGR of 13%, the nutraceutical industry in India is anticipated to reach US$2731 million in 2016 (Biospectrum Asia Edition, March 16, 2012). According to a poll conducted by the corporate research and consulting firm Frost & Sullivan, functional foods would experience the fastest growth

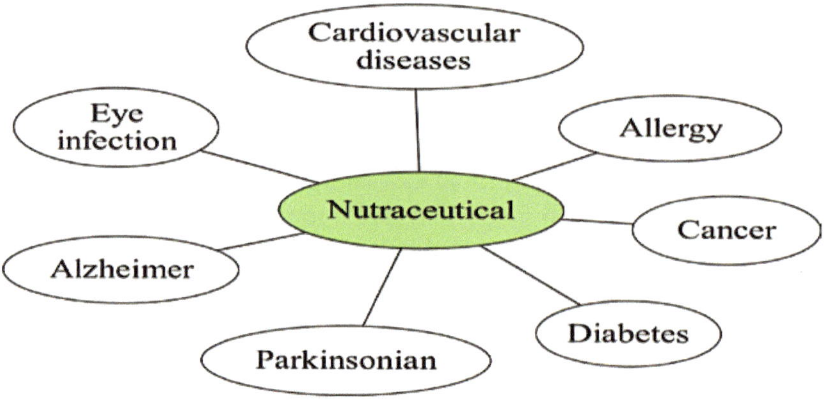

FIGURE 12.2 Nutraceuticals in disease management.

through 2015, followed by dietary supplements. The survey revealed the most lucrative business for nutraceutical companies would be dietary supplements, particularly herbal and dietetic supplements. A food or food product is known as a nutraceutical, which is a mix of the words "nutrition" and "pharmacy," if it includes health and medical claims, such as the prevention or treatment of illness. According to the report, the market for nutraceuticals is currently driven by dietary supplements, which make up 64% of sales. Generally speaking, the sale of vitamin and mineral supplements drives the nutraceuticals industry. It is the primary driver of this market, claims the report. The researchers estimated that the global nutraceutical market was worth US$ 150 billion approx. in 2011 and that over 93% of the pressures on the market came from the United States, Europe, and Japan. According to the report, industries are looking at developing nations like India and China as significant development locations as these markets mature because of the high per capita spending on nutraceuticals. Increased cases are of obesity in the population, as well as an increase in diabetes and cardiovascular disease cases. Additional variables may contribute to the expansion of dietary supplements in India. Due to growing worries about food security and the need for greater nutrition, the government is also pitching in by sponsoring vitamin fortification initiatives in India. The Global Nutraceuticals Market Report from Variant Market Research predicts that the global nutraceuticals market will reach $340 billion by 2024 with a compound annual growth rate (CAGR) of around 7.3% from 2016 to 2024. The market for dietary supplements is projected to increase at a Compound Annual Growth Rate (CAGR) of 8.3%, or roughly to $722.49 billion by 2027, according to Grand View Research's Global Nutraceuticals Market Report.

Figure 12.1 shows the growth of the worldwide nutraceutical market from 2015 to 2019.[25] According to international standards, India contributed around 2% of the world market in 2015 and brought in \$2.8 billion. Due to the country's substantial population, rising urban density, and growing awareness, this share is anticipated to rise to around 3% by 2022. The market for nutraceuticals in India had a value of over \$4 billion in 2017. By 2022, it is anticipated to increase at a CAGR of 21% and reach \$10 billion. Dietary supplements are the sector in India that is expanding the fastest. The top three firms in the world's nutraceuticals market are from the United States, Germany, and France. The market shares for Indian nutraceuticals from 2015 to 2022 are presented in Figure 12.3.

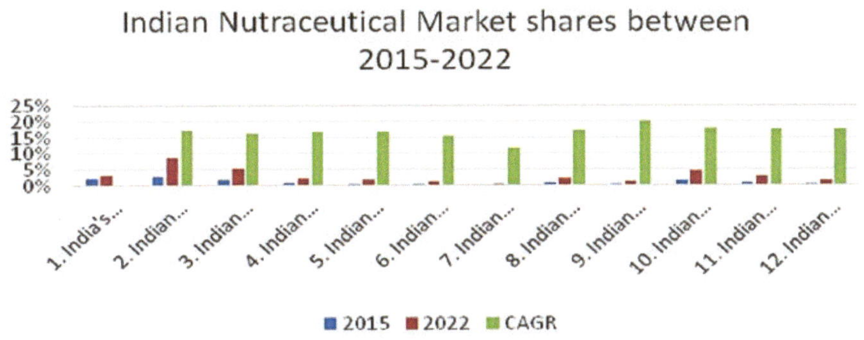

FIGURE 12.3 Indian nutraceutical market shares in 2015–2022.

12.2.1 THE US MARKET

1. The nutraceutical market in the United States was worth \$50.4 Billion in 2010, making it the largest worldwide.
2. Dietary supplements are rapidly approaching maturity, while functional foods and beverages are catching up.
3. Dietary supplements grew at a rate of around 3.1%, while functional foods and beverages grew at a rate of 5.6%.
4. Consumers in the United States are health-conscious and demand active ingredients in nutraceuticals, requiring nutraceutical customization for each target group.

Due to increased customer demand for all-natural, nonmodified functional ingredients, companies in the United States are looking to diversify their product line and are shifting their emphasis more and more toward natural nutraceutical ingredients.

12.2.2 THE EUROPEAN MARKET

1. In 2010, the European nutraceutical industry was worth US $ 35 billion.
2. The nutraceutical sector in Europe is undergoing significant consolidation, with the most recent examples being DSM's acquisition of Martek Biosciences and Dupont's acquisition of Danisco.
3. The creation of new products and innovation are the nutraceutical industry's main priorities in Europe, which has resulted in increased R&D spending, which has increased from 0.24% of industry revenue in 2004 to between 0.8 and 1% in 2010.
4. Product and ingredient innovation, according to European companies, is the way forward for the nutraceutical business. Germany, The Netherlands, and Sweden are now considered to be Europe's leading nutraceutical innovation centers, with the United Kingdom and Spain serving as crucial test markets for new products.

12.2.3 THE JAPANESE MARKET

1. Japan is the world's second-largest market for nutraceutical products by country.

12.2.4 MARKET IN THE UNITED STATES

1. Its functional food business is worth USD27.7 billion, according to estimates.
2. Functional foods with recognized health claims, or FOSHU (Foods for Specified Health Use), and those that may provide health benefits have been approved by the Japanese Government (without any health claims).

12.2.5 THE SOUTHEAST ASIAN MARKET

1. Taiwan, Thailand, The Philippines, Vietnam, Malaysia, North Korea, South Korea, and Singapore are all large, growing international markets for nutraceuticals in Southeast Asia.
2. The Southeast Asian nutraceuticals market has significant growth potential, with a market share of 7% of the global market (USD 8 billion) in 2008, and is predicted to grow to 10% of the global market in 2009 (USD 10.96 billion).

12.2.6 THE CHINESE MARKET

1. In 2008, China's nutraceutical business was estimated to be worth around USD 15 billion.
2. Productive capacities in functional foods, beverages, nutritional supplements, and pharmaceuticals are upgraded and diversified as a result of strong economic expansion.
3. Government policies have eased limitations on the selling of health foods.
4. This is also assisting in the growth of nutraceutical product demand in China.
5. Medical costs are rising, and people are becoming more concerned about their health, which is propelling the Chinese nutraceuticals industry forward.

India is overtaking the United States in terms of the size of the global neutraceutical market, which is growing at a rate of 21% annually. Table 12.3 lists some of the most popular neutraceuticals on the global market.

TABLE 12.2 Most Popular Neutraceuticals on the Global Market.

S.N.	Product name	Category	Use	Name of companies
1	Omega women	Vitamins and antioxidant	Immune booster	Wassen, UK
2	Calcirol D-3®	Vitamins and calcium	Calcium dietary supplement	Cadilla-Healthcare Limited, India.
3	EFAGoldTM Super Lignan capsule	Lignan and Vitamin D	Cardiovascular drugs	Nature's Way, USA
4	Proteinex®	Predigested/sproteins	Protein-supplement	Pfizer Limited
5	5-Hour energy	Caffeine, Taurine	Energy drink	Living essentials, USA
6	Revital®	Minerals and vitamins	Daily health augmentation	Ranbaxy Lab Ltd
7	Muscle Optimeal®	Dietary fiber, protein, vitamins	Substitute drink for meals	Jarrow Formulas, USA
8	Weight smartTM	Trace elements and vitamins	Nutritional supplement	Bayer Corporation, USA
9	WelLife®	Granulated-L-glutamine	Amino acid supplement	Daesang America Inc., Hackensach, NJ, USA
10	GRD®	Proteins, vitamins	Nutritional supplement	Zydus Cadila Ltd. Ahmedabad, India

TABLE 12.2 *(Continued)*

S.N.	Product name	Category	Use	Name of companies
11	Rox®	Taurine, Caffeine	Energy drink	Rox America, Spartanburg, SA, USA
12	Coral calcium	Calcium and trace minerals	Calcium supplement	Nature's answer, Hauppauge, NY, USA
13	ChaserTM	Activated calcium carbonate	Hangover supplement	Living Essentials, Walled Lake, MI, USA
14	Proplus®	Soy proteins	Nutritional supplement	Campbell Soup Company, Camden, NJ, USA
15	PNer plusTM	Vitamin	Neuropathic pain supplement	NeuroHelp, San Antonio, Texas, USA
16	OlivenolTM	Natural antioxidant	Dietary supplement	Cre Agri, Hayward, CA, USA
17	Snapple-adayTM	Vitamins and minerals	Meal replacement beverage	Snapple Beverage Group, USA
18	TOZAL Eye Health Formula	Zinc, antioxidants Omega-3-fatty acids, and lutein	Improved vision	AmeriSciences, USA
19	Ferradol Food® powder	Niacinamide, carbohydrate, protein	Dietary supplement for both adults and children	Pfizer Limited, India
20	BrainSpeed Memory®	Vitamins, minerals	Boost mental health	Natrol, USA

12.3 RATIONALE FOR USE OF NUTRACEUTICALS

Chronic disease onset, development, morbidity, and death are all regulated by dietary factors. 40–50% of cardiovascular diseases, 35–50% of cancers, and 20% of osteoporosis cases are caused by dietary factors. It is not a new concept to use food as therapy to treat and prevent various illnesses. For a long time, iodine supplementation of table salt and iron/folic acid supplementation of wheat flour have been used to prevent iodine deficiency goiter and anemia. Vitamin A-fortified foods have also been found to be a cost-effective and practical way to combat vitamin A deficiency. Nutraceutical products, which include a diverse range of bioactive chemicals, are used to treat and manage a number of disorders. Several frequently applied bioactive components are listed in Table 12.3.

TABLE 12.3 Diseases and Bioactive Compounds Having Neutraceuticals.

S.N.	Diseases	Bioactive compounds having neutraceuticals
1	Cardiovascular disease	Tannins, anthocyanins, and octacosanol are n-3 PUFAs
2	Eye disorder	Green tea, vitamin C, vitamin E, astaxanthin, lycopene, and -carotene
3	Alzheimer's	Turmerin, Lycopene, Lutein, and Curcumin
4	Diabetes	Psyllium, omega-3 fatty acids, and isoflavones
5	Cancer	Biochanin, Daidzein, Lycopene, and β-carotene
6	Allergy	Quercetin
7	Immunity problems	Garlic, Echinacea angustfolia, and astragalus
8	Inflammation	Vitamin C, glucosamine, and chondrointin
9	Obesity	Capsaicin, Psyllium fiber, green tea, mahuang guarana
10	Miscellaneous	Caempherol, B-carotene, and Moringa oleifera Chitosan, Lam, Saponins, Terpenes, and Curcumin
11	Toxicity	The vast majority of medicinal plant extracts [Cynara scolymus]

12.4 REGULATION OF NUTRACEUTICAL INDUSTRY IN INDIA

It has become vital to regulate nutraceuticals as their use has increased. In India, many laws have been created to control and develop nutraceuticals. The sheer quantity of laws perplexed stakeholders. By switching from multi-level, multidepartmental management to a single chain of command, the Food Safety and Standard Authority of India (FSSAI) was founded under a single point of contact for all complaints regarding food safety and standards were established by the Food Safety and Standard Act, 2006. Following its publication on May 5, 2011, in the Indian Gazette, the Food Safety and Standard Rules came into force on August 5, 2011. The FSSAI's main goals were to create standards based on science, control the production, sale, import, processing, and distribution of food, and guarantee the accessibility of safe food for human use.[5] This study focuses on the laws governing import, labeling, claims, registration, and licensure of nutritional supplements. This article attempts to highlight some of the benefits and drawbacks of the Indian nutraceutical industry as well as several FSSAI initiatives. This paper also offers some recommendations for improving India's market for nutraceuticals as well as their quality and safety. The formation of the Food Safety and Standard Authority of India (FSSAI) to certify nutraceuticals in India was permitted under the Food Safety and Standard Act of 2006.[15] The Food Safety and Standard Rules of 2011 entered into effect on August 5, 2011, following their publication in the Indian Gazette on May 5, 2011. Standards and food safety are involved (Licensing and Registration of Food Businesses). The lone law and rule pertaining to the licensing and registration of food businesses selling nutraceuticals are

1. Regulations from 2011 Concerning Food Safety and Standards (Licensing and Registration of Food Businesses).
2. Food Additives and Product Standards (Food Safety and Standards) Regulations, 2011.
3. Food Safety and Standard (Prohibition and Restriction of Sales) Regulations, 2011.
4. Food Safety and Standard (Packaging and Labeling) Regulations, 2011
5. Regulations for Food Safety and Standards from 2011 related to (Contaminants, Toxins, and Residues).
6. Regulations on Food Safety and Standards (Laboratory and Sampling Analysis), 2011.
7. Regulations for Food Safety and Standards, 2016 (Food or Health Supplements, Nutraceuticals, Food for Special Dietary Purpose, Functional Food, and Novel Food).
8. Regulations on Food Safety and Standards (Food Recall Procedure), 2017.
9. Food Safety and Standard (Import) Regulations, 2017.
10. Approval of Non-Specific Food and Food Ingredients is underneath the Food Safety and Standards Regulations.
11. Regulations on Food Safety and Standards (Organic Food), 2017.
12. Regulations on Food Safety and Standards (Alcoholic Beverages), 2018.
13. Food Safety and Standard (Food Fortification) Regulations, 2018.
14. Regulations for Food Safety and Standards (Food Safety Auditing), 2018.
15. Food Safety and Standard (Recognition and Notification of Laboratories) Regulations, 2018.
16. Food Safety and Standard (Advertising and Claims) Regulations, 2018.
17. Regulations for Food Safety and Standards (Packaging), 2018.

12.4.1 LICENSING AND REGISTRATION

The Food Safety and Standards (Licensing and Registration of Food Businesses) Regulations took effect on August 5, 2011. All food business operators in the nation are required to be registered or licensed in accordance with the Food Safety and Standards (Licensing and Registration of Food Businesses) Regulations 2011.[17] Petty food producers must register with the commissioner if their annual sales are less than 12 lakhs. A manufacturer with annual revenue of more than 12 lakhs is required to apply for a license with the FSSAI office. A manufacturer's registrations/licenses should be converted to FSSAI registrations/licenses. There must be an additional Central License for the Head/Registered Office if the food business operates in more than one state.[18] Registration for small-scale food production: Before launching his or her business, a petty food manufacturer must register with the commissioner. The following papers must be submitted to register: (1) An application for Schedule 1. (b) Rs. 100 in annual fees. (c) Address verification (government documents like voter ID cards). (d) Photograph of the applicant.[19]

12.5 CHALLENGES GROWING FOOD SEGMENTS

Despite the fact that nutraceuticals have a significant market demand in India, a variety of barriers prevent it from realizing its full market potential in both the global and the domestic markets. Ref. [33] is one of them.

12.5.1 REGULATORY CHALLENGES

12.5.1.1 LACK OF CLEAR UNIFIED DEFINITION

Nutraceuticals are defined differently in different nations based on the laws that govern them. In the United States, nutraceuticals are classed as dietary supplements, but in Europe, they are classified as food supplements. Nutraceuticals still need a universally agreed-upon definition. As a result, international regulatory cooperation will be required. As a result of their adherence to standardized regulations, the ASEAN (Association of Southeast Asian Nations) nations are currently setting the bar high.

12.5.1.2 COMPLICATED REGULATORY LANDSCAPE

While three agencies—FSSAI, AYUSH (Ayurveda, Yoga and Naturopathy, Unani, Siddha, and Homeopathy), and (Central Drug Standard Control Org.)—monitor nutraceuticals based on claims made by products, other countries may regulate products defined as nutraceuticals in India as dietary supplements, traditional medicines, food supplements, and other products. This presents a challenge for Indian manufacturers and exporters to comply with other importing country regulations.

12.5.1.3 CLAIMS ALLOWED

Different countries allow different health claims based on their rules, making it challenging to reach a product for the worldwide market. For example, the FSSAI policy prohibits preventative claims, although other nations let them.

12.5.1.4 NEED FOR MORE LEGAL CLASSIFICATION

Additional legal classification is necessary to stop different participants from assigning different categories to various products. The items are thought to be of a medical nature on the one hand, but they are also intended to keep people healthy. Identifying whether nutraceuticals are preventive or curative is the problem.

12.5.2 DIFFICULTIES FACED BY NEW PLAYERS

Because there is no regulatory framework for setting up manufacturing units, it is challenging for new businesses to enter the nutraceutical sector and obtain subsidies. Nutraceuticals are classified as either foods or medications; therefore, quality and price become issues.

12.5.3 ADULTERATION

12.5.3.1 ECONOMIC ADULTERATION

Customers are put at risk when producers' full or partially substitute listed ingredients with the less costly and simpler to acquire ingredients that are not mentioned by the producer in the claims.[10]

12.5.3.2 PHARMACEUTICAL ADULTERATION

Active chemical medicinal components were added to the nutraceutical to increase its effectiveness, but they were not listed as an ingredient on the label. There have been several recorded instances of unregistered active pharmaceutical ingredients being used in nutraceutical goods worldwide.[10]

12.5.4 ONLINE PHARMACY

Since most of the products are nonprescription (OTC), regulatory scrutiny for nutraceuticals in e-retail pharmacies is significantly less because no prescription uploads are necessary. Consumers are more likely to buy products with higher concentrations because they are uninformed of the legal limitations.[45]

12.5.5 LOWER CONSUMER KNOWLEDGE ABOUT MICRONUTRIENT DIETARY SUPPLEMENTS

Even if Indian people are becoming more health-conscious and pursuing a healthy lifestyle, there is a lack of awareness about certain nutritional deficiencies, such as daily requirements of macro- and micronutrients that are not satisfied through a conventional diet.[45]

12.5.6 FORMULATING A NUTRACEUTICAL DOSE FORM PRESENTS DIFFICULTIES

12.5.7 CHALLENGES IN ANALYSIS

1. Because nutraceuticals are a collection of chemical entities, identifying and quantifying all of the chemicals in the products is difficult.
2. Defining and recognizing contaminants, as well as guaranteeing that they are not dangerous to consumers.
3. A structural study of each entity in a formulation is challenging.

12.5.8 FORMULATION CHALLENGES

12.5.8.1 SOLID DOSAGE FORM

Botanicals are intricate, with up to 50 active chemicals and several chemical components making about 70–90% of the composition. Excipients or active ingredients are absent from this product. Particle size, flow, compressibility, moisture sensitivity, ingredient interaction, content homogeneity, and quality control (QC) testing are challenges with nutritional components. Because botanicals and extracts differ depending on the region, season, and other factors in which they were grown.

The dose size of the active constituent is big, leaving little place in the final formulation for excipients, and the amount of each ingredient is sufficient to enable the proper distribution of the favorable characteristics. In comparison to pharmaceutical formulations, nutritional formulations frequently have more active ingredients at higher concentrations. Excipient space is frequently constrained due to limitations on dosage size. Nutraceutical formulations comprise 70–90% excipients and 10–30% active compounds compared to traditional pharmaceutical formulations, which contain the opposite ratio. Some desirable outcomes, such as disintegration time, hardness, and friability, when choosing the right tools, are important to carefully consider the tablet's shape and form. It is challenging to read many nutraceutical tablets since they are made with neutral colors like brown and gray that appear mottled, gritty, or granular.

The processing components are harmed by the natural nutraceuticals ingredients, which are frequently coarse, abrasive, caustic, and hard.

12.5.9 PSYCHOLOGICAL CHALLENGES

Manufacturers of dietary supplements must first segregate the products and handle dietary supplements and functional meals separately.

1. Modifying products is to suit the preferences and tastes of domestic customers such as vegetarianism, halal, or Hindu diets, conventional treatments, and taste. Examples of this include a failure to recognize the functional benefits found in everyday foods as well as preferences for specific nutritional formulations that represent social and cultural diversity. Choosing this study in population might be challenging (depending on age, sickness, and other factors).[8]

12.5.10 OTHER DIFFICULTIES

1. It is necessary to provide sufficient information and scientific data to demonstrate that the product is safe, repeatable, and therapeutically effective, as well as if it provides these effects over a specific time period, such as 2 or 3 years.

2. It is necessary to develop a method to demonstrate that product quality is repeatable, and this mechanism must be supported by strong, scientific evidence that can be demonstrated experimentally using a reliable technique. Certification requirements are typically applicable to excipients and active ingredients. Specific nation and product compliance for WADA (World Anti-Doping Agency) requires free of GMOs Halal KosRegistration category/classification. The formula may fall into several categories depending on the claims and substances used in each jurisdiction.

12.5.11 INITIATIVES THAT ARE TAKEN BY FSSAI

The Food Safety and Standards Authority of India (FSSAI) have started a number of projects to enhance India's food safety culture. Here are a few notable projects:

1. On December 31, 2019, the FSSAI finalized the rules for the approval of rapid testing instruments, kits, and procedures to improve the environment for food testing in the country. The FSSAI has approved a total of 30 rapid food testing kits/devices in accordance with these regulations. Real-time food testing is now "faster, better, and less expensive" thanks to these gadgets. These are projected to have a substantial impact on the food industry's regulatory, monitoring, and quality assurance systems.[19]
2. The nation's food regulator has chosen to have food testing done through consumer organizations to gain consumer trust. If a customer sample does not pass inspection, the food standard rule allows for the repayment of testing costs. However, due to practical issues, this clause has never been put into practice. On December 13, 2019, the FSSAI made a decision to allow trustworthy Voluntary Consumer Organizations (VCOs) to help clients obtain these tests in an effort to address these problems. In some circumstances, testing costs will be covered even if the sample passes.[19]
3. The Eat Right India programme, which attempts to increase consumer awareness of the need of eating wholesome food, includes the Swasth Bharat Yatra as a component. The second iteration of the "National Eat Right Mela" was held on December 29, 2019, according to the FSSAI. The honorable minister will also announce "Network for Scientific Co-operation for Food Safety

and Applied Nutrition," a collaboration of academic institutions working on diverse topics. They are responsible for gathering information and data on current and potential threats to food safety in their respective fields, conducting surveys, conducting research, and other related tasks, exchanging testing equipment and procedures, etc.

4. The Hon'ble Minister will also present "The PURPLE Book," a recipe book for meals tailored to particular illnesses. The book[31] provides hospitals with detailed instructions on appropriate meals for serious medical conditions like diabetes, hypertension, cancer, gastrointestinal disorders, and more in a model format [written and approved by professionals in the food and nutrition industry].
5. Heart Attack Rewind aims to remove trans-fat from India by 2022.[31] This effort was promoted across India.
6. FSSAI-CHIFSS (Confederation of Indian Industry (CII) and Hindustan Unilever (HUL) Initiative on Food Safety Sciences) is a partnership between FSSAI and CII-HUL designed to encourage food safety collaboration between industry, the scientific community, and academia.[31]
7. The Food Safety and Standards Authority of India (FSSAI) launched the Food Product Identity Verification System (FPIVS) and the "eplatform for Comments" on November 5, 2018."
8. FPIVS enables FBOs to determine whether a product is subject to applicable food regulations or whether product approval is necessary. The "eplatform for Comments" will make it easier for stakeholders to submit their comments online for upcoming food restrictions and requirements.[23,34]
9. The presence of color-coded food safety display boards on the premises of food outlets will be compulsory. It must include the FSSAI license number, safety and food hygiene recommendations, and a customer feedback from Ref. [23,34].
10. The unification of food standards to remove ambiguity, the improvement of food testing facilities, and cooperation with the food industry to encourage self-regulation.[23,34] These FSSAI projects are all a part of the ongoing effort to revitalize the FSS Act. It is the first step toward the FSSAI's wider goal of establishing a national food safety strategy in India.

12.6 A RECENT REPORT ON NONCONFORMITIES

Food adulteration is not the only problem with food safety. Despite the fact that the government has taken numerous initiatives to ensure food safety, the number of nutraceutical nonconformities has increased recently. In comparison to 2018, nonconformities have increased by 25% in 2019. According to the Food Enforcement report, 1,06,459 samples in total were evaluated over the 2018–2019 fiscal years. Only 3.7% of the samples were determined to be dangerous, whereas 15.8% were found to be subpar, and 9% had labeling issues.

In 2018–2019, there were 7% more samples tested than there were in 2017–2018. 25% more samples than the previous year were found to be nonconforming. The number of civil lawsuits filed has risen by 36%, while the number of cases where penalties have been imposed has risen by 67%. When compared to the prior year, the amount of penalties levied increased by 23% in 2018–2019. During the fiscal year 2018–2019, a total of Rs. 32.58 crore was realized. Criminal cases have seen an 86% increase in the number of filed cases. The states and union territories with the highest rates of unsafe food were Tamil Nadu (12.7%), Assam (8.9%), Jharkhand (8.8%), West Bengal (7.6%), and Odisha (6.7%).[38] In recent years, the number of samples, nonconformities, and tainted food has also increased (Table 12.1).

12.7 RECOMMENDATIONS

India's nutraceutical industry is growing as a result of factors like rising disposable income, greater health awareness, the development of specialist nutraceuticals, and others. As a result, appropriate rules for ensuring the safe and effective use of nutraceuticals are required. Despite the numerous hurdles, the already available FSSAI has come a long way toward providing unique and transparent regulations. Numerous food safety projects are being worked on by the FSSAI, including FSSAI-CHIFSS, Heart Attack Rewind, FPIVS, and the e-Platform for Comments. Despite the FSSAI's substantial progress, a number of challenges still exist, including the regulatory environment's complexity, the lack of a precise, all-encompassing definition and classification, adulteration, and dangerous food items. According to the food safety enforcement report, out of 1,06,459 samples examined, 3900 were determined to be dangerous and 16,870 to be subpar (2018–2019). FSSAI will therefore need to take a few further actions to address the issues.

The following proposals should be considered by regulatory bodies to increase nutraceutical quality, safety, and market:

1. The harmonization of global regulatory bodies facilitates the creation of consistent definitions and rules for nutraceuticals; it is similar to the Global Harmonization Task Force (GHTF) for medical devices and the International Regulatory Cooperation for Herbal Medicines (IRCH) for herbal medicine.
2. The manufacturer should back up their claims with data on safety, effectiveness, and mechanism of action.
3. To solve the issue of adulteration, more severe rules are required.
4. In order to prevent unregulated nutraceutical items from being sold on e-commerce websites, more stringent restrictions should be implemented that require merchants to submit their FSSAI license code before entering the platforms.
5. The FSSAI should adopt new programmes to enhance awareness, boost R&D, and improve product development skills.

12.8 THE PROSPECT OF NUTRACEUTICALS

The bulk of the world's population is concerned with maintaining good health, and nutraceuticals play an important role in this regard. Furthermore, the negative side effects of medications on the market have boosted their need for such items. Today's global market offers a wide range of neutraceuticals formulations with different bioactives, including vitamins and plant bioactives. Based on evidence that their use can help avoid serious illnesses like cancer, diabetes, and cardiovascular disease, the list of nutraceuticals products is continuously expanding. The fewer health risks compared to synthetic medications account for the high consumer acceptance.

Customers are interested in different neutraceutical types for treating oxidative stress, decreasing stress, boosting energy, enhancing mental clarity, preventing metabolic illnesses, and other uses. To succeed and compete in the global market, neutraceutical businesses strive to create new products and advertise them in print and online media to draw clients. Neutraceutical research is being massively funded by nations including the United States, Germany, China, and the United Kingdom with the hopes of making huge profits in the future, when important problems like antibiotic resistance will rule and these products may only be short-term fixes. In affluent nations like the United States, more than half of the population use dietary supplements to maintain daily health. Newer neutraceuticals are being developed using technological developments such as neutrigenomics and real-time cellular imaging techniques. Such technological advancements in the creation of nutraceutical formulations and the related food business will have a significant positive impact on health care management. Multiple chemicals can now be contained in a single capsule thanks to advancements in encapsulation technology.

Encapsulation in liquid Pellets, oils, granules, and powders can all be encapsulated in hard gelatin capsules thanks to advancements in neutraceutical formulation technology. This method could also be used to separate and purify neutraceuticals. The future of highly effective next-generation nutraceuticals is being influenced by innovative delivery systems such as nanoparticles, phytosomes, liposomes, and lipid nanoparticles. Enzymes are a unique form of nutraceutical with a bright future ahead of them. Future enzyme-consisting nutraceuticals, made possible by cutting-edge technologies like nanotechnology, are expected to account for a sizeable portion of the industry. For the enhancement of nutraceutical products, novel and enhanced characterization methods based on genomics, proteomics, and metabolomics are being researched. Thorough pharmacokinetics investigations and to guarantee such medications' biological activity, anticipate their mode of action, and verify their health claims, preclinical and clinical tests are crucial. Additionally, to meet regulatory standards and guarantee product safety, manufacturers must carefully review the toxicity profiles of all nutraceuticals. In conclusion, even though nutraceuticals products are widely accepted, more technological development and scientific validation are needed to ensure their viability and widespread use in clinical settings.

12.9 CONCLUSION

Currently, the global market for nutraceuticals is expanding. At the moment, Indian businesses are exporting and importing nutraceuticals. Concentrated nutrients, nutritional supplements, and special diets that include GM foods, herbal items, and processed foods like cereals, soups, and drinks are referred to as nutraceuticals. Supplements known as nutraceuticals provide all of the vital elements that people should take. The body receives energy and nutrient supplements from nutraceuticals, which are crucial for optimal health. The food and pharmaceutical industries both use a lot of nutraceuticals. In addition to treating restlessness and insomnia, nutritional supplements can help maintain prostate health. Chondroprotective qualities of nutraceuticals like glucosamine and chondroitin sulfate have been demonstrated in the prevention of joint injury. Despite its flaws, the existing FSSAI has made significant progress in creating distinct and unambiguous regulations. FSSAI has a number of food safety projects, including FSSAI-CHIFSS, Heart Attack Rewind, FPIVS, and the e-Platform for Comments, to name a few. The FSSAI has taken a number of efforts; however, there are still some issues that need to be handled. Producers that want to export their goods to other nations might save time and effort by harmonizing the laws governing nutraceuticals. Nutraceutical safety is still a major concern for Indian health regulators. Product analysis, development, and authentication of extracts, as well as the adoption of strict norms in nutraceutical manufacturing, marketing, and availability, can all be done using scientific methods and collective expertise.

KEYWORDS

- **nutraceuticals**
- **dietary supplements**
- **food segment**
- **nutrients**
- **functional food**
- **market**
- **regulations**
- **challenges**
- **bioactive compounds**

REFERENCES

1. Verma, A. Nutraceutical Industry in India: Current Scenario and Future Trends. Nuffoods Spectrum, Jan 27, 2017 [Online]. https://nuffoodsspectrum.in/column/26/1688/nutraceutical-industryin-india-current-scenario-and-future-trends.html (accessed Aug 31, 2019).
2. Benkouider, C. In *Functional Foods and Nutraceuticals*, 2005; vol 44, pp 8–11.
3. Borkar, N.; Saurabh, S. S.; Rathore, K. S.; Pandit, A.; Khandelwal, K. R. An Insight on Nutraceuticals. *Pharma Tutor* **2015,** *3* (8), 13–23.

4. Brower V. Nutraceuticals: Poised for a Healthy Slice of the Healthcare Market. *Nat. Biotechnol.* **1999,** *16*, 728–731.
5. Bureau Report. Industry Leaders Share their Views on the Growth Path for the Indian Nutra Market in 2015. Nuffoods Spectrum, January 12, 2015 [Online]. https://www.nuffoodsspectrum.in/column/33/1179/industry-leaders-share-theirviews-on-the-growth-path-for-the-indian-nutra-market-in-2015.html (accessed Aug 14, 2019).
6. Cardenas, D. Let not thy Food be Confused with thy Medicine: The Hippocratic Misquotation. *e-SPEN J.* **2013,** *8*, 260–262.
7. Chand, P. Survey Shows that 49% of Indians use the Internet for Health Information. PR Moment India, Mar 3, 2105. Available from: https://www.prmoment.in/pr-news/survey-shows-that-49-ofindians-use-the-internet-for-health-information (accessed Aug 18, 2019).
8. Chaudhari, S. P.; Powar, P. V.; Pratapwar, M. N. Nutraceuticals: A Review. *World J. Pharm. Pharm. Sci.* **2017,** *6* (8), 681–739.
9. Consumers can Now Get Food Tested Through Consumer Organizations. FSSAI, Dec 13, 2019.
10. Desai, H.; Sav, A.; Amin, P. Formulation and Evaluation of Mucoadhesive Anti-Infective Solution Containing Solubilised Tea Tree Oil for Vaginal Infections. *Int. J. Adv. Pharm. Biol. Chem.* **2013,** *2* (2), 385–389.
11. Dillard, C. J.; German, J. B. Phytochemicals: Nutraceuticals and Human Health. *J. Sci. Food Agric.* **2000,** *80*, 1744–1756.
12. Economic Survey 2016: Treatment Cost at Private Hospitals 4 Times than Government Ones. *The Economic Times,* Feb 26, 2016. . https://economictimes.indiatimes.com/industry/healthcare/biotech/healthcare/economicsurvey-2016-treatment-cost-at-private-hospitals-4-times-than governmentones/articleshow/51156935.cms?utm_source=contentofinterest&utm_medium=text&utm_campaign=cppst (accessed Aug 4, 2019).
13. McClements, D. J.; Zou, L.; Zhang, R.; Salvia-Trujillo, L.; Kumosani, T.; Xiao, H. Enhancing Nutraceutical Performance Using Excipient Foods: Designing Food Structures and Compositions to Increase Bioavailability Compr. Rev. Food Sci. Food Saf. **2015,** *14*, 824–847.
14. Eskin, N. A. M.; Tamir, S. In *Dictionary of Nutraceuticals and Functional Foods*; CRC Press: Boca Raton, USA, 2006.
15. Food Safety and Standard (Advertising and Claims) Regulations. FSSAI, 2018 [Online]. https://archive.fssai.gov.in/home/fss-legislation/fss-regulations.html (accessed Nov 19, 2019).
16. Food Safety and Standard (Import) Regulations. FSSAI, 2017 [Online]. https://archive.fssai.gov.in/home/fsslegislation/fss-regulations.html (accessed Oct 8, 2019).
17. Food Safety and Standard (Licensing and Registration of Food Businesses) Regulations. FSSAI, 2011 [Online]. https://www.fssai.gov.in/upload/uploadfiles/files/Licensing_Regulations.pdf (accessed Oct 26, 2019).
18. Food Safety and Standard Act, 2006 [Online]. https://archive.fssai.gov.in/home/about-us/introduction.html (accessed Oct1, 2019).
19. Food Safety and Standard Authority of India, FAQs on Enforcement. FSSAI [Online]. https://fssai.gov.in/cms/compliancefaq.php (accessed Oct 31, 2019).
20. Food Safety and Standard Authority of India, Food Safety and Standard Regulations. FSSAI.

21. Food Safety and Standard Authority of India, Licensing and Registration System [Online]. https://foodlicensing.fssai.gov.in/index.aspx (accessed Oct 12, 2019).
22. Food Safety and Standard Authority of India [Online]. https://archive.fssai.gov.in/home/about-us/introduction.html (accessed on Jan 2, 2020).
23. FSSAI Role in Food Quality and Consumer Safety is Essential [Online]. https://www.fssaifoodlicense.com/fssai-role-infood-quality (accessed on Jan 3, 2020).
24. FSSAI's 2nd Eat Right Mela Highlights Healthy Diets for a Better Life. FSSAI, Dec 23, 2019.
25. Global Nutraceuticals Market Will Reach USD 317.3 Billion by 2024. Zion Market Research, June 25, 2018 [Online]. https://www.globenewswire.com/news-release/2018/06/25/1528968/0/en/Global-Nutraceuticals-Market-WillReach-USD-317-3-Billion-by-2024-Zion-Market-Research.html (accessed July 6, 2019).
26. Health Supplements to India Trends and Opportunities. Australian Government Trade and Investment Commission [Online]. https://www.google.com/url?sa=i&url=https%3A%2F%2Fwww.austrade.gov.au%2Faustralian%2Fexport%2Fexport-markets%2Fcountries-and-economies%2Findia%2Findustries%2Fhealth supplements-toindia&psig=AOvVaw26Q3D19FDLtz1hS7wiU- (accessed July 18, 2019).
27. https://archive.fssai.gov.in/home/fss-legislation/fss-regulations.html# (accessed Oct 1, 2019).
28. https://fssai.gov.in/upload/press_release/2019/12/5df499f9a0c50Press_Release_NetCOFaN_14_12_2019.pdf (accessed Jan 2, 2020).
29. https://fssai.gov.in/upload/press_release/2019/12/5e0084fa8b3e5Press_Release_Eat_Right_Mela_23_12_2019. pdf (accessed Jan 2, 2020).
30. https://www.biovoicenews.com/nutraceuticals-market-in-india-challenges-and-trends/ (accessed Aug 18, 2019).
31. Important Developments and Major Achievements During the Month of November, 2018. FSSAI, Nov 03, 2018 [Online]. https://www.fssai.gov.in/upload/monthly_update/2018/12/5c0612adde19dMonthly_Achievements_November_
32. Improved Food Safety Enforcement: Data for 2018-19 Released [Online]. https://fssai.gov.in/upload/press_release/2019/11/5ddce90b2867fPress_Release_Enforcement_Data_26_11_2019.pdf (accessed Jan 3, 2020).
33. Indian Nutraceuticals Industry. Smergers Investment Banking For SMEs [Online]. https://www.smergers.com/industrywatch/indian-nutraceuticals-industry/ (accessed July 6, 2019).
34. Initiatives Announced by FSSAI to Promote Food Safety Culture [Online]. https://www.fssaifoodlicense.com/fssaiinitiatives/ (accessed Jan 3, 2020).
35. International Food Information Council. Functional Foods Fact Sheet: Omega-3 Fatty Acids. http://www. foodinsight. org/Functional_Foods_Fact_Sheet_Omega_3_Fatty_Acids.
36. Hugenholtz, J.; Smid, E. J.; Ladero, V.; Pascal Hols. Metabolic Engineering of Lactic Acid Bacteria for the Production of Nutraceuticals. *Antonie Van Leeuwenhoek* **2002,** *82*, 217–235.
37. Kalra, E. K. Nutraceutical-Definition and Introduction. *AAPS PharmSci.* **2003,** *5*, 27–28.
38. Kastor, A.; Mohanty, S. K. Disease and Age Pattern of Hospitalisation and Associated Costs in India: 1995–2014. *BMJ Open* **2018,** *8* (1), e016990.
39. Keservani, R. K.; Sharma, A. K.; Ahmad, F.; Baig, M. E. Nutraceutical and Functional Food Regulations in India. In *Nutraceutical and Functional Food Regulations in the*

United States and Around the World, 3rd ed.; Debasis, B., Ed.; Academic Press, 2014; pp 327–342.

40. Keservani, R. K.; Kesharwani, R. K.; Vyas, N.; Jain, S.; Raghuvanshi, R.; Sharma, A. K. Nutraceutical and Functional Food as Future Food: A Review. *Der Pharmacia Lettre* **2010a,** *2* (1), 106–116.
41. Keservani, R. K.; Sharma, A. K.; Kesharwani, R. K. An Overview and Therapeutic Applications of Nutraceutical and Functional Foods. In *Recent Advances in Drug Delivery Technology*, 2017; pp 160–201, ISBN: 9781522507543.
42. Keservani, R. K.; Sharma, A. K.; Kesharwani, R. K. Eds.; In *Nutraceuticals and Dietary Supplements: Applications in Health Improvement and Disease Management*; CRC Press, 2020, ISBN: 9781771888738.
43. Keservani, R. K.; Kesharwani, R. K.; Sharma, A. K.; Vyas, N.; Chadoker, A. Nutritional Supplements: An Overview. *Int. J. Curr. Pharm. Rev. Res.* **2010b,** *1* (1), 59–75.
44. Kulkarni, S. A Brief Report on Nutraceutical Products in India. Corporate Crystal India Pvt. Ltd. 2015. http://nutra-vita.com/wp-content/uploads/2017/04/Nutraceutical-Products-in-India.pdf. (accessed Sep 3, 2019).
45. Kumar, A. K.; Choudhary, R. K. Determination of Antibacterial, Antifungal Activity and Chemical Composition of Essential Oil Portion of Unani Formulation Kulzam. *Int. J. Green Pharm.* **2011,** *5* (1), 28–33.
46. Montrose, D. C.; Floch, M. H. Probiotics Used in Human Studies. *J. Clin. Gastroenterol.* **2005,** *39* (6), 469–484.
47. Nelson, N. J. Purple Carrots, Margarine Laced with Wood Pulp? Nutraceuticals Move into the Supermarket. *J. Natl. Cancer Inst.* **1999,** *91*, 755–757.
48. Nutraceutical Market Size Worth $722.49 Billion By 2027 | CAGR: 8.3%. Grand View Research, April 2020 [Online]. https://www.grandviewresearch.com/press-release/global-nutraceuticals-market (accessed July 4, 2020).
49. Nutraceutical Market Size, Share & Trends Analysis Report by Product (Dietary Supplements, Functional Foods, Functional Beverages), By Region, And Segment Forecasts, 2020 – 2027. Grand View Research, April 2020 [Online]. https://www.grandviewresearch.com/industry-analysis/nutraceuticals-market (accessed July 4, 2020).
50. Nutraceuticals Market in India – Challenges and Trends. Bio Voice News Desk, Nov 18, 2019.
51. Nutraceuticals: Challenges of a Fast Growing Food Segment in India. Food Safety Helpline, Sept 24, 2014 [Online]. https://foodsafetyhelpline.com/nutraceuticals-challenges-fast-growing-sector-india/ (accessed July, 2019).
52. Nutraceuticals: Global Markets to 2023. BCC Research, July 2018 [Online]. https://www.bccresearch.com/marketresearch/food-and-beverage/nutraceuticals-global-markets.html (accessed July 5, 2019).
53. Nutraceuticals: Global Markets. Research and Markets, March 2017 [Online]. https://www.researchandmarkets.com/reports/4090190/nutraceuticals-global-markets (accessed July 5, 2019).
54. Rapid Food Testing Devices and Kits set to Redefine Food Safety in 2020. FSSAI, Dec 31, 2019 [Online]. https://fssai.gov.in/upload/press_release/2019/12/5e0aef7e1cb3ePress_Release_Rapid_Kits_31_12_2019.pdf (accessed Jan 2, 2020)
55. Rise in Health Awareness to Take Nutraceuticals Market to $ 8.5 bn by 2022. Business Standard, Apr 9, 2017 [Online]. https://www.business-standard.com/

content/b2b-pharma/rise-in-health-awareness-to-take-nutraceuticalsmarket-to-8-5-bn-by-2022-117041900637_1.html (accessed July, 2019).

56. Santini, A.; Cammarata, S. M.; Capone, G.; Ianaro, A.; Tenore, G. C.; Pani, L.; Novellino, E. Nutraceuticals: Opening the Debate for a Regulatory Framework. *Br. J. Clin. Pharmacol.* **2018,** *84*, 659–672.
57. Kulkarni, S. Regulations for Nutraceuticals Products in India. *FnB News*, Aug 16, 2018 [Online]. http://www.fnbnews.com/FB-Specials/regulations-for-nutraceuticals-products-in-india-44998. (accessed July 5, 2019).
58. Urbanisation in India. In *Wikipedia*, June 2016 [Online]. https://en.wikipedia.org/wiki/Urbanisation_in_India (accessed Sep 16, 2019).
59. Vouloumanou, E. K.; Makris, G. C.; Karageorgopoulos, D. E. Probiotics for the Prevention of Respiratory Tract Infections: A Systematic Review. *Int. J. Antimicrob. Agents* **2009,** *34*, e1–e10.
60. White Paper On 'Indian Nutraceuticals Industry – Building Global Momentum Amid Regulatory Challenges' At ASSOCHAM – 5th National Symposium On Nutraceuticals. Sathguru, Sept 30, 2019 [Online]. https://www.sathguru.com/news/2019/09/30/white-paper-on-indian-nutraceuticals-industry-building-globalmomentum-amid-regulatory-challenges/ (accessed Sep 9, 2019).
61. Whitman, M. Understanding the Perceived Need Forcomplementary and Alternative Nutraceuticals: Lifestyle Issues. *Clin. J. Oncol. Nurs.* **2001,** *5*, 190–194.
62. Zeisel, S. H. Regulation of Nutraceuticals. *Science* **1999,** *285*, 185–186.

CHAPTER 13

CHALLENGES FOR INCORPORATION OF NUTRACEUTICALS IN FOOD

PRIYA SEN, PREETI KHULBE, MANISH GUPTA, and RIYA THAPA

Gyan Vihar School of Pharmacy, Suresh Gyan Vihar University Jaipur, Jagatpura, Jaipur, Rajasthan, India

ABSTRACT

Nutraceuticals are food products that are part of a regular diet but have health benefits. The words "nutrient" and "pharmaceutical" are combined to form the term "nutraceutical." Nutraceuticals are often foods or components of meals that significantly affect and maintain human biological functions. The current population-based trends are the primary players of the nutraceutical market's global expansion. Dietary fiber, prebiotics, probiotics, polyunsaturated fatty acids, antioxidants, and other herbal/natural foods are examples of food products utilized as nutraceuticals. India provides a number of distinct obstacles to market newcomers, in addition to a diverse range of categories and ingredients. Tailoring products to home tastes and preferences is one of them. Some of the biggest hurdles includes putting the fundamentals in place, maximizing unlimited potential, domestic players, and raw material inequality. Pharma markets now dominate the nutraceutical market, and this trend is likely to continue because there appears to be very less information available to prospective players interested in investing in this area. The main focus of this chapter is on the issues of incorporating nutraceuticals into the food.

Formulations, Regulations, and Challenges of Nutraceuticals.
Tingirikari Jagan Mohan Rao, Rajesh K. Kesharwani, Raj K. Keservani, Anil Kumar Sharma (Eds.)

13.1 INTRODUCTION

Stephen Defelice introduced the word nutraceuticals in 1989, "Nutraceuticals" are food product or components of foods that provide medicinal or medical benefits,[5,21] such as preventing infection and/or therapy, according to Brian C. Defelice. "Let food be your medication," urged Greek physician HIPPOCRATES (considered to be the "Father of Medicines"). "Focus on prevention" is the principle underlying it. Nutritional supplements, food ingredient, multifunctional food, and other terms are also used in this area. Functional foods are conventional foods that contain elements or compounds that provide a therapeutic or health advantage along with a dietary effect.[37]Nutraceuticals are natural products that are used to promote health, delay ageing, prevent infection, and maintain the human body's proper function. Nutraceuticals are presently garnering a lot of significance because of their nutritional and medicinal benefits. They are classified as nutritional supplements and botanical bioactive substances based on their origins.[35]

In 2014, the worldwide economy for nutritional vitamin components would grow by 6% yearly to approximately $1300 million. In India, a considerable number of people are reliant on organic and alternative medications due to inconsistencies and flaws in the official medical delivery system. The growing popularity of Indian Ayurvedic medicines expands the market for formulations containing Withaniasomnifera, Curcuma longa, Zingiber officinale, Ocimum tenuiflorum, and other herbs.[30–32,36] Due to increased scientific proof of swine flu, melanoma, and other preventative health benefits, Vit. D will see the largest surge in popularity. The global economy for dietary supplement ingredients is expected to rise by 6% each year to $13 billion in 2014. Due to irregularities and problems in the approved medical delivery system, a large number of individuals in India depend on natural and unconventional medications.[37]

Nutraceuticals[19,20] is a method of combining nourishment with pharmaceuticals. They are foodstuffs or parts of diets that aid in the maintenance of various physiological processes, as well as the prevention and treatment of several acute or chronic disorders. Herbal, minerals, and nutritional supplements are the 3 primary types of nutraceuticals. Various plants are used to make a variety of herbal supplements. Vitamins, minerals, and α-aminolike micro nutrients having known nutritional benefits are necessary for good health. Roughage, plant fiber, high fiber food containing beneficial bacteria stearidonic acid, antioxidants,[39] spices, and other types of herbs and natural foods are examples of food products utilized as nutraceuticals. Probiotics are used for their health benefits in the gastrointestinal tract as well as in systemic areas such as the liver, central nervous system, vaginal canal, and blood circulation.[38]

Nutritional supplements are components intended to improve the nutritional quality that endure or consist some or all of the following dietary ingredients: a multis, native element, an crude or other natural product, an amino, and roughage material can be consumed by persons to support the edibles by enhanced overall daily consumption, according to the Dietary Supplement Health and Education Act (DSHEA). It is available in pill capsules, tablets, and liquid form. It is not intended to be based on traditional cuisine or as the sole component of a diet. It is a "nutritional supplement," according to the label. A nutritional supplement's producer is able to make sure that the material is healthy before it is advertised under the Act (DSHEA), 1994.[40]

Nutraceuticals are presently garnering a lot of significance since their nutritional and medicinal benefits. They are classified as nutritional supplements and natural phytochemical compounds based on their origins. Natural nutraceuticals aid in the maintenance of good health and the promotion of lifespan and standard of living. Nutritional supplements have shown promising outcomes in the treatment of a variety of ailments, including tumors, brain disorders, heart disorders, and others. An introduction of numerous bioactive substances that serve as nutraceuticals is included in this review like carbohydrates, lipids, different plant parts, alkaloids, herbal plants, etc. (Tyagi, P. FSSAI Regulations for Nutraceuticals).

Medical Advantages:

- Stay rid of unwanted effects.
- It is probable that the medical benefits will be amplified.
- It is possible to have found naturally nutritional supplement with no adverse effects.
- It has the ability to improve people's health, nutrition, and health status.
- It is possible that it will be easily obtainable and reasonably priced.

This treatment is built on the concept that meals might provide therapeutic qualities in addition to providing nourishment. It accomplishes this purpose, as per nutraceutical and nutritional therapy theory, by utilizing the efficiency of these nutraceuticals in detoxifying, avoiding vitamin and mineral shortages, and restoring good digestion and eating habits. Macronutrients with specific systemic response in support of human wellness are known as phytonutrients.

The phytochemicals perform in the following manner:

1. Enzymatic reaction substrates.
2. Enzyme-catalyzed reaction cofactors.
3. Enzyme antagonists.
4. Intestinal absorbents which adhere to and remove unwanted constituents.
5. Enhance critical nutrient uptake and/or stabilization.
6. Therapeutic bacteria-specific cell activation.
7. Healthy bacteria synthesis substrates.
8. Specific inhibition of pathogenic microorganisms in the intestine.
9. Active or hazardous chemicals scavengers.
10. Agonizing or antagonistic compounds for cell membrane or intracellular sites.[34]

As an addition to standard treatment, diet offers a consistent approach to managing and suppressing Osteoarthritis. In recent years, nutraceutical supplements such as chondroitin sulfate (CS), glucosamine sulfate (GS), and Methyl sulfonyl methane (MSM) have been used to control and help to treat Osteoarthritis (OA). Nutraceuticals are supplements that contain a bulk amount of a biological element that was originally separated from food but now exists in a non-food substrate and is used to maintain or improve health in proportions greater than those available from regular foods. Nonetheless, there is no consensus on whether the terms "nutraceutical" or "nutritional supplements" should be used. As seen by the growing sales of micronutrients, "active ageing" is a primary goal of nutritional supplements. In people with Osteoarthritis, nutritional active combinations seemed to be effective in improving clinical symptoms and reducing inflammatory indicators. Currently, 69% of people with Osteoarthritis are given dietary supplements to help them manage their condition (Tyagi, P. FSSAI Regulations for Nutraceuticals).

Nutraceutical categories:

Roughage including phytology:

- Vitamins, native elements, acetyl-COA Q, levo-carnitine
- Panax Gingsing, maidenhair tree, hypericum perforatum, serenoa repens

Nutraceuticals:

A food item that is commonly consumed but offers health benefits that go beyond those provided by standard nutrition.

Examples:

- Yogurt contains probiotics, which are beneficial to gut function.
- Boosted foods, grains, and snacks with dietary fiber, nutrients, and elements.
- Milk containing docosahexaenoic acid may help to avoid cardiovascular disorders.
- Rapeseed oil with reduced triglycerides is good for lowering cholesterol.
- Cardiovascular disease and colorectal tumor: cereals, miller's bran, ispaghula, and monolignols.
- Oligofructose, a prebiotic, is used to manage intestinal bacteria.
- Inhibition of cholesterol adsorption by stanols (Benecol).

Products with Therapeutic Properties:

- Bars with additional medicines.
- Lactoferrin and transgenic cows for immunological enhancement.
- Oral vaccination against infectious illnesses with transgenic plants.[37]

13.2 PHARMACEUTICALS VERSUS NUTRACEUTICALS

Nutraceuticals currently lack a separate definition from other food-derived dietary supplements, crude treatments, prescription drugs and probiotics, functional food ingredients, and so on. They do, however, have a health-promoting therapeutic activity.[33]

In general, nutraceuticals are foods or parts of foods that play a substantial role in changing and sustaining proper metabolic function in humans. The present demographic and health trends are the primary drivers of the nutritious market's global expansion.[33]

The best part is that we are not forced to make a decision. Both medicines have a role to play in our overall health. Pharmaceuticals, in a nutshell, are the results of clinical trials targeted at curing specific ailments. Naturopathic remedies are food-based substances that are used to diagnose and reduce diseases. They are often derived from Eastern herbal and pharmaceutical formulations.

Both could be able to aid you, depending on your problem. Consider carefully about most popular areas people seek healthcare and how both are utilized to properly tell the difference between medications and health supplements.

Unfortunately, most of us have spent some duration in the health center at a certain stage in life. The purpose of emergency department personnel is to stabilize your health condition and, more importantly, to save you from death. Medicines will be given to you in these urgent instances.

Hospitals employ them pre, though, and then after surgery because they are FDA-approved. In general, these are medicines that are created in a lab and then examined to see if they may cure certain conditions.

When the hospital determines that you are no longer in urgent danger, you are usually sent home or to a physician for treatment. Nutraceuticals are rarely used, save in the case of anticipated future nutritional guidelines.

In truth, no two humans are alike; thus, everyone will respond differently to medications and health supplements. So, there is space for both in a society like ours, when new health concerns emerge each day. These variations can be shown: utilize drugs only when essential to cure serious conditions and prolong life. Use nutraceuticals and Eastern medicine benefits to treat ailments and improve your quality of life.[29]

Many investigations have already been carried out to look into their security, effectiveness, and regulatory, since they are gaining increasing commercial and research interest, with the goal of distinguishing them from other market-available nutrition products that claim to get a positive effect on humans.

The major difficulty for nutraceuticals to be recognized as a preventive or therapeutic useful method in approaching a few pathogens is to recognize the possible mode of action of therapeutically active compounds present in nutritional supplements that may enhance health and lessen the chance of disease states while enhancing overall well-being.

Expert opinions: It is critical to have a clear and unambiguous definition of nutraceuticals as well as a unified regulatory framework. Nonetheless, there is an absence of clarity information, and stated health advantages are frequently not adequately verified by safety and efficacy in vivo evidence, which can lead to misleading assumptions and lose possession for a product to be successful, as promised.[33]

13.3 HEALTH IMPLICATIONS

Nutraceuticals have a variety of health benefits, from general well-being to chronic illness therapy. Nutraceuticals promote health through boosting nutritional uptake, purification, and digestive tract microbial maintenance.[27]

Fiber, prebiotics bacteria, probiotics bacteria, polyunsaturated fatty acids, antioxidants, and other medicinal herbs are examples of food products utilized as nutraceuticals. These nutraceuticals aid in the therapy of a few of the decade's most serious medical conditions, including overweight, heart disorders, melanoma, osteoarthritis, arthritis, diabetes mellitus, and cholesterol. Overall, the term "nutraceutical" has ushered in an era of science and treatment, where the food sector has evolved into a research-driven market.[26]

13.3.1 MEDICAL ADVANTAGES OF NUTRACEUTICALS

We are also conscious that our lifestyles and eating habits have deteriorated, so few people depend on fast food with empty calories. It is required for a small number of people since these supplements assist them meet their dietary requirements, thereby boosting our immunity.

13.3.1.1 ENHANCES FUNDAMENTAL WELL-BEING

Nutraceuticals aid in the enhancement of our general health. As we previously discussed, a well-balanced diet is beneficial to the body. It provides our bodies with nourishing substances and promotes improvement, including both emotional and physical well-being.

13.3.1.2 IMPROVEMENT IN ENERGY LEVEL

There are a variety of nutritional supplements that can help us enhance our activity levels, such as all vitamins, specific in Vit. B class. The Vit. D class also aids in keeping you energetic. Vegetarian and vegans' diets should aim to up their Vit. B-12 intake since it is mostly found in meat and dairy product, which they do not eat.

13.3.1.3 PREVENTS THE FORMATION OF CHRONIC DISORDERS

Nutraceuticals have been considered in the medication of dangerous diseases such as type I and type II diabetes and elevated blood pressure. Fiber supplements help to maintain the arteries free from clotting. Diabetes mellitus can also be managed by weight control and eating a balanced diet.

13.3.1.4 SLEEPING ROUTINE IS IMPROVED

Supplements are thought to enhance sleeping habits. Numerous natural supplements, including Valeriana officinalis and other micronutrients like mg, may promote restful sleep. It also helps us to improve our sleeping patterns.

13.3.1.5 REDUCES STRESS

Certain vitamins, such as vit. D, play a critical role in determining our emotions. Vit B unquestionably increases stress and depression. Vit. B2, often known as riboflavin, is a vital nutrient found primarily in meat. As a result, vegans might take multivitamin supplements.

13.3.2 ADVERSE EFFECTS OF NUTRACEUTICALS

Certain components in supplements can have effect on our bodies. There are a variety of fortified goods on the market that include vitamins and minerals. Combining nonprescribed health supplements in your diet with fortified foods or drugs you are already taking can lead to complications. Here are some possible nutraceutical adverse effects.

13.3.2.1 HEADACHE

Vit. A deficiency is linked to headaches. In addition to nausea and dizziness, a single big dose can induce. In various persons, it might range from moderate to severe. As a result, it is critical to follow the physician's recommendations for supplementation.

13.3.2.2 VERTIGO

Vit. A excess can also cause vertigo and genetic abnormalities. To counteract this negative impact, consume the proper diet and consume health supplements in the recommended dosage. As a sick person, it is important to take suggestion from your doctor for the proper dosage before taking any medicines.

13.3.2.3 HYPERTENSION

Some nutrients may also result in hypertension, which can be damaging to our health. It has the potential to be lethal for people who have heart problems. Nutritional supplements are used by a high percentage of the total population. Medical practitioners should be aware of this, and as a result, they should ask patients about their supplement consumption.

13.3.2.4 GUT DISORDERS

Overconsumption of some nutrients might result in gastrointestinal issues such as emesis, loose stools, and constipation. Constipation is frequently linked to an excess of vit. C. The term "Nutraceutical" directly translates to "pharmaceutical nutrition for the body." Nutraceuticals are supplements to the food we eat that provide additional nutrition to our bodies. It works on the preventive premise. A well-nourished, healthy body is capable of fighting ailments on its own.[28]

13.4 AYURVEDIC LEGACY

Ayurveda, Siddha, and Unani are three historic Indian medicinal systems. Medical systems are described in the Vedas and other ancient religions. Between 25th century and 5th century before Christ in India, the Ayurvedic philosophy arose and developed. Because considering the relationship between man and his illness is central to the traditional Indian medical system. Ayurveda literally means "life science." It has been said that individuals in excellent health have balanced metabolisms. The ancient Indian medical system emphasizes concepts of man and his illness. Because it provides a full system for living a long and healthy life, it offers diet and health supplements regimens to aid in the renewal of the body. For many different sorts of common illnesses, including food intolerances, it offers treatment choices. Food is the primary source of sustenance, but as the world becomes more sophisticated, certain ancient practises are being abandoned. As a result, current eating habits have an impact on a well-balanced diet. There is an ever-widening nutritional gap, and normal living is no longer stable as a result. However, with increase in growth of working population across the world, people are unable to spend time and money on their health and well-being. These are the some of the reasons which is encouraging individuals to consider about their health.[9]

13.4.1 AYURVEDIC POSTERITY

Recent studies suggest that the market is outgrowing, but is not abandoning its traditional herbal and ayurvedic medicinal heritage.

Because some Indian customers may be unfamiliar of newer constituents like omega-3 fatty acids, reliance on current products may dilute demand for developing nutraceuticals. Natural and herbal treatments offer a wealth of regional ingredients as a basis for health supplements.

Herbal honey, or chyavanprasam, has become an immensely popular. Traditional remedies also reflect a culture of natural goods and self-care that aligns with nutraceutical developments worldwide. India's pharma industry is mostly oriented on the health supplements industry. With a few exceptions such as Bournvita, Complain, Gritzo, and Quista are the medical nutritional health drinks with added supplements which can provide long term health benefits in children's, and adults.

Leading national participants in the functional-foods space also involve ITC, Britannia, etc., as well as Indian enterprises selling ayurveda-based functional nutritional supplements (e.g., Gynoveda, Prorganiq, Neuherbs, Kapiva, Capsoft-Healthcare, Sycap Pharma, Nutramarck Life Sciences).

13.4.2 UNIQUE CHALLENGES

India poses a number of distinct difficulties for market newcomers, in addition to a large variety of categories and components. Among these are: modifying products to suit domestic customers' interests and inclinations. Vegan and non-vegan, dietary habits, traditional treatments, flavor with choices for design that take social and cultural variety, or a failure to acknowledge functional advantages in basic foods just are a few examples.

In India, for example, yoghourt is a part of the diet and is frequently prepared at home. Premium-priced nutraceutical companies will benefit from educational campaigns on diet, nutrition, and illness risk reduction. Pharmaceutical companies looking to bring on the dietary supplements sector may seek to use joint ventures, research alliances, or acquisitions to gain access to these abilities.

The appropriate pricing points: The class of the society in cities in India may be able to purchase premium nutraceutical prices, but the rural poor will not.

It will necessitate a greater focus on scientific data and professional backing to support medical claims. A split-level strategy with specialized branding, formulation, positioning, cost, and distribution is required for a balanced approach to the Indian market.

Known for its commitment to social responsibility, in collaboration with nongovernment groups or government-fortified initiatives, there may be potential to access underprivileged areas of the market.

A regulatory environment is unpredictable. The mainstream industry claims that most of the Indian markets lack transparency and standardization on problems like product details, approvals, health claims, and manufacturing norms; many products are fake or essentially unregulated.

Nutraceuticals are now regulated by the Food Safety and Standards Act. They are recognized as food items, with manufacturing/quality regulations and limitations on illness prevention or reduction claims.

Market participants allege that the lack of specifically stated nutraceuticals rules causes uncertainty and expediency. Even if it simply makes structure-function assumptions, any dietary supplement in a "drug-like" oral format is considered a medication, with appropriate licencing procedures.

These provide forth the regulations for nutraceutical manufacturing, formulation, approval, and sale, as well as health claims, RDAs, scientific evidence, and labeling. There are lists of additions, vitamins, minerals, amino acids, plant substances, and other compounds that are approved. The Food Safety and Standards Authority of India said in March 2016 that the new proposals will apply to any nutraceuticals produced in India after 2011 till the regulations were put into effect.

As these challenges indicate, genuine competitors in India's nutraceuticals sector must bring tenacity, adaptability, sensitivity, and patience to the table. The Indian government recently declared, along with purpose of making India a global leader in nutrition research and innovation, center for "health" foods in the long term. India is a spectacular new front in a global market where a point of attraction is shifting more to emerging economies for food and pharmaceutical enterprises with innovation and staying ability.[13]

13.5 NUTRACEUTICALS INGREDIENTS USED AS FOOD

Nutraceuticals food-derived product provides additional health advantages beyond the fundamental nutritional value present in meals. "Nutraceutical" is a concept that combines the words "nutrient" (a healthy dietary component) and "pharmaceutical" (a medical drug). Nutraceuticals are based on the idea of preventive medicine, as described by the Greek physician Hippocrates (known as the "Father of Modern Medicine"), who said, "Let food be your medicine." One of the most essential fields of research is their significance in human nutrition, which has far-reaching effects for consumers, medical experts, regulators, food producers, and distributors. Nutraceuticals have created a lot of attention throughout the years because of their dietary supplements, safety, and medicinal benefits. Aurea Biolabs offers a vast selection of Nutraceutical Ingredients with specific health advantages. See the list below for a complete list of similar items.

Amla polyphenols are extracted from the dried fruit of Phyllanthus Emblica Linn. using aqueous alcohol extraction. The product is brown in color that has an amla-like flavor and odor. Amla polyphenols are used in nutraceutical compounds, antidiabetic products, antioxidants, antimicrobial, anti-inflammatory, antiemetic properties. The polyphenol content of the product is regulated to a maximum of NLT 25%. The dried leaves of Centella Asiatica are ethanolically extracted to form Asiatic acid. The Asiatic acid concentration of the greenish powder is specified to be between 20% and 25%. The ingredient is used in brain supplements for memory improvement, cognitive wellness, and other applications. Boswellia Serrata is extracted in a solvent and then filtered to produce Bio-AKBA. The powder is off-white to cream in color and has a unique Boswellia aroma. It is optimized to produce 3-o-acetyl-11- keto-beta-boswellic acid (AKBA) with a purity of 20–25%. Nutraceutical supplements and anti-inflammatory formulations contain the ingredient. Solvent extraction of powdered dried fruits of Capsicum Annum L. yields Capsaicin Crystals (95%). It is a white to light-yellow crystal with a pungent chilli aroma. Nutraceutical supplements and topical treatments such as pain ointments contain it. Cassia polyphenols are collected from dried cassia bark (Chinese cassia) via aqueous alcohol extraction. Antioxidant and nutraceutical supplements have included the component. The product is regulated to contain no more than 20% total polyphenols.[22]

The dietary supplements sector is categorized by health benefits, category, application, form, and region, according to the survey.

By dietary supplements ingredients type

- Probiotic proteins and amino acid, phytochemical and plant extracts, fibers and specialty carbohydrates, polyunsaturated fatty acid, vitamin, prebiotic, carotenoids, minerals

  By nutraceutical ingredients application

- Food, beverages, personal care, animal nutrition, dietary supplements

By nutraceutical ingredients health benefits, cognitive health, gut health, heart health, bone health, immunity nutrition, weight management

By nutraceutical ingredients form, dry, liquid

By Region, North America

- According to medical web, a nutraceutical might be a dietary food that has therapeutic properties, or a specific portion of a cuisine. Examples include garlic, polyunsaturated fatty acid (found in salmon fish, mackerel fish, etc.), glycine max, zingiber officinale, essential nutrients, supplements, nutritional food, hydrolyseda-amino, etc.

Components with additional nutritional supplements such as essential nutritional fibers, minerals, α-aminoacids, vitamins, and fatty acids, respectively. Following are some health advantages:

- Vit. A contains essential antioxidant for growth and development as well as the treatment of several skin conditions.
- Vit. E contains essential antioxidant, supports the development of muscle, lung, nerve, and blood cells, and strengthens the immune system.
- Vit. K is essential for blood clotting.
- Vit. C contains essential antioxidant.
- Vit. B1 performs neurological functions and converts food into energy.
- Vit. B2 produces energy and maintains the skin, eye, and nervous system.
- Vit. B3 maintains proper neurological function.
- Vit. B6 contains genetic information of cells, produces RBCs, maintains CNS, and synthesizes amino amido, fats, protein, and carbohydrates.
- Vit. B9 produces RBCs, genetic materials and prevents birth defects.
- Ca maintains the bone strength and teeth and is important for the nerve, muscle, and glandular function.
- Fe produces energy and carries oxygen to the tissues.
- Mg helps to prevent premenstrual syndrome (PMS).
- P helps in the formation of genetic material and energy production.
- Cu produces hemoglobin and collagen, which is beneficial for our heart.
- I helps in proper function of thyroid (Nutraceutical Ingredients Market by Type, Application, Region).[25]

13.6 NUTRACEUTICAL INCORPORATION STRATEGIES

Several approved compounds that can be used in functional nutrients are designed to alleviate the risks. The capability of nutraceuticals to affect circulatory system and atheroma vascular disease in a positive way should be considered a huge opportunity for disease prevention and treatment. The dietary, medical, and pharmaceutical sectors have a compelling need for edible delivery methods that encapsulate, preserve, and release bioactive lipids. Because these delivery methods must be edible, the variety of materials and processing activities that may be employed to make them are limited. The primary biological activated tri-glycerides lipids that need to be distributed within the food business highlight the major obstacles that they face in their current integration into products. The delivery systems were manufactured by simple processing methods from food-grade (GRAS) materials (such as tri-glycerides, α-amino, polysaccharides, surface active agents, and minerals) (for example, extrusion and drying).[7]

Excipients used in nutraceutical preparations

This research examines the market for excipients in nutraceutical oral solid formulations in the two most important regional segments. It includes a variety of excipient categories and provides a detailed dive into particular excipient markets, making it much easier to evaluate market trends and spot market possibilities.

13.6.1 REPORT BENEFITS

This study is a valuable resource for developers and marketers of unique active ingredients to nutraceutical oral solid dosage producers. This report specifically aids subscribers by supplying:

- A very accurate and unprejudiced estimate of the excipients market in each category and function
- Extensive knowledge of industry developments in terms of customer perception, technology, and products
- Some in examination of the marketing viewpoint and predictions at the active ingredient level.

13.6.2 NUTRACEUTICALS TOXICITY

The inaccurate notion that nutraceuticals are secure because they may have foods or food derivatives is leading worldwide use of nutraceuticals upward.

However, adverse effects and toxicity have been observed repeatedly, not only as a result of the nutraceutical's consumption, but also there is a chance of contamination. Pesticides, other hazardous plants, metals, herbicides, or even intentional adulteration with chemical compounds might contaminate nutraceuticals. Furthermore, numerous studies have indicated the occurrence of drug–herbal interaction after coadministration due to the complex form of natural remedies and the presence of various phytochemicals. As a result of this interaction, the medications may be misused, increasing the chance of toxicity and severe adverse effects. It might potentially impair pharmacological effectiveness and result in poor treatment response.

A large number of regularly used nutraceuticals have such a connection to serious side effects and toxicity. Maidenhair tree extracts are strong in flavonoids and are frequently recommended to older individuals to help with vascular endothelium. Maidenhair tree extracts are also used to treat dementia, Alzheimer's disease, schizophrenia, and cerebrovascular inadequacy. However, it has been shown that long-term exposure to large dosages of Ginkgo biloba (Maidenhair tree) extracts is correlated to spontaneous bleeding. Ginkgo biloba shows to be carcinogenic in a lot of animal examinations.

Furthermore, catechins, which have anticancer effects, are found in green tea extracts. Green tea extracts are also used to prevent possible overweight and metabolic abnormalities. Yet, excessive exposure can cause renal toxicity, liver toxicity, and endocrine disruption. Furthermore, caffeine use has been related to stress, tachycardia, osteoporosis, and reproductive problems, especially in early age. Many other regularly used herbal items, such as cardamom, leaf extract, and St. John's wort, have been known to cause cancer, hepatotoxicity, teratogenic effects, and mutagenicity, among several other things.

13.6.2.1 NUTRACEUTICAL CONTAMINANTS

Among the most dangerous aspects of utilizing nutraceuticals is the existence of nutraceutical contaminants. Pesticides are the most harmful nutraceutical pollutants because of the extreme toxicity they cause when consumed. The symptoms of toxicity can extend from a little rash to severe breathing, neurological, and reproductive problems. Pesticides are required to keep the therapeutic plants in good condition. As a result, pesticide residue limitations must be strictly observed. Organochlorines are the most persistent insecticides in herbal products and have the greatest hazardous effects after long-term exposure.

Heavy metals, such as Pb, Cd, Hg, and As, are also common pollutants in nutraceuticals. The World Health Organization (WHO) has established guidelines and restrictions for the presence of environmental contaminants in finished herbal products. Acute exposure to high concentrations of heavy metals or long-term contact to low concentrations of heavy metals can both cause heavy metal poisoning. Carcinoma, heart intoxication, neurodegenerative disorders, renal, and liver failure are all side effects of metal poisoning.

Pyrrolizidine alkaloids are secondary metabolites of plants that are naturally occurring and shown in animal research to cause developmental toxicity, carcinogenicity, genotoxicity, and liver toxicity. Meanwhile, liver damage has been reported as a symptom of their acute toxicity. Because of their significant alkylating properties, these alkaloids have a hazardous effect. They cause cellular malfunction and tissue necrosis by reacting with cellular proteins[16] and DNA. Pyrrolizidine alkaloids are found in approximately 5000–6000 plant species that 1105 samples of food derivatives acquired from online portals and supermarkets across Europe. According to the study, pyrrolizidine alkaloids were found in 60% of the dietary supplement items.

13.6.2.2 NUTRACEUTICALS–DRUG INTERACTION

Several studies have recently shown numerous nutraceutical–drug interactions. Interfering with the medication's metabolic route or altering drug transporters might be the nature of the interaction. It is hard to estimate the incidence of nutraceutical–drug interactions due to the complex nature of nutraceuticals and the insufficiency of pharmacokinetic, pharmacodynamics, and safety investigations.

Some of the observed interactions may be dangerous and even fatal. Aspirin and several nonsteroidal anti-inflammatory drugs, for example, interact with ginkgo, curcumin, Allium sativum, Panax quinquefolium, Linn. chamomile, and in herbal preparations. As a result of the suppression of platelet aggregation ability, these identified interactions raised the risk of bleeding in the liver and small intestine, the C-P450 enzymes are often expressed metabolic enzymes. They also contribute to the metabolism of a number of different pharmaceuticals. The herbal extract of Saint John's Wort usually reported nutraceuticals that activates a wide variety of CYP, including CYP1A2, CYP2C9, etc. The induction of these enzymes has the effect of speeding up the metabolism of the medicines eaten. As a result, the medication is rapidly removed from the body, and its action is lost. Treatment failure of antidepressant medicines and loss of oral contraceptive function were also described as side effects of Saint John's Wort.

When combined with P-glycoprotein inhibitor phytochemicals, the bioavailability of fexofenadine, a P-glycoprotein receptor substrate, is greatly enhanced. P-glycoprotein phytochemical inhibitors include naringenin, genistein, and quercetin, which are flavonoids.[8,14,17] As a result, combining P-glycoprotein inhibitor-containing nutraceuticals with powerful substrate medications like digoxin, or cyclosporine might cause intolerable toxicity.

Another significant transporter in drug assimilation and disposal is the Organic Anion Transporting Polypeptide receptor (OATP). Iijima et al. examined 98 widely used medicinal herbs in Japan to see whether there was any interaction between nutraceuticals and pharmaceutical medications. The activity of the intestinal 2B1 (OATP2B1) receptor was shown to be reduced by less than 20% in 12 herbal species. Moreover, seven herbal species improved the action of the binding site by more than 150%.

As a result, taking these herbal items together has a significant impact on the absorption rate of substrate drugs including bosentan, penicillin G, and other pharmaceuticals.

13.6.2.3 NUTRACEUTICALS ADVANCED DELIVERY SYSTEMS TOXICITY

The integration of nutraceuticals with nano distribution techniques has grown in popularity in recent years. These innovative delivery technologies are employed to increase nutraceutical absorption and bioavailability, offer a targeted drug release profile, and improve the stability of phytochemicals that are easily degraded. Because of their specific, chemical, and physical characteristics following nanonization, nano-sized nutraceuticals boost the efficacy of nutraceuticals. The same favorable physicochemical features of nanoparticles, on the other side, are the primary source of human toxicity.

The nano delivery system's safety is controversial. Furthermore, there is a scarcity of knowledge and research on the absorption, metabolism, distribution, excretion, and toxicity of nanomaterials. Understanding the molecular interactions of nanomaterials with biological systems is difficult to understand their mechanism. Respiratory illnesses, cardiovascular ailments, carcinogenicity, and lower life expectancy are some of the most widely documented negative impacts of nanomaterials. The majority of these negative effects are believed to be caused by oxidative stress and inflammation caused by nanomaterials at the molecular level.

Zhang et al. produced AU, Hg, Pd, and palladium nanoparticles to load Ascorbic acid, ()-epigallocatechin gallate, and gallic acid into three distinct nutritional supplements. The manufactured systems were evaluated to scavenge hydroxyl radicals. Surprisingly, they discovered that the nanoparticles reduced the antioxidant phytochemicals' capacity to scavenge hydroxyl radicals and created a Reactive Oxygen Species (ROS).

EFSA list of parameters have included:

- The size of the molecules in nanoparticles.
- The physical characteristics of nanoparticles.
- The shape and composition of nanoparticles.
- Weight and molecule concentration.
- The surface chemistry of nanoparticles.
- The ionic strength of nanoparticles.
- Partition and solubility properties.
- Dustiness, viscosity, density, and pour density.
- The chemical reactivity of nanoparticles.
- The higher energy level to form chemical reaction activity in nanoparticles.

The development of prediction models and high throughput methods are currently receiving more attention to support the early evaluation of the toxicity of nanosystems. This idea, called "safety by design," will make it easier to design nano delivery systems and enable the creation of safer systems. The reduction of the chance of a medicine being withdrawn from the market due to safety concerns from 50% to 20% was made possible by the use of the safety by design approach in drug creation.[8]

13.7 CHALLENGES WITH NUTRACEUTICALS IN FOOD

Supplements and macronutrients are now the most prominent nutraceutical products in India, followed by probiotics and docosahexaenoic acid which helps in development and functioning of brains in infants and adults. However, preventative health products have the greatest commercial penetration potential. Obesity rates and heart diseases mortality rates have both increased by 20% in India. Diabetes affects a large portion of the population.

As a result, nutraceutical companies must work on creating products for the Indian market based on data for the prevention of disease.

The nutraceutical food sector in India has a lot of promise because the products are so diverse. Nutraceuticals include supplements, cow milk, gluten-free diet meals, bifidus, potable, Phytomedicines, dieting products, and others.

- The Indian nutraceutical industry has always been in profit in exports.
- The sector is now completely accessible to the local market because of the growth in per capita revenue and public awareness campaigns.

However, changing people's perceptions of the nutraceutical sector as a gray area is its greatest obstacle. A more formal category would stop diverse competitors from using different criteria for specific items. The items are believed to be from a medical character on one end while maintaining health on the other. The issue is deciding if nutraceuticals are required to be therapeutic or preventative.

The difference in local manufacturers and raw material prices

In India, pharmaceutical firms that also produce allopathic medications in almost the same facilities are primarily responsible for the production of nutraceutical items. The Indian pharmaceutical manufacturer is dominated by a large number of powerful companies who have complete influence over all elements of the sector, and middle- and small-sized businesses follow the patterns they create. It is a significant task to integrate all of these businesses into a single, regulated, and standardized sector, especially given that the market is predicted to expand between 18% and 20 %.

Due to the abundance of raw resources, India and China lead the Asia-Pacific nutraceutical market.

- Unfortunately, due to insufficiency of legal criteria, raw material quality varies widely.
- Numerous nutraceuticals did not exhibit good quality, rather the quality of many formulations was consistently low, according to studies conducted by local investors.
- Whole food sources exhibit a comparable disparity since their elements have varying amounts.

This shows that there is a clear need for well spelt out regulations for raw materials as well as for manufacturers of nutraceuticals.

13.7.1 PHARMACEUTICALS VERSUS NUTRACEUTICALS

Pharmaceutical firms have a systemic focus on pharmaceuticals and internal marketing control, whereas nutraceuticals do not. Pharma authorities frequently see dietary supplements and drugs as having distinct marketing strategies. The medication market is often conservative, taking into account all medical and regulatory requirements.

Similarities cannot be drawn with the marketing of nutraceuticals, where inflated claims regarding their efficacy are common. As a result, there is a great deal of discrepancy in the material's promotion and selling since the claims are interpreted differently. Another issue is that firms try to advertise their products through doctors because the Indian customer is focused on prescriptions. This strategy does not produce growth; therefore, businesses should search for a more direct path to the customer.

13.7.2 DIFFICULTIES FACED BY NEW PLAYERS

Pharma corporations already control a large portion of the nutraceutical market, and this trend is likely to continue as there does not appear to be much information available to new businesses looking to engage in this area.

Additionally, it is challenging for new businesses to join the uncharted market because there is no regularized method for setting up production factories for nutraceutical items. They find it tough to access subsidies due to their insufficiency. Nutraceuticals are also classified as either foods or medications, which raises questions about both quality and price.

Since India is ideally positioned to offer ingredients for the global nutraceutical sector, particularly in the case of plant extracts and phytochemicals, new entrepreneurs would like to enter the nutraceutical area. If the laws are enforced as soon as feasible and there is clarity for new participants, this might result in international investment. Without a defined legal framework and the appropriate distribution network, the importer of nutraceuticals is now having difficulty bringing in nutraceutical foods or components.

13.7.3 LACK OF DATA OR INFORMATION

Foods that are fortified, supplemented, or improved with nutraceuticals offer more health advantages than just vitamins and minerals. The relationship between the use of nutraceuticals and health claims should be supported by reliable scientific data. It is difficult for them to import nutraceutical foods or components without a defined regulatory framework and the appropriate distribution network.

- Despite the fact that nutraceuticals are currently a trendy issue in nutrition, there is not enough information available on suggested intakes and efficacy.
- Nutraceuticals are merely dietary supplements and cannot be used as a replacement for real food.
- There is no science-based literature available to inform consumers about nutraceuticals as their usage rises.

13.7.4 RESEARCH-BASED AS OPPOSED TO PERCEPTION-BASED

Medicinal plant has a lengthy history in ancient India. There is an increasing need to consider the utility and safety of nutraceuticals based on sufficient research as they start to control the preventive health sector. Concerns exist on how private-sector research will compare to the traditional Ayurvedic herbal medicine formulations because this industry is still developing. Some suggestions demand that nutraceuticals be made in compliance with the standards of the Indian Pharmacopoeia. There is an expectation that nutraceutical producers would then adhere to its safety and quality requirements, making India a renowned provider of nutraceuticals.[4]

Due to deficiency of authoritative oversight, several issues regarding the development of nutritional supplements are frequently disregarded. These difficulties include determining the genuine source of the raw supplies, compound purity, the existence of additional active compounds, quality, a lack of scientific proof, deceptive advertising, heavy metal contamination, and interactions between supplements and medications. A popular plant called "ginseng," for instance, comes in many distinct kinds, including different species of Panax Ginseng, like Chinese ginseng, Korean ginseng, etc. Although all of them are marketed as ginseng, only the Korean ginseng (P. ginseng), South China ginseng (P. notoginseng), and Indica americana are really members of the genus Panax (P. quinquefolius). Some star anise variants contain several hundred times as much anisatin, a neurotoxin, as the original star anise (Illicium verum). The possibility of inadvertent contamination with germs (pathogenic microbes), chemicals, mycotoxin (aflatoxin), toxic substances (seaweeds), Zn, Cd, Pb, and Ca exists for supplements that are not made under stringent GMP guidelines. To increase effectiveness, certain supplements, such as those frequently used for weight reduction, muscle growth, and sex enhancement, are laced with illegal substances. A substantially comparable plant, which might or might not contain the active components, is another strategy employed by certain producers. For instance, Yellow Snakeroot or mountain grape, which may have little or no berberine/hydrastine concentration, is frequently used in place of Yellow Puccoon, which is utilized for berberine/hydrastine content.

Additionally, because phytochemistry is inherently unstable owing to seasonal and regional to control the quality of nutraceuticals is difficult. Measuring final goods, maintaining uniformity, and limiting undesired components are difficult tasks. Another major issue is how herbal supplements may interact with prescription medications. As an effective antidepressant, (hypericinherb) is utilized; however, it also activates numerous cyp450 isoenzymes, rendering many pharmaceuticals useless. The absence of scientific support is among the main issues with utilizing nutraceuticals. Contrary to medicines, most nutraceuticals are not subjected to "randomized controlled clinical research," and some are never studied under carefully monitored experiment.[5]"

13.7.5 NUTRACEUTICAL MARKET

According to one research, it is projected that the nutraceuticals industry will reach $319,000.6 million by 2025 with a predicted Growth of 6.70%, and $722,000.49 million by 2027, according to another research report. The nutraceuticals manufacturer is dominated by a number of large corporations, such as Bionova Life Sciences, Lacto-nova, Nutripharm, and Maxnova Heath Care, but there are also some intriguing smaller firms, such as Akmed Biotech.

13.7.6 HEALTH IMPLICATIONS

The rise of chronic diseases, which leads many consumers to turn to their diets to control symptoms, is a key factor driving the nutraceuticals industry. For instance, Monash Food Innovation reports that the number of probiotic products increased from 100 in 2002 to 1800 in 2016 as a result of increased awareness of chronic GI disorders and revelation that a healthy gut microbiome is crucial for central nervous system, cardiovascular, and psychological wellness. Nutraceutical application in the management of chronic illnesses is a developing field of research.

According to the Centres for Disease Control and Prevention (CDC), six out of ten Americans have at least one persistent illness, and these conditions have a significant negative social, economic, and psychological effect. Cardiovascular, neurological, intestinal, and autoimmune illnesses, in certain nations, more than 50% are using nutraceutical products.

13.7.6.1 CARDIOVASCULAR DISORDERS

One of the main causes of mortality worldwide is cardiovascular disease (CVD). Nutraceuticals can be used to decrease blood pressure, reduce plaque and cholesterol, and other conditions.

Polyunsaturated fatty acids are proven to be important for cardiovascular, brain, and eye health as well as lowering CVD risk. Omega-3 fatty acids are made by a large number of businesses. Fatty acids have the ability to prevent disease, but certain products can also help break down the fibrinogen in blood clots through enzymatic means. Natto kinase is an enzyme produced by Bacillus subtilis and obtained from the Japanese fermented meal known as natto. By rendering the plasminogen inhibitor-1 inactive, this enzyme exhibits potent fibrinolytic action.

Coenzyme Q10 supplementation is thought to be a CVD prevention strategy. Natural antioxidants like coenzyme Q, or ubiquinone, protect cells and organ systems against reactive oxygen species. According to reports, coenzyme Q10 is among the most popular high-protein diets in the United States and has yearly sales of $100 million worldwide. Ubiqsome by Indena is one instance of a bioavailable coenzyme Q10 formulation.

13.7.6.2 NEURODEGENERATIVE DISEASES

Numerous neurodegenerative diseases are characterized by inadequate gluconeogenesis. Companies are creating novel remedies that nourish the brain through the ketone pathway to address this issue. Tricaprylin, a triglyceride found in cerecin, can trigger ketosis and improve mitochondria metabolism. Phase 2 studies are presently taking place. Tricaprilin is used to treat different neurodisorders, Parkinson's disease, and so on.

Alzheimer's disease is predicted to affect 220 lakh individuals by 2025. One of our greatest opportunities to stop this epidemic may be through the use of nutraceuticals, among other factors including age, heredity, habitat, and diet. Nutraceuticals like Ginger root and Huperzia serrata extracts, as well as phosphatidylserine, acetyl carnitine, and coenzyme Q10, are noteworthy in preventative research.

13.7.6.3 GASTROINTESTINAL DISEASES

The nutraceutical business now prioritizes GI health, and by 2023, the probiotics market is projected to grow to $53,000 million. The term "probiotic" should be used when referring to fermented foods that include microbes that have been scientifically demonstrated to offer health advantages. The concept of fermented foods was recently agreed upon by scientists.[25]

Here are some interesting collaborations and projects on probiotics:

- GanedenBC30, also known as Bacillus coagulants GBI-30, 6086, is a naturally occurring probiotic that supports immunological and digestive health and may also have some effect on protein utilisation. Compared to the typical Lactobacillus and Bifidobacterium probiotics, it has a naturally tough protective shell that makes it pH resistant and gives it a longer shelf life.
- Probiotics and other microbial treatments are being developed by Holobiome to aid those suffering from depression and mental illnesses. In 2019, a team of researchers presented findings demonstrating that two distinct bacterial species—which have an impact on our neurons and brain—are missing from the guts of persons with depression. They discovered that some bacterial species, including Bacteroides, generate gamma-aminobutyric acid (GABA), a neurotransmitter that lowers brain activity.

13.7.7 NEXT-GENERATION PROBIOTICS (NGP)

Due to the simplicity of sequencing the microbiota, (NGPs) are now available on the market. NGPs have the potential to be classified as live biotherapeutic agents. They are used as drugs and used to address certain health issues. NGPs are being researched for their potential to be used in the cure of chronic illnesses, such as Streptococcus spp. for the prevention of the advancement of chronic renal disease and Akkermansia muciniphila for the improvement of the metabolism in obese and diabetic individuals.

13.7.8 FUTURE ASPECTS

Further clinical studies on nutraceuticals are now required to distinguish the best from the others. Organizations like Vizera and Nizo, etc., support the design of clinical trials for nutraceutical businesses, among other things. For instance, Atlantia is hiring for a number of clinical studies on food, including ones looking at how different meals may affect prediabetes, memory loss, and constipation.

Numerous nutraceutical goods need additional in-depth scientific research and adaption depending on geography, population, dietary preferences, and other pertinent factors. According to a review article in the British Journal of Clinical Pharmacology, the product can be classified as a nutraceutical rather than a nutritional supplements if clinical proof is provided. It should have a well-established mechanism of action, as well as pharmacokinetics and pharmacodynamics. Finally, appropriate clinical studies should be used to demonstrate the product's effectiveness and safety.

Consumers need to be protected from fake nutritional supplements; hence, the industry needs more stringent regulation. Regulation of goods at the intersection of pharmaceuticals and nutrition is urgently needed. Diverse nations have quite different regulations. The Dietary Supplement Health and Education Act of 1994, which mandates that the maker must confirm the product's safety before it is promoted, defines "nutraceuticals" as a word recognized by the FDA in the United States and separates them from traditional food and medications.

The European Food Safety Authority (EFSA), which does not distinguish both food supplements and nutraceuticals, must certify the product before it may be sold in the European Union. Before a food or nutritional product may be sold in the EU, EFSA must give its detailed approval for any health advantages claimed. The Food Safety and Standards Authority of India (FSSAI), whose regulations are less strict than those for pharmaceutical goods and which lists eight categories of such items, is responsible for regulating health supplements and nutraceuticals in India. Every country in the world, including China, Japan, and many others, has its own procedures for registering functional foods and nutraceuticals.

The security, efficiency, and quantity of items being produced might be impacted by this absence of standard laws. Due to an upsurge in bogus claims of health advantages, India's FSSAI issued new directions in 2019 asking all food safety commissioners to monitor all items falling under their purview. In an effort to combat the rising risk of fraudulent claims and unlawful goods, the FDA also took a position in 2019 to amend the regulations on nutritional supplements. Countries should set new guidelines to direct the creation of new nutraceutical goods. Strict oversight can guarantee that fraudulent promises are not flooding the consumer market.6

13.8 FSSAI AND NEED FOR REGULATIONS

The Food Safety and Standards Authority of India (FSSAI) Regulations, 2016 were published in the Indian Gazette on December 23, 2016 for high protein diet, Nutritional supplements, and Dietary food.

These regulations must be followed by any Food Business Operator (FBO) that desires to manufacture, import, or sell these items. These rules will apply to foods that fall into the following categories:

i. Nutrition
ii. Nutraceuticals
iii. (FSDU) Food for Special Dietary Use
iv. (FSMP) Food for Special Medical Purpose
v. Probiotic substances added to food
vi. Prebiotic substance added to food
vii. Botanical or Herbal Ingredients in Specialty Foods
viii. Novel Food (Need prior approval as per FSS (Approval of nonspecified food and food ingredients Regulations, 2017).

FBOs could use two or more materials and additions from the numerous schedules listed in such regulations for each category.

For example, in the nutraceutical category, at least one component from Schedule VI is necessary, while substances from other schedules that are permissible in the nutritious group are either or not utilized. Furthermore, these items may contain substances that may be standardized or approved for use in the manufacturing of other standardized foods, as defined by the FSS (Food Products Standards and Food Additives) Regulations of 2011. Because these chemicals can be found naturally in ordinary food products for mass consumption, such foods should not be regarded supplements, nutraceuticals, or probiotic or prebiotic foods. As a result, if an article of food has still not considered in either way but is appropriate for use in a special dietary routine due to its natural constituents, the food may be labeled with the statement "this food is by its nature X" (where "X" refers to the fundamental distinctive characteristic as demonstrated by commonly accepted scientific data) to notify consumers.[1]

A nutraceutical is a food or food-related component that is intended to provide health advantages, including therapeutic applications. Many people use nutritional supplements, such as phytomedicines and health products, as an alternative to traditional medicine. Nutraceuticals are in high demand around the world. In India, the annual increase of nutritional supplements was estimated to be 25%, which is comparable to that of Japan which is a developed country. In this context, plenty of new enterprises have sprung up in India to produce health supplements. The growth of the Indian nutraceuticals sector was mostly due to pharma corporations and FMCG (fast moving consumer goods) suppliers. Beneficial foods and beverages, as well as dietary supplements, make up the Indian nutritional supplements market. Multivitamins make for about 64% of the nutritional supplements in Indian economy. The Indian nutraceuticals business, which grew from about $1000 million in 2008 to $1800 million in 2013, broke the $2000 million milestone in 2014 and is predicted to reach $4000 million by 2018. The FSS Act, 2006, Rules and Regulations, 2011 classifies nutraceuticals as foodstuff. The FSSAI has created rules regarding food manufacturing licenses and permits, production, packaging and labeling, and packaged food requirements, among other things. The FSSAI has established regulatory criteria for the certification of nutritional supplements in the Indian economy. This Act contains 21 sections, and the 4th Article, or Article 22, of the Act, deals with health supplements, nutritional supplements, and different functional foods, and these items can be created, promoted (sold), or supplied (imported) by any corporation. The FSSAI stated in the Nutraceuticals Regulations, 2016, under the "Basic Criteria," that (21) simple combinations of micronutrients created in pills, capsule, or syrup formats will not be recognized in any of these regulations' classes unless micronutrients are added. FSS Regulations, published in 2011, provided guidelines on food company license and insurance, packaging and labeling, packaged food standards and food additives, sales prohibitions and restrictions, pollutants, allergens, and leftovers, and testing and sampling analysis. In 2015, the Food and Drug Administration (FDA) released regulations for nutritional supplement, nutraceuticals, foods for supplements, therapeutic purposes, functional ingredients, and innovative foods. Any claimed novel foods must provide the following information:

a. Biochemical make-up of genetically modified foods
b. Surface chemistry/alteration
c. Diameter of the basic granules
d. Dissolution
e. Bioavailability
f. Quantity of nanoparticle in the packaged food

Nutraceuticals must consist of some of the components listed in Schedule I of the Food Act: vitamins and minerals. Necessary α-amino and other nutrition (Schedule II) List of flora and botanic substances (Schedule IV) Nutraceutical Substances List (Schedule VI) Appendix VII List of probiotic strains List of prebiotic chemicals (Schedule VIII). No component besides those listed in Schedule VI may be utilized as a multivitamin, with standardization to the marker chemicals listed and at the daily consumption levels listed. Nutraceuticals would only include health supplements or extracts of ingredients listed in Schedule IV. Schedule IV and Schedule VI ingredients with a herbal origin may be used in the prescribed form or as an extraction, according to the extraction proportions in relation to the current consumption value. Only compounds listed in Schedule VA, Schedule VE, or Schedule VF must be utilized in the formulation of nutritional supplements. On December 31, 2018, the FSSAI issued a notice prohibiting the use of numerous substances in nutraceutical foods. The following is of the 3rd notice on the topic of Health supplement Foods Regulatory.[41]

i. Until the planned revisions to the Nutritional supplement Regulations are implemented and notified, the ingredient "Vitamin D3" from fungus (Cladoniarangiferina) as a vegetable origin may be used in goods specified under Nutraceutical Regulations.
ii. Mustard powder may remain to use it as a component in goods regulated by the health supplement Regulations as long as compliance is maintained.
iii. Certain substances indicated in the notification should be discontinued.
iv. Allow food business operators (FBOs) to market current formulations including basic combinations of minerals and vitamins in medicaments such as pills, capsule, and syrups, up to 1 recommended daily allowance.

This is in force for 3 months from the current notification's date, or until new instructions are announced, whichever comes first. The FSSAI has ordered Health supplement Regulations to cease using 14 components for which there is no scientific evidence of their safety. The nation's top food authority ordered FBOs to stop using of p-hydroxylbenzyl acetone, silica, female ginseng, guarana, serenoarepens, Chinese ginseng, pinus maritima extracted to Monterey pine, pinus maritima extracted from maritime pine, Vitamin D3-veg, inonotusobliqws extract, Gram-negative bacteria, melaleuca oil. FBOs were also ordered to stop using Cynara cardunculus, Brassica oleracea L. var. acephala, chia seed, Anacardium occidentale, Passiflora edulis, Actinidia extracts, broccoli, and enzymes, including pectinase and xylanase, were also requested to stop selling them as health supplements. Their inclusion in products was not restricted, but FBOs could not claim the item. Although it was not forbidden, FBOs could not describe the goods as supplements or nutraceuticals. No more items containing these substances will be allowed to be manufactured until the authority has reviewed and authorized them. FBOs are also required to provide information and data on these components within 1 month. For the risk of overconsumption, nutritional supplements goods should have a "Nutraceuticals" Recommended Usage Warning on the label. For Nutraceuticals Products, prohibitions there must be restrictions on labeling claims. Treatment of disease claims, such as "prevents bone fragility in postmenopausal women," has indicated remedies for disease claims via visuals, vignettes, or symbols. Increased demand for nutraceuticals products in India demonstrated its potential for illness prevention and treatment. As a result, product quality and safety regulatory agencies are required to reduce adverse outcomes, cytotoxicity, impurity, abuse, and excess use of drug during human utilization. Maintaining food safety and quality standards as per numerous regulatory criteria issued by the respective governments becomes increasingly crucial as food goods travel from one country to another. To preserve nutraceutical quality and safety issues, all new and current enterprises should follow the FSSAI regulatory guidance for optimal utilization.[2]

The (FSSAI) has established some regulatory standards for the certification of nutritional supplements in the Indian economy. The Act is divided into 21 chapters, with the fourth article discussing nutraceuticals and other functional foods.

The (21) simple combination of mineral and vitamins formulated in capsules, tablets, or syrup layouts will not be encapsulated in any of classifications of these standards anticipated where essential nutrients are added, according to the FSSAI's Nutraceuticals Regulations, 2016, under the General Requirements.

Note: The Food Safety and Standards Act of 2011 classifies dietary supplementation as food.

The FSS Regulations of 2011 established comprehensive licencing and registration requirements for;

- FBOs (food business operators)
- Food ingredients and quality standards
- Labeling and packing

- Contaminants and wastes
- Toxins
- Testing in the lab and collection

In 2015, the FSSAI issued guidelines for vitamins and supplements, multivitamins, functional ingredients, herbal supplements, and novel foods.

There must be any of the components listed in the Food Act

Schedules:

- Schedule I contains fat-soluble vitamins, water-soluble vitamins, and minerals.
- Schedule II contains necessary protein and other nutrients.
- Schedule IV contains herbal and biological substances.
- Schedule VI contains substances as health supplements.
- Schedule VII contains bacterial species as probiotics.
- Schedule VIII contains essential prebiotic substances.

No ingredients other than those listed in Schedule VI may be used as nutritional supplements, with standardization to the marker chemicals listed and at the daily consumption levels specified. Nutraceuticals shown in Schedule IV are the only ones that can be used in supplements.

Components of phytochemical composition or plants included in Schedules IV and VI can be used in the form provided or as an extract, subjected to extraction ratios corresponding to the everyday usage requirement. In the formulation of nutritional supplements, only additives listed in Schedule VA, VE, or VF should be used.

The FSSAI issued a notification on December 31, 2018, prohibiting the company from using numerous components in nutraceuticals.

The Nutraceutical Foods Regulation had received its third notice:

- Permitted the use of the component "Vitamin D3" from lichen as a vegetable source in the Nutraceutical Regulations-required items until the planned Health supplement Regulations or amendments are finalised.
- No restrictions were placed on the consumption of mustard seed powder in the foods listed in the Nutraceutical Regulations.
- Ensured that certain components mentioned in the warning were no longer used.
- Permitted FBOs to sell the current composition, which contains a mixture of essential vitamins and minerals, in medicaments such as capsules, pills, and syrups, only to one recommended daily intake.

This will be in existence for 12 weeks from the date of this announcement or until further instructions are issued, whichever comes first. The (FSSAI) has changed nutritional supplement standards to prohibit producers from using 14 substances that are potentially harmful to human health. Furthermore, the FSSAI has asked the FBOs to guarantee that the following components are never used:

- Silica dioxide
- Ras ketone
- Chinese ginseng
- Guarana
- Serenia repens
- Danggui
- Bark extracted from maritime pine
- Bark extracted to Monterey pine
- Phytavail iron
- Oxalobacterformigenes (Gram-negative bacteria)

In addition to the aforementioned goods, the FSSAI has instructed FBOs to stop utilizing the following components as health supplements.

- Melaleuca oil
- Leaf cabbage powder
- Actinidia deliciosa extract
- Pectolyase
- Brassica oleracea and enzymes
- Mexican chia
- Passiflora edulis

The consumption of the ingredients listed above requires prior permission from the authority; otherwise, the maker may be fined.

Furthermore, the FBOs are needed to submit information about these chemicals within a month. Nutraceuticals-based products should have the word "Nutraceuticals" on the label.

Labeling for nutraceuticals goods must represent the stated therapeutic advantages.[3]

The FSSAI has a bigger role to play in creating guidelines that will simplify the overall nutritional manufacturing process. Consumers need to address the issues they are worried about.

- A good standard of raw resources
- A substance that is made in a secure environment

- Health-related allegations
- Categorization
- Transporting and preserving

The FSSA of 2006 governs the production, preservation, transportation, marketing, and export of nutraceuticals in India. The FSSA has unified a number of previous legislations dealing with food products and nutraceutical safety and standards. However, there are no specific restrictions in place, and the administration is still looking for a number of designs.

Rules relating to quality and security, according to nutraceutical businesses, will lead to economic growth and keep uncontrolled activities under check. The fragmented nature of the legislation now makes it impossible for players to cooperate fully or make the required expenditures.[4]

13.9 CONCLUSION

In India, there is a large market for nutraceuticals due to their effectiveness in the treatment and prevention of diseases. As a result, FSSAI is required to prevent adverse effects, abuse, impurity, toxicity, and overdose in humans. The market growth for nutritional supplements in India is forcing regulatory authorities to strengthen applicable regulations in order to limit the availability of substandard products.[3]

Nutraceuticals have the ability to change the way we think about food and medicine. They allow food medications to be more diverse, but they also necessitate thorough investigation, digital innovations, clinical studies, creative product formulations, and stricter, more standard supervision.

Despite the uncertainty, the nutraceuticals industry is constantly expanding and overflowing with fresh concepts. Nutraceutical innovations with fewer side effects and demonstrable medical benefits are always decided to seek items for the treatment of illness issues.

Nutraceutical use is fast increasing over the world, owing to their stated safety and efficacy. The use of sophisticated nanosized drug delivery systems as a platform for nutraceutical formulation has recently gained traction. The encouraging results gained in terms of bioavailability, safety, targeting, and durability of the nutritional supplements examined have affected this. As a result, nanotechnology, nanostructured lipid carriers, dendrimers, phytosomes, nano emulsions, and other drug delivery methods were used.[6]

Nutraceutical formulation, on the other way, poses numerous hurdles in terms of efficacy, safety, and regulatory compliance. Concerned institutions were unable to come up with a universal definition. Regulatory agencies to govern and regulate nutraceuticals have yet to be established around the world.[8]

KEYWORDS

- **disorders**
- **nutraceuticals**
- **pharmaceuticals**
- **phytochemicals**
- **probiotics**
- **vitamins**

REFERENCES

1. Ansari, S.; Chauhan, B.; Kalam, N.; Kumar, G. Current Concepts and Prospects of Herbal Nutraceutical: A Review. *J. Adv. Pharm. Technol. Res.* **2013,** *4* (1), 4.
2. Baradaran, A. Lipoprotein (A), Type 2 Diabetes and Nephropathy; the Mystery Continues. *J. Nephropathol.* **2012,** *1* (3), 126–129.
3. Daliu, P.; Santini, A.; Novellino, E. From Pharmaceuticals to Nutraceuticals: Bridging Disease Prevention and Management. *Expert Rev. Clin. Pharmacol.* **2018,** *12* (1), 1–7.
4. Das, L.; Bhaumik, E.; Raychaudhuri, U.; Chakraborty, R. Role of Nutraceuticals in Human Health. *J. Food Sci. Technol.* **2011,** *49* (2), 173–183.
5. Ebert, A. W. Sprouts and Microgreens—Novel Food Sources for Healthy Diets. *Plants* **2022,** *11* (4), 571.
6. Excipients For Nutraceutical Oral Solid Dosage Forms: Market Analysis And Opportunities [Online]. https://klinegroup.com/reports/excipients_for_nutraceutical_osdf/ (accessed June 21, 2022).
7. FSSAI [Online]. https://www.fssai.gov.in/cms/food-safety-and-standards-regulations.php (accessed June 12, 2022).
8. Hardy, G. Nutraceuticals and Functional Foods: Introduction and Meaning. *Nutrition* **2000,** *16* (7-8), 688–689.
9. Helal, N. A.; Eassa, H. A.; Amer, A. M.; Eltokhy, M. A.; Edafiogho, I.; Nounou, M. I. Nutraceuticals' Novel Formulations: The Good, the Bad, the Unknown and Patents Involved. *Recent Pat. Drug Del. Formul.* **2019,** *13* (2), 105–156.
10. https://corpbiz.io/learning/fssai-regulations-for-nutraceuticals/ (accessed June 10, 2022).
11. https://www.prescouter.com/2021/09/nutraceuticals-the-promises-and-challenges-of-medicinal-foods/ (accessed June 21, 2022).

12. https://www.sciencedirect.com/science/article/pii/B9780128202845000113#:~:text=Nutraceuticals%20provide%20multiple%20health%20benefits (accessed June 21, 2022).
13. Kalra, E. K. Nutraceutical-Definition and Introduction. *AAPS PharmSci* **2003,** *5* (3), 27–28.
14. Keservani, R. K.; Sharma, A. K. Flavonoids: Emerging Trends and Potential Health Benefits. *J. Chin. Pharm. Sci.* **2014,** *23* (12), 815.
15. Keservani, R. K.; Kesharwani, R. K.; Vyas, N.; Jain, S.; Raghuvanshi, R.; Sharma, A. K. Nutraceutical and Functional Food as Future Food: A Review. *Der Pharmacia Lettre* **2010a,** *2* (1), 106–116.
16. Keservani, R. K.; Sharma, A. K.; Jarouliya, U. Protein and Peptide in Drug Targeting and its Therapeutic Approach. *ArsPharmaceutica* **2015,** *56* (3), 165–177.
17. Keservani, R. K.; Sharma, A. K.; Kesharwani, R. K. Nutraceutical and Functional Foods for Cardiovascular Health. In *Food Process Engineering*; Apple Academic Press, 2016a; pp 291–312; CRC Press, 2016a; pp 257–278, ISBN: 9781771884020.
18. Keservani, R. K.; Sharma, A. K.; Kesharwani, R. K. Medicinal Effect of Nutraceutical Fruits for the Cognition and Brain Health. *Scientifica* **2016b,** Article ID 3109254, 10. DOI: 10.1155/2016/3109254.
19. Keservani, R. K; Sharma, A. K.; Kesharwani, R. K. An Overview and Therapeutic Applications of Nutraceutical and Functional Foods. In *Recent Advances in Drug Delivery Technology*, 2017; pp 160–201, ISBN: 9781522507543.
20. Keservani, R. K.; Sharma, A. K.; Kesharwani, R. K. Eds.; In *Nutraceuticals and Dietary Supplements: Applications in Health Improvement and Disease Management*; CRC Press, 2020; ISBN: 9781771888738.
21. Keservani, R. K.; Kesharwani, R. K.; Sharma, A. K.; Vyas, N.; Chadoker, A. Nutritional Supplements: An Overview. *Int. J. Curr. Pharm. Rev. Res.* **2010b,** *1* (1), 59–75.
22. Maurya, A. P.; Chauhan, J.; Yadav, D. K.; Gangwar, R.; Maurya, V. K. Chapter 11 - Nutraceuticals and their Impact on Human Health.
23. Nasri, H.; Baradaran, A.; Shirzad, H.; Rafieian-Kopaei, M. New Concepts in Nutraceuticals as Alternative for Pharmaceuticals. *Int. J. Prev. Med.* **2014,** *5* (12).
24. Nicoletti, M. Nutraceuticals and Botanicals: Overview and Perspectives. *Int. J. Food Sci. Nutr.* **2012,** *63 Suppl 1*, 2–6.
25. Nutraceutical Ingredients Market by Type, Application, Region - Global Forecast 2025. MarketsandMarkets [Online]. https://www.marketsandmarkets.com/Market-Reports/nutraceutical-ingredient-market-1319.html (accessed June 12, 2022).
26. Pandey, M. M.; Rastogi, S.; Rawat, A. K. S. Indian Traditional Ayurvedic System of Medicine and Nutritional Supplementation. *Evid. Based Complementary Altern. Med.* **2013,** *2013*, 1–12.
27. Pradhan, S. Nutraceuticals: The Promises and Challenges of Medicinal Foods. PreScouter - Custom Intelligence from a Global Network of Experts.
28. Putta, S. *FSSAI Guidance and Notification on Nutraceuticals -an Insight*, 2020.
29. Ramirez, D.; Abellán-Victorio, A.; Beretta, V.; Camargo, A.; Moreno, D. A. Functional Ingredients from Brassicaceae Species: Overview and Perspectives. *Int. J. Mol. Sci.* **2020,** *21* (6), E1998.
30. Rane, B. R.; Bharath, M. S.; Patil, R. R.; Keservani, R. K.; Jain, A. S. Novel Approaches in Nutraceuticals, Chapter 11. In *Enhancing the Therapeutic Efficacy of Herbal Formulations through Novel Drug Delivery Systems*; Kesharwani, R. K., Keservani, R.

K., Sharma, A. K., Ed.; IGI Global International Publisher: Pennsylvania, USA, 2020b; pp 241–266, ISBN13: 9781799844532.

31. Rane, B. R.; Patil, A. S.; Keservani, R. K.; Jain, A. S. Novel Approaches in Herbal Formulation, Chapter 2. In *Enhancing the Therapeutic Efficacy of Herbal Formulations through Novel Drug Delivery Systems*; Kesharwani, R. K., Keservani, R. K., Sharma, A. K., Ed.; IGI Global International Publisher: Pennsylvania, USA, 2020a; pp. 43–68, ISBN13: 9781799844532.
32. Rane, B. R.; Tadavi, S. A.; Keservani, R. K. Naturopathy, Chapter 12. In *Herbal Product Development*; Sharma, A. K., Keservani, R. K., Gautam, S. P., Ed.; Apple Academic Press, CRC Press, Taylor & Francis Group, 2020c; pp 321–347, ISBN: 9781771888776.
33. Sachdeva, V.; Roy, A.; Bharadvaja, N. Current Prospects of Nutraceuticals: A Review. *Curr. Pharm. Biotechnol.* **2020,** *21* (10), 884–896.
34. Santini, A.; Cammarata, S. M.; Capone, G.; Ianaro, A.; Tenore, G. C.; Pani, L.; Novellino, E. Nutraceuticals: Opening the Debate for a Regulatory Framework. *Brit. J. Clin. Pharmacol.* **2018,** *84* (4), 659–672.
35. Serafini, M.; Stanzione, A.; Foddai, S. Functional Foods: Traditional Use and European Legislation. *Int. J. Food Sci. Nutr.* **2011,** *63* (sup1), 7–9.
36. Sharma, A. K.; Keservani, R. K.; Gautam S. P. Herbal Product Development, Apple Academic Press, CRC Press, Taylor & Francis Group, 2020; ISBN: 9781771888776.
37. Shinde, N.; Bangar, B.; Deshmukh, S. Nutraceuticals: A Review on Current Status. *Res. J. Pharm. Technol.***2014,** *7* (1), 110–113.
38. Siddiqui, R. A.; Moghadasian, M. H. Nutraceuticals and Nutrition Supplements: Challenges and Opportunities. *Nutrients* **2020,** *12* (6), 1593.
39. Singh, P.; Kesharwani, R. K.; Keservani, R. K. Antioxidants and Vitamins: Roles in Cellular Function and Metabolism. In *Sustained Energy for Enhanced Human Functions and Activity*; Academic Press, 2017; pp. 385–407, ISBN:9780128054130.
40. Varzakas, T. The Incorporation of Nutraceuticals in Strategic Reformulation of Food Products for Better Health, Nutrition and Disease Prevention: The Case Studies of LCPUFA, Probiotics, Prebiotics, Phytosterol Esters, Andbeta;-Glucan Addition. *Appl. Food Sci. J.* **2020.**
41. Whitman, M. Understanding the Perceived Need for Complementary and Alternative Nutraceuticals: Lifestyle Issues. *Clin. J. Oncol. Nurs.* **2001,** *5* (5), 190–194.

INDEX

A

Achilles tendon ruptures (ATR), 177
Artificial general intelligence (AGI), 113
Artificial intelligence (AI), 2, 105
  ACM SigCHI meeting, 110, 111
  artificial general intelligence (AGI), 113
  business intelligence (BI)
    commonplace measurements, 131
    exploration views (EV), 134
    frameworks, 131
    human memory, restrictions, 135
    human-focused plan, 133
    measure of information, 136
    OLAP inquiries, 131
    primary objective, 133
    SAP Research, 132
  digital hearing
    comparative perceptions, 123
    Dicta-Sign Wiki, 125
    human–PC cooperation plan, 120, 121
    intelligent sonifications, 121
    k-Nearest Neighbors Algorithm (k-NN), 125
    Sign Wiki model, 125
    signal-based gadgets, 125
    sound items, 124
    unpredictable human activities, 122
  digital humanities (DH)
    advanced humanities, 126
    advanced philology, 127
    challenge, 130
    computational philology, 127
    enormous tent, 127
    human–PC connection, 129
    interfaces, 130, 131
    rural American 1950s, 128
  Edified experimentation, 112
  flexibility, 109
  Hadoop with, 112
  healthcare systems
    computerized diagnostics, 115
    digitalization, 120
    DMAIC (define, measure, analyze, improve, and control), 116
    DSS and CDSS, 115
    EMR, 116, 117
    EMR quality improvement, 114
    HCI professionals, 118
    MOOCs, 119
    portable innovations, 119
    Subjective Mental Effort Question (SMEQ), 116
    UI augmentation, 115
  HospiSign stage, 106
  human-PC communication stage, 106
  logical and designing disciplines, 108
  man-made consciousness, 108
  market investigation
    hyper-combined foundation market, 138
    hyper-converged infrastructure market, 139
    Hyper-Converged Infrastructure (HCI) Market, 136
    hyper-intermingling framework, 137
    hyper-joined foundation market, 138
    hyper-merged foundation market, 137
    hyper-united foundation market, 138
    monetary firms, 137
  Ordinary AI (GOFAI), 110, 111
  rationalistic reaction, 112
  track plans, 107
  two-way course, 110
Atrial fibrillation, 78
  materials and methods
    ADAM solver, 81–82
    classifier performance enhancement, 82–83
    ECG data description, 78, 79, 81
    time-frequency features, 83–84
Attack target scale (ATS), 288
Attacker scale (AS), 288
Autism spectrum disorder (ASD), 180
Automatic segmentation, 41

B

Binary BAT Algorithm (BBA), 245
Biomedical image processing
challenges in, 24
heterogeneity of images, 25
robustness of algorithms, 25
unknown delineation of objects, 25
Blood oxygen level-dependent (BOLD), 183
Blood–brain barrier (BBB), 181
Brain surface extractor (BSE), 43
Brainsuite tool, 67
Business intelligence (BI)
commonplace measurements, 131
exploration views (EV), 134
frameworks, 131
human memory, restrictions, 135
human-focused plan, 133
measure of information, 136
OLAP inquiries, 131
primary objective, 133
SAP Research, 132

C

Cache management error, 295, 296
Cancer cells, 256
Cephalometric, 202
Cervical cancer, 148
detection of, 150
feature extraction, 153
literature survey
general diagnosis, 150
techniques for, 150–152
local binary pattern (LBP), 154
directive pixel area, 155
GLCM medium, 156–158
performance evaluation parameter, 159
reckoning operation, 155–156
methodology
materials and methods, 152
methods, 153
Cervix cancer, 148
Channeled hotelling observer (CHO), 102
Collusion attack strength (CAS), 289
Computer-aided automatic detection, 147
feature extraction, 153
literature survey
general diagnosis, 150
techniques for, 150–152
local binary pattern (LBP), 154
directive pixel area, 155
GLCM medium, 156–158
performance evaluation parameter, 159
reckoning operation, 155–156
methodology
materials and methods, 152
methods, 153
Computer-aided design (CAD), 102
Contrast Limited Adaptive Histogram Equalization (CLAHE), 256
Cooperation attack frequency (CAF), 287

D

Dental images
literature survey
computed tomography (CT), 203
deep learning model, 203
semiautomatic model, 203
methodology
Haung thresholding, 204
mandible region in sample 1 and 2, 204
Otsu thresholding, 204
Panoramic Images, 204
results and discussion, 204
edge detection technique, 211, 212
grayscale segmentation, 205
Haung threshold method, 211
original grayscale image, 209
segmented mandible length, 206, 207
segmented mandible region, 205, 206, 210
threshold value, 207, 208, 209
Dermoscopy
methods and materials
classifier explanation, 232
NSCT algorithm, 231–232
result and discussion
normal and abnormal images, 232
Design detecting
materials and methods
classification accuracy, 250
CSO + K means, 251, 252
Cuckoo search optimization, 247
extraction formula table, 248–249
image description, 245
K-Means is clustering, 247–248
median filter, 246

segmentation accuracy comparison, 251
SVM classifier, 249, 250
Diabetes mellitus (DM), 215
biomedical datasets, 216
healthcare professionals, 217
linear discriminant analysis (LDA), 218
machine learning (ML), 219
Naive Bayes (NB), 217
National Health and Nutrition Survey (NHNS), 218
Pima dataset, 218
principal component analysis (PCA), 218
proposed method, 219
classification, 220–221
results and discussion
dataset, 221–225
support vector machine (SVM), 217
training methods, 218
Dicta-Sign Wiki, 125
Diffusion kurtosis imaging (DKI), 183
Diffusion tensor imaging (DTI), 179
Digital hearing
comparative perceptions, 123
Dicta-Sign Wiki, 125
human–PC cooperation plan, 120, 121
intelligent sonifications, 121
k-Nearest Neighbors Algorithm (k-NN), 125
Sign Wiki model, 125
signal-based gadgets, 125
sound items, 124
unpredictable human activities, 122
Digital humanities (DH)
advanced humanities, 126
advanced philology, 127
challenge, 130
computational philology, 127
enormous tent, 127
human–PC connection, 129
interfaces, 130, 131
rural American 1950s, 128

E

Edge detection technique, 211
Edified experimentation, 112
Electroencephalography (EEG), 88
Electronic pictures, 301–302
Exploration views (EV), 134

F

Fast Discrete Curvelet Transform, 266
Feedback
accuracy of, 282
collusion assaults, 282–283
discrimination attacks, 283
intoxication attacks, 283
newcomer attacks, 284
storage and evaluation, 281–282
TMS, 284–285
T-RBAC scheme, 284
Fluid attenuated inversion recovery (FLAIR), 178
Full width half maximum (FWHM), 31
Functional magnetic resonance imaging (fMRI), 68
blood oxygen level-dependent (BOLD), 183
brain functional activities, 181–182
brain seizure images
brain parts, 71–73
emotional changes, 183
neuro-developmental disorder, 182–183
post-traumatic stress disorder (PTSD), 182
Fused images, 266
Fuzzy Local Information C-means Clustering, 262

G

Gaussian filter, 202
Genetic meta analysis
GWA datasets, 5
HapMap loci, 4
image processing, 5–6
meta analysis risk, 5–6
Gradient echo imaging (GRE), 179

H

Healthcare systems
computerized diagnostics, 115
digitalization, 120
DMAIC (define, measure, analyze, improve, and control), 116
DSS and CDSS, 115
EMR, 116, 117
EMR quality improvement, 114
HCI professionals, 118

MOOCs, 119
portable innovations, 119
Subjective Mental Effort Question (SMEQ), 116
UI augmentation, 115
High angular resolution diffusion imaging (HARDI), 183
Histogram equalization (HE), 101
HospiSign stage, 106
Human Papilloma Virus (HPV), 148
Human-computer interaction (HCI), 105
ACM SigCHI meeting, 110, 111
artificial general intelligence (AGI), 113
business intelligence (BI)
commonplace measurements, 131
exploration views (EV), 134
frameworks, 131
human memory, restrictions, 135
human-focused plan, 133
measure of information, 136
OLAP inquiries, 131
primary objective, 133
SAP Research, 132
digital hearing
comparative perceptions, 123
Dicta-Sign Wiki, 125
human–PC cooperation plan, 120, 121
intelligent sonifications, 121
k-Nearest Neighbors Algorithm (k-NN), 125
Sign Wiki model, 125
signal-based gadgets, 125
sound items, 124
unpredictable human activities, 122
digital humanities (DH)
advanced humanities, 126
advanced philology, 127
challenge, 130
computational philology, 127
enormous tent, 127
human–PC connection, 129
interfaces, 130, 131
rural American 1950s, 128
Edified experimentation, 112
flexibility, 109
Hadoop with, 112
healthcare systems
computerized diagnostics, 115
digitalization, 120
DMAIC (define, measure, analyze, improve, and control), 116
DSS and CDSS, 115
EMR, 116, 117
EMR quality improvement, 114
HCI professionals, 118
MOOCs, 119
portable innovations, 119
Subjective Mental Effort Question (SMEQ), 116
UI augmentation, 115
human-PC communication stage, 106
logical and designing disciplines, 108
man-made consciousness, 108
market investigation
hyper-combined foundation market, 138
hyper-converged infrastructure (HCI) market, 136, 139
hyper-intermingling framework, 137
hyper-joined foundation market, 138
hyper-merged foundation market, 137
hyper-united foundation market, 138
monetary firms, 137
Ordinary AI (GOFAI), 110, 111
rationalistic reaction, 112
track plans, 107
two-way course, 110
Hyper-converged infrastructure (HCI) market, 136

I

Image quality assessment (IQA), 268

K

K-Nearest Neighbors Algorithm (k-NN), 125

L

Larmor frequency, 25, 26
Linear discriminant analysis (LDA), 218
Local binary pattern (LBP), 154
directive pixel area, 155
GLCM medium, 156–158
performance evaluation parameter, 159
reckoning operation, 155–156
Long short-term memory (LSTM), 78

Lung cancer, 236
convolutionary neural network (ConvNet/CNN)
architecture, 238
detection using CT images
bar graph-comparison, 262
data sets, 261
removal of bone region, 257–258
late diagnosis, 236, 237
methods and materials
pre-processing, 237
PET images, detection
bar graph, 265
data sets, 265
FLICM detection, 262–264
Fuzzy Local Information C-means Clustering, 262
image fusion, 266
random forest classifier, 239
result and discussion
Fast Discrete Curvelet Transform, 266
fused images, 266
image quality assessment (IQA), 268
KNN classifier, 239
multi model imaging fusion, 268
Non Sub-sampling Contourlet Transform (NSCT), 267
peak signal-to-noise ratio (PSNR), 268, 269
performance analysis for, 268
random forest classifier, 239
RMSE, 270–271
segmented images, 239
SIM, 270
spatially weighted Fuzzy C means clustering (SWFCM)
algorithm, 259–260

M

Machine learning (ML), 219
architecture, 98
active learning, 100
deep learning, 100–101
evolutionary learning, 100
reinforcement learning, 100
semisupervised learning, 99
supervised learning, 99
unsupervised learning, 99
artificial intelligence (AI), 310
clinical test and exploration, 313–314
conditions and opinion, 312
crowdsourced information collection, 314
drug discovery and producing, 312
grounded behavioral revision, 313
higher radiotherapy, 314
irruption prediction, 315
medical imaging designation, 312
personalized drugs, 312–313
sensible health records, 313
use, 308
digital image processing
defined, 302
key phases of, 301
electronic pictures, 301–302
image process, 300–301
man-made intelligence applications, 308–310
medical imaging techniques, 303
advantages of, 306
channeled hotelling observer (CHO), 102
computer-aided design (CAD), 102
histogram equalization (HE), 101
history of, 98
KARDIO, 102
meaning, 305–306
modified histogram-based contrast enhancement with homomorphic filtering (MHFIL), 101
OS-EM rebuilding method, 103
processing, 303–304
technologies, 304–305
and types, 97
uses, 305
working, 306–307
Magnetic resonance images (MRI)
applications
Achilles tendon ruptures (ATR), 177
autism spectrum disorder (ASD), 180
blood–brain barrier (BBB), 181
cardiovascular disease diagnosis, 181
colorectal cancer, 177
diffusion kurtosis imaging (DKI), 183
diffusion tensor imaging (DTI), 179
fluid attenuated inversion recovery (FLAIR), 178

gliomas, 177
gradient echo imaging (GRE), 179
hematologic marrow diseases, 177
high angular resolution diffusion imaging (HARDI), 183
lung cancer, 178
mapping cerebral connectivity, 179–180
out-of-field recurrence (OFR), 177–178
SPIN echo magnetic resonance imaging, 179
stroke, automated quantitative segmentation, 178
susceptibility-weighted imaging (SWI), 183–184
concept of
Larmor frequency, 25, 26
longitudinal magnetization, 27
longitudinal relaxation, 27
out-of-phase condition, 26
transversal relaxation, 27, 28
functional magnetic resonance imaging (fMRI)
blood oxygen level-dependent (BOLD), 183
brain functional activities, 181–182
emotional changes, 183
neuro-developmental disorder, 182–183
post-traumatic stress disorder (PTSD), 182
noise in
Rician noise, 24
SNR, 23–24
zero-mean Gaussian-distributed noise, 23
quality metrics
SSIM, 38
T1 and T2 images, 38
results and discussions, 31
deformation field, 34, 35
DICOM format, 33
joint histogram, 34, 36
Nifti images, 31, 32, 33, 35
segmented brain parts, 36, 37
spatial preprocessing, 29
coregister, 30
full width half maximum (FWHM), 31
normalization, 30–31
realignment (REALIGN), 30
segmentation, 31
smoothing, 31
T1-weighted *versus* T2-weighted images, 29
types
T1-weighted images, 176
T2-weighted images, 176
Market investigation
hyper-combined foundation market, 138
hyper-converged infrastructure (HCI) market, 136, 139
hyper-intermingling framework, 137
hyper-joined foundation market, 138
hyper-merged foundation market, 137
hyper-united foundation market, 138
monetary firms, 137
Medical imaging techniques, 21, 22
clustering methods
classification of, 192
hard computing clustering, 193
hybridsegmentation, 196
clustering techniques, 197
MATLAB software, 197–198
image segmentation strategies, 191–192
MRI brain image, 188
automatic segmentation, 189–190
manual segmentation, 189
MRI scan, 22–23
MRI technique, 190–191
segmentation, classification, 191
soft computing clustering
Fuzzy C-means algorithm, 194–195
K-means clustering, 193–194
spatial Fuzzy C-means algorithm, 195–196
Melanoma skin cancer, 229
materials and methods
classification accuracy, 250
CSO + K means, 251, 252
Cuckoo search optimization, 247
extraction formula table, 248–249
image description, 245
K-Means is clustering, 247–248
median filter, 246
segmentation accuracy comparison, 251
SVM classifier, 249, 250
Microarray data, 2

Modified histogram-based contrast enhancement with homomorphic filtering (MHFIL), 101

N

National Health and Nutrition Survey (NHNS), 218
Non Sub-sampling Contourlet Transform (NSCT), 267

O

Odds ratio (OR), 6
Open source tools
  Doccano, 171
  NLTK, 171
  Spark NLP, 171
  TextBlob, 171
Ordinary AI (GOFAI), 110, 111
Out-of-field recurrence (OFR), 177–178

P

Papanicolaou smear test, 149
Partial volume classifier (PVC), 44
Peak signal-to-noise ratio (PSNR), 268, 269
Pediatric brain
  cortical surface extraction
    brain regions for different datasets, 48–59
    brain surface extractor (BSE), 43
    cerebrum labeling, 44
    gray matter (GM), 59
    nonuniformity correction, 43–44
    outputs of, 45–49
    partial volume classifier (PVC), 44
    skull stripping, 43
    surface and volume registration (SvReg), 44–45
    tissue classification, 44
    topology correction, 44
    volume estimation, 45
    white matter (WM), 59
    WISP removal, 44
  materials and methods
    human brain's important parts and functions, 42–43
    real-time datasets, 42
Pima dataset, 218
Post-traumatic stress disorder (PTSD), 182
Principal component analysis (PCA), 218

Q

Quality metrics
  SSIM, 38
  T1 and T2 images, 38

R

Radial Basis Function Network (RBFN), 245
Realignment (REALIGN), 30
Receiver-running feature (ROC), 6–7
Registration process, 65–66
Regression, 66
Risk ratio (RR), 6

S

Seizure
  genetic meta analysis
    GWA datasets, 5
    HapMap loci, 4
    image processing, 5–6
    meta analysis risk, 5–6
  hemosiderin, 64
  image preprocessing, 64
    intensity normalization, 65
    motion, 65
    registration process, 65–66
    regression, 66
    and slice time correction, 65
    smoothing spatial filter, 66
  materials and methods
    Brainsuite tool, 67
    clustering, method, 3–4
    MedCalc, 3
  meta-analysis, 4
    nonlinear regression, 7
    odds ratio (OR), 6
    receiver-running feature (ROC), 6–7
    risk ratio (RR), 6
  results and discussion, 67
    analysis techniques, 73–74
    cerebrum labeling, 69
    cluster analysis, 8
    continuous measure, 11–12
    correction of topology, 69
    correlation, 13
    cortical thickness estimation, 70
    functional magnetic resonance imaging (fMRI), 68

generic meta analysis, 10–11
hemosiderin, 68
identification of cortex, 69
MedCalc, 7
medical care system, 73
non-linear regression method, 15–16
pial area, 70
risk ratio and difference, 9, 13–14
scalp and skull processing, 69
scrub mask, 69
seizures, 74
splitting of hemisphere, 70
stripping of skull, 68
surface thickness of cortical area, 70
surface volume registration, 70
tissue, 69
WISP and DEWISP process, 70
Sentiment analysis, 164
approaches for
hybrid-based approach, 167
Lexicon-based approach, 166
machine learning techniques, 166–167
classification
levels, 165
limitation of, 172
machine learning techniques
ADA boost algorithms, 170
classification algorithms, 168
data preprocessing, 168
decision tree, 169
dimensionality reduction algorithms, 169
feature selection, 168
feature vector construction, 168
gradient boosting, 170
K-means clustering, 169
KNN, 169
linear regression, 168–169
logistic regression, 169
Naive-Bayes, 169
random forest algorithm, 169
support vector machine (SVM), 169
open source tools
Doccano, 171
NLTK, 171
Spark NLP, 171
TextBlob, 171
social media analytics
architecture, 170–171
Sexually Transmitted Infections (STIs), 149
Short time Fourier
materials and methods
EOG artifacts, 90
implementation using MatLab, 90–93
Matlab *transformSTFT,* 90
STFT implementation, 88–89
Sign Wiki model, 125
Social media, 164
Spatial preprocessing, 29
coregister, 30
full width half maximum (FWHM), 31
normalization, 30–31
realignment (REALIGN), 30
segmentation, 31
smoothing, 31
Spatially weighted Fuzzy C means clustering (SWFCM), 255
algorithm, 259–260
Subjective Mental Effort Question (SMEQ), 116
Support vector machine (SVM), 169, 217
Surface and volume registration (SvReg), 44–45
Surgery, 148
Susceptibility-weighted imaging (SWI), 183–184

T

Total dermoscopic value (TDO), 244
T-RBAC scheme, 284
Trust evaluation
design methodologies
trust evaluation system (TES), 279–281
existing system
proposed system, 278–279
Service-Level Agreements (SLAs), 277
system architecture, 278
feedback
accuracy of, 282
collusion assaults, 282–283
discrimination attacks, 283
intoxication attacks, 283
newcomer attacks, 284
storage and evaluation, 281–282
TMS, 284–285
T-RBAC scheme, 284
security and accuracy, comparison, 289

actual availability, 294
availability model experiments, 295–296
collusion attacks, 293
credibility model, 291–292
experimental evaluation, 290–291
Peaks behavior model, 294
reallocation effectiveness, 296
structural rigidity, 294
tolerable caching error rate, 296
simulation results
attack target scale (ATS), 288
attacker scale (AS), 288
collusion attack frequency, 287
collusion attack strength (CAS), 289
collusion set (CS), 288
cooperation attack frequency (CAF), 287
feedback frequency, 287
trust model, 285, 286

U

UI augmentation, 115

V

Visual interpretation, 2

X

X-ray image, 202
literature survey
computed tomography (CT), 203
deep learning model, 203
semiautomatic model, 203
methodology
Haung thresholding, 204
mandible region in sample 1 and 2, 204
Otsu thresholding, 204
Panoramic Images, 204
results and discussion, 204
edge detection technique, 211, 212
grayscale segmentation, 205
Haung threshold method, 211
original grayscale image, 209
segmented mandible length, 206, 207
segmented mandible region, 205, 206, 210
threshold value, 207, 208, 209

Z

Zero-mean Gaussian-distributed noise, 23